High Performance Permanent Magnet Materials

MATERIALS RESEARCH SOCIETY SYMPOSIA PROCEEDINGS

ISSN 0272 - 9172

Volume 25—Thin Films and Interfaces II, J. E. E. Baglin, D. R. Campbell, W. K. Chu, 1984, ISBN 0-444-00905-1

Volume 26—Scientific Basis for Nuclear Waste Management VII, G. L. McVay, 1984, ISBN 0-444-00906-X

Volume 27—Ion Implantation and Ion Beam Processing of Materials, G. K. Hubler, O. W. Holland, C. R. Clayton, C. W. White, 1984, ISBN 0-444-00869-1

Volume 28—Rapidly Solidified Metastable Materials, B. H. Kear, B. C. Giessen, 1984, ISBN 0-444-00935-3

Volume 29—Laser-Controlled Chemical Processing of Surfaces, A. W. Johnson, D. J. Ehrlich, H. R. Schlossberg, 1984, ISBN 0-444-00894-2

Volume 30—Plasma Processsing and Synthesis of Materials, J. Szekely, D. Apelian, 1984, ISBN 0-444-00895-0

Volume 31—Electron Microscopy of Materials, W. Krakow, D. A. Smith, L. W. Hobbs, 1984, ISBN 0-444-00898-7

Volume 32—Better Ceramics Through Chemistry, C. J. Brinker, D. E. Clark, D. R. Ulrich, 1984, ISBN 0-444-00898-5

Volume 33—Comparison of Thin Film Transistor and SOI Technologies, H. W. Lam, M. J. Thompson, 1984, ISBN 0-444-00899-3

Volume 34—Physical Metallurgy of Cast Iron, H. Fredriksson, M. Hillerts, 1985, ISBN 0-444-00938-8

Volume 35—Energy Beam-Solid Interactions and Transient Thermal Processing/1984, D. K. Biegelsen, G. A. Rozgonyi, C. V. Shank, 1985, ISBN 0-931837-00-6

Volume 36—Impurity Diffusion and Gettering in Silicon, R. B. Fair, C. W. Pearce, J. Washburn, 1985, ISBN 0-931837-01-4

Volume 37—Layered Structures, Epitaxy, and Interfaces, J. M. Gibson, L. R. Dawson, 1985, ISBN 0-931837-02-2

Volume 38—Plasma Synthesis and Etching of Electronic Materials, R. P. H. Chang, B. Abeles, 1985, ISBN 0-931837-03-0

Volume 39—High-Temperature Ordered Intermetallic Alloys, C. C. Koch, C. T. Liu, N. S. Stoloff, 1985, ISBN 0-931837-04-9

Volume 40—Electronic Packaging Materials Science, E. A. Giess, K.-N. Tu, D. R. Uhlmann, 1985, ISBN 0-931837-05-7

Volume 41—Advanced Photon and Particle Techniques for the Characterization of Defects in Solids, J. B. Roberto, R. W. Carpenter, M. C. Wittels, 1985, ISBN 0-931837-06-5

Volume 42—Very High Strength Cement-Based Materials, J. F. Young, 1985, ISBN 0-931837-07-3

Volume 43—Fly Ash and Coal Conversion By-Products: Characterization, Utilization, and Disposal I, G. J. McCarthy, R. J. Lauf, 1985, ISBN 0-931837-08-1

Volume 44—Scientific Basis for Nuclear Waste Management VIII, C. M. Jantzen, J. A. Stone, R. C. Ewing, 1985, ISBN 0-931837-09-X

Volume 45—Ion Beam Processes in Advanced Electronic Materials and Device Technology, B. R. Appleton, F. H. Eisen, T. W. Sigmon, 1985, ISBN 0-931837-10-3

Volume 46—Microscopic Identification of Electronic Defects in Semiconductors, N. M. Johnson, S. G. Bishop, G. D. Watkins, 1985, ISBN 0-931837-11-1

MATERIALS RESEARCH SOCIETY SYMPOSIA PROCEEDINGS

Volume 93—Materials Modification and Growth Using Ion Beams, U. Gibson, A. E. White, P. P. Pronko, 1987, ISBN 0-931837-60-X

Volume 94—Initial Stages of Epitaxial Growth, R. Hull, J. M. Gibson, David A. Smith, 1987, ISBN 0-931837-61-8

Volume 95—Amorphous Silicon Semiconductors—Pure and Hydrogenated, A. Madan, M. Thompson, D. Adler, Y. Hamakawa, 1987, ISBN 0-931837-62-6

Volume 96—Permanent Magnet Materials, S. G. Sankar, J. F. Herbst, N. C. Koon, 1987, ISBN 0-931837-63-4

Volume 97—Novel Refractory Semiconductors, D. Emin, T. Aselage, C. Wood, 1987, ISBN 0-931837-64-2

Volume 98—Plasma Processing and Synthesis of Materials, D. Apelian, J. Szekely, 1987, ISBN 0-931837-65-0

MATERIALS RESEARCH SOCIETY CONFERENCE PROCEEDINGS

VLSI-I—Tungsten and Other Refractory Metals for VLSI Applications, R. S. Blewer, 1986; ISSN: 0886-7860; ISBN: 0-931837-32-4

VLSI-II—Tungsten and Other Refractory Metals for VLSI Applications II, E.K. Broadbent, 1987; ISSN: 0886-7860; ISBN: 0-931837-66-9

TMC—Ternary and Multinary Compounds, S. Deb, A. Zunger, 1987; ISBN:0-931837-57-x

MATERIALS RESEARCH SOCIETY SYMPOSIA PROCEEDINGS VOLUME 96

High Performance Permanent Magnet Materials

Symposium held April 23 and 24, 1987, Anaheim,California, U.S.A.

EDITORS:

S. G. Sankar
Carnegie-Mellon University, Pittsburgh, Pennsylvania, U.S.A.

J. F. Herbst
General Motors Research Laboratories, Warren, Michigan, U.S.A.

N. C. Koon
Naval Research Laboratory, Washington, DC, U.S.A.

MRS MATERIALS RESEARCH SOCIETY
Pittsburgh, Pennsylvania

This work was supported in part by the Air Force Office of Scientific Research, Air Force Systems Command, USAF, under Grant Number AFOSR 85-0355.

Published by:

Materials Research Society
9800 McKnight Road, Suite 327
Pittsburgh, Pennsylvania 15237
Telephone (412) 367-3003

Library of Congress Cataloging in Publication Data

High performance permanent magnet materials.

   (Materials Research Society symposia proceedings, ISSN 0272-9172 ; v. 96)
   Includes bibliographies.
   1. Magnets, Permanent—Congresses.   2. Magnetic alloys—Congresses.
   3. Magnetic materials—Congresses.   I. Sankar, S. G.   II. Herbst, J. F.   III. Koon,
   N.C.   IV. Materials Research Society.   V. Series.
QC757.9.H54   1987   620.1'1297   87-20371
ISBN 0-931837-63-4

Manufactured in the United States of America

Manufactured by Publishers Choice Book Mfg. Co.
Mars, Pennsylvania 16046

Dedication

René Pauthenet
(1925-1987)

Dedication

René Pauthenet was born in 1925 in Jura, France. He obtained an Engineering Diploma in 1948 from the Institut Polytechnique de Grenoble and began his professional career at C.N.R.S., the French National Center of Scientific Research. He was awarded the Doctor of Engineering in 1951 for his thesis on the magnetic properties of spinel ferrites. His study constituted the first experimental proof of Néel's theory of ferrimagnetism.

In 1957, Pauthenet was awarded the French "Docteur d'Etat" for his famous thesis on the rare-earth iron garnets. This "classical" study won him the 1970 Grand Award of the French Academy of Sciences, which he shared with E.F. Bertaut and J. Forrat. His very precise measurements of the weak ferromagnetism in a single crystal of hematite, $\alpha-Fe_2O_3$, are the basis of Dzialoshinski's interpretation of a new type of interaction, the vectorial Dzialoshinski-Moriya interaction.

He became a professor in 1961. In 1971, he was named Director of the National School of Electrical Engineering and in 1976, Vice-Chairman of the Institut Polytechnique de Grenoble. He was instrumental in establishing the Service National de Champs Intenses (S.N.C.I.) and became the first director, a post he held until 1980. He developed the collaboration contract between S.N.C.I. and the "Max Planck Gesellschaft" and in 1979, initiated the hybrid magnet project common to the French S.N.C.I. and the German Hochfeldlab.

Professor Pauthenet is the author of many publications covering such topics as magnetic properties and structures of compounds of oxides, sulfides and alloys of 3d transition elements, with rare earths and actinides; the experimental check of Neel's theory of the superparamagnetism of fine size powders; permanent magnets, and magnetic properties under high fields. Recently, his interest focused on plasma confinement and optimizing the configuration of permanent magnets and he contributed to a new type of stripped ion source, the "micromafios". His last patent was on the use of permanent magnets in NMR imaging. For his scientific contributions, he was awarded the silver medal of the C.N.R.S. and the Blondel medal in 1961.

He was responsible for organizing several national and international conferences and co-edited the Journal of Magnetism and Magnetic Materials. He was a member of the Administration Council of Ugimag, the largest French producer of permanent magnets, and since the end of 1985 served as Chairman of C.E.A.M., the Concerted Action on Permanent Magnets of the European Economic Community. He was Chevalier du Merite National and Chevalier de la Legion d'Honneur.

Professor Pauthenet is survived by his wife and three sons. We have lost a great scientist, a friend.

Contents

*Invited Paper

*Invited Paper

Preface

High Performance Permanent Magnet Materials play a key role in the advancement of modern technology. The recent discovery of a new class of permanent magnet materials using relatively abundant light rare earth elements with iron and boron has generated a great deal of excitement in scientific and technological communities. Advances in the synthesis of permanent magnet materials, the design of new magnet structures, and utility of these magnets in free electron lasers for medical applications, for strategic defense research, and in the development of NMR imaging units are expected to enhance the pace of research in this important materials field. The possibilities of designing several novel hybrid structures employing permanent magnets coupled with the newly discovered high T_c oxide superconductors will provide innumerable scientific challenges and technological opportunities in the next few years.

In an effort to assess the state-of-the-art in this rapidly growing subject, a two-day symposium was organized as a part of the 1987 Spring Meeting of the Materials Research Society (MRS) at Anaheim, California. The symposium covered topics of interdisciplinary nature, ranging from fundamental aspects, such as alloy synthesis, characterization and crystal field effects, through technical issues such as magnet fabrication, coercivity mechanism and magnet design considerations. Applications such as nuclear magnet resonance imaging and free electron lasers were also addressed.

Eight invited papers and seventeen contributed papers were presented in four sessions at the symposium. Participants profited enormously from the exchange of ideas.

This book on High Performance Permanent Magnet Materials should serve as a valuable resource of information and reference to students and researchers who wish to update their knowledge with the most recent developments.

S.G. Sankar
Carnegie Mellon University

J.F. Herbst   N.C. Koon
General Motors Research Laboratories   Naval Research Laboratory

June 1987

Acknowledgments

I wish to extend appreciation for the support and coopera-
tion of co-chairs Jan F. Herbst and Norm C. Koon in organizing
this symposium. It is a great pleasure to acknowledge the
financial support provided by Harold Weinstock, Air Force Office
of Scientific Research, Boiling AFB. I am grateful to Graham
Hubler, Naval Research Laboratory, who served as a Spring Meeting chair, for helping me in many ways with the organization of
the symposium. The staff of MRS, especially Mary Kaufold, has
been very helpful in arranging this symposium and in preparing
this proceedings.

A large measure of credit for the success of the symposium
goes to the invited speakers and the session chairs. I am
grateful to K.H.J. Buschow, K. Halbach, S. Hirosawa, J.
Laforest, H.A. Leupold, F. Pinkerton, W.E. Wallace, and C.
Williams for delivering very illuminating invited talks, and to
session chairs E.B. Boltich, K.H.J. Buschow, J. Fidler, W.J.
James, S. Hirosawa, J. Laforest, A. Tauber, and H. Weinstock for
running efficient, highly informative sessions.

The symposium noted with sadness the death of Professor
René Pauthenet just a week prior to the meeting. He had made
several significant contributions in the area of permanent
magnets and had been scheduled as an invited speaker. I am
especially grateful to Drs. E.F. Bertaut, J. Laforest and their
colleagues at the Centre National de la Recherche Scientifique,
Grenoble, France for preparing the biography of the late
Professor Pauthenet which appears in this proceedings.

A number of reviewers, who will remain unnamed, have ex-
erted meticulous care to enhance the quality of the papers
published in this volume.

I wish to thank my wife, Veni and our children, Meena and
Kami, for their love and understanding.

S.G. Sankar

Pittsburgh
June 1987

PHYSICAL PROPERTIES OF TERNARY RE-Fe BASE ALLOYS AND THEIR RELATIONSHIP TO PERMANENT MAGNET APPLICATIONS.

K.H.J. BUSCHOW
Philips Research Laboratories
P.O.Box 80000, 5600 JA Eindhoven, The Netherlands

ABSTRACT

Several ternary compounds of tetragonal structure and with the composition $R_2Fe_{14}B$ have found an application as starting materials for permanent magnets. Several physical properties of $R_2Fe_{14}B$ compounds will be discussed in relation to the modification in properties that can be obtained by substituting other elements for R, Fe or B. Also investigated is the extent to which it will be possible to find alternative materials differing in structure and composition from $R_2Fe_{14}B$. For this purpose the results obtained on 3d-rich ternaries based on well-known structure types will be reviewed. Included in this survey will be novel ternary intermetallics made by crystallization from the amorphous state. These compounds, though stable enough for applications, are actually metastable. They are not found in the corresponding equilibrium diagrams.

INTRODUCTION

Permanent magnets based on Nd-Fe-B can be manufactured in two different ways. One of the methods consists of powder metallurgy of cast Nd-Fe-B alloys and includes liquid phase sintering [1[. In the other method the starting material consist of melt-spun Nd-Fe-B which is then hot pressed in one or two different stages [2]. Both manufacturing processes lead to magnet bodies with favourable values of the energy product $(BH)_{max}$. The intrinsic coercive force $_jH_C$ and B_r would hardly justify the enormous scientific activity which has taken place during the last few years [3], since it is not to be expected that there will be any major improvements in these magnetic properties. However, these magnetic materials also have some less favourable properties, such as a relatively low corrosion resistance and a relatively high temperature coefficients of the magnetization and coercive force. This poor thermal stability of the magnets comes into play when these magnets are applied in electrical machines requiring a recoil line coincident with their demagnetization characteristic over a fairly large temperature range. In the new Nd-Fe-B magnets this temperature range is rather limited, so that an increase in temperature leads to irreversible demagnetization by armature reaction fields.

The research efforts devoted to 3d-rich ternary rare earth intermetallics comprise three distinct directions. One of these directions encompasses experimental and theoretical studies of the exchange interaction and crystal field effects in the ternary compounds in relation to similar types of interactions found in the large class of binary rare earth 3d intermetallics. The two other directions aim at finding materials in which the deficiencies mentioned above are less serious or are absent altogether. One way of looking for such materials is to investigate the modifications in properties that can be obtained by substituting one or more elements for the constituent elements in $R_2Fe_{14}B$. In a somewhat more ambitious approach the research activities are focussed on the possibility of finding novel types of materials. In the present paper a survey will be given of the physical properties of various ternary 3d-rich compounds based on rare earths and either boron or other non-magnetic components. The physical properties of these materials and their modifications obtained by substitutions will be discussed in terms of the requirements set by permanent magnet applications. Included in the survey will be the properties of various series of ternary intermetallics of different structure, and materials made by crystallization from the amorphous state. These latter compounds, though stable enough for applications, are actually metastable and are not found in the corresponding equilibrium diagrams.

MODIFICATION OF PROPERTIES OBTAINED BY SUBSTITUTIONS

The problem of changing the intrinsic properties of the $R_2Fe_{14}B$ compounds by substitution of one or more different elements was addressed by quite a number of authors. One of the less favourable properties of the $R_2Fe_{14}B$ compounds is their relatively low Curie temperature and the prime goal of investigations has been to raise the Curie temperature. Results of magnetic measurements of the pseudobinary systems $R_2(Fe_{1-x}Co_x)_{14}B$ were reported by many different authors (see Ref. 3 and references cited therein). There is satisfactory agreement between the data reported in all these investigations. They show that the Curie temperature increases considerably for small Co concentrations while the effect on T_c tends to saturate at the highest Co concentrations. This behaviour is partly due to a preferential site occupation of the Co(Fe) atoms. Experimental evidence for this can be obtained relatively easily by means of ^{57}Fe Mössbauer spectroscopy and neutron diffraction [4,5]. It follows from these measurements (see for instance Fig. 1) that the Fe atoms show a pronounced preference for occupying the $8j_2$ site. Moreover, there are indications from Mössbauer spectroscopy that the Fe atoms are reluctant to occupy the $16k_2$ site. This means that there is a strong preference of Co to occupy the $16k_2$ site when Co is substituted into the Fe sites of $Nd_2Fe_{14}B$. The $16k_2$ sites are characterized by a relatively small interatomic distance between the 3d atoms. Such small interatomic distances are known to favour an antiferromagnetic interaction which may be responsible for the relatively low Curie temperatures of the $R_2Fe_{14}B$ compound. The preferential substitution of Co into the $16k_2$ sites may therefore explain why T_c shows a relatively strong initial increase with x in the series $R_2Fe_{14-x}Co_xB$.

A shallow maximum located at a Co concentration corresponding to $x \approx 0.2$ is frequently observed when the room temperature magnetization of the above mentioned pseudobinary compounds is plotted versus x. This maximum may originate in part from the increase of T_c with x. But it may also be indicative of a maximum in the concentration dependence of the 3d sublattice magnetization. In favour of the latter interpretation are results of measurements of the magnetization made at 4.2 K, in which similar maxima were observed [6,7]. However, not all of the investigations cited agree as to the occurrence of a maximum in the low-temperature concentration dependence of the magnetization [8]. Recent high-field measurements made on aligned powders of $Pr_2(Fe_{1-x}Co_x)_{14}B$ at 4.2K do not point to the presence of a maximum [9].

Spin polarized electronic structure calculations were made by Zong-Quan Gu and

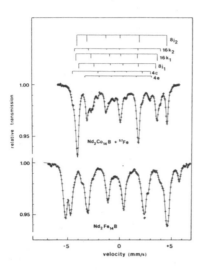

Figure 1.
^{57}Fe *Mössbauer Spectrum in* $Nd_2Co_{14}B$ *doped with* ^{57}Fe *(top part) and in* $Nd_2Fe_{14}B$ *(bottom part). Note the strongly increased intensity of the* $8j_2$ *site in the former compared to the latter* [4].

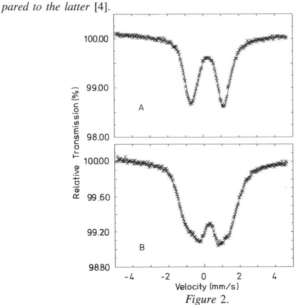

Figure 2.
^{155}Gd *Mössbauer spectrum measured in magnetically aligned* $Gd_2Co_{14}B$. *(A) Measurement without external magnetic field. (B) Measurement with an external field of 7T. Taken from Ref.* [11].

Chin, who also investigated the effects of Co substitution into the various Fe sites in this compound [10]. It follows from these band structure calculations that the moments of the (unsubstituted) Fe atoms are almost unchanged and that the Co moments in $Y_2(Fe_{1-x}Co_x)_{14}B$ are substantially smaller than those in $Y_2Co_{14}B$. However, the moment of Co when residing at the 4c site (3.28 μ_B in $Y_2Co_{14}B$) is equal to 2.14 μ_B which is much larger than 1.60 μ_B calculated for Fe atoms when occupying this site. The authors argue that the large differences in local moments found by them may be in concord with a maximum in the concentration dependence of the saturation magnetization only if the Co substitution proceeds by means of preferential site occupation.

The effect of Co substitution has a beneficial effect on T_c and on the temperature coefficient of the magnetization. However, the temperature coefficient of the coercive force may be of equal importance in permanent magnet applications. Since the coercive force is strongly related to the magnetocrystalline anisotropy, several authors have included in their investigation the effect of Co substitution on the anisotropy. This anisotropy originates from crystal field effects. Conclusive information regarding crystal field parameters can be obtained from [155]Gd Mössbauer spectroscopy. The [155]Gd Mössbauer spectrum of $Gd_2Co_{14}B$ is shown in Fig. 2 together with the best fit to the spectrum [11]. From the corresponding values of the electric field gradient the second order parameters of crystal field experienced by the 4f electrons may be obtained. The second order parameters found for $Gd_2Co_{14}B$ can be compared with those obtained for $Gd_2Fe_{14}B$ in Table I.

Table I.

Second-order crystal field parameters derived from [155]Gd Mössbauer spectroscopy on $Gd_2Co_{14}B$ (A). The values given under B are those obtained by Boge et al. in $Gd_2Fe_{14}B$ (Ref. 12), while those given in column C pertain to a refit of the $Gd_2Fe_{14}B$ powder data, as discussed in Ref. 11. All values are given in units of K/a^2.

R-site	A		B		C	
	A_2^0	$|A_2^2|$	A_2^0	$|A_2^2|$	A_2^0	$|A_2^2|$
(1)	580	354	680	414	570	234
(2)	406	800	661	1298	525	1030

The results given in Table I suggest that the rare earth sublattice anisotropy in $R_2Co_{14}B$ is almost of the same magnitude as that in the corresponding compound of the series $R_2Fe_{14}B$. This is true at temperatures higher than room temperature in particular, where higher order crystal field terms can be neglected.

Experimentally it is found that the room temperature anisotropy field (H_A) in $Nd_2Co_{14}B$ is substantially lower than that of $Nd_2Fe_{14}B$ [13-15]. By contrast, the room temperature anisotropy field of $Pr_2Co_{14}B$ is considerably larger than its Fe counterpart [7-9]. These facts and also the occurrence of minima in the concentration dependence of H_A for $Nd_2(Fe_{1-x}Co_x{14}B$ and $Pr_2(Fe_{1-x}Co_x)_{14}B$ (see Fig. 3) are difficult to understand. Whereas the room temperature anisotropy is mainly determined by K_1, it

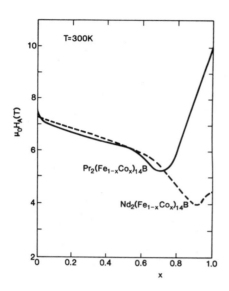

Figure 3.
Concentration dependence of the room temperature anisotropy field in
$Nd_2(Fe_{1-x}Co_x)_{14}B$ *and* $Pr_2(Fe_{1-x}Co_x)_{14}B$. *Taken from Ref.* [9].

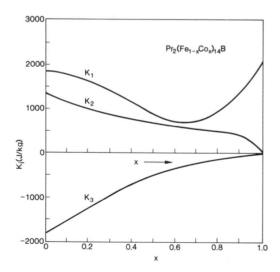

Figure 4.
Concentration dependence of the anisotropy constants K_n at 4K in $Pr_2(Fe_{1-x}Co_x)_{14}B$.
Taken from Ref. [9].

becomes necessary to consider higher order terms when going to lower temperatures. The concentration dependence of K_1, K_2 and K_3 for $Pr_2(Fe_{1-x}Co_x)_{14}B$ is shown in Fig. 4 . The minimum of K_1 still corresponds to the minimum of H_A shown for this series in Fig. 3.

The slightly reduced values of the room temperature anisotropy fields when relatively small amounts of Fe in $R_2Fe_{14}B$ are replaced by Co (see Fig. 3) are not the only reason why Co substitution does not lead to improved hard magnetic properties. This is true for the temperature coefficient of the coercive force in particular. A probable reason for this is that not only H_A itself, but also its temperature dependence, becomes less favourable upon Co substitution. This may be inferred already from a comparison of the temperature dependences of H_A in the pure ternaries $Nd_2Fe_{14}B$ and $Nd_2Co_{17}B$ shown in Fig. 5, where it can be seen that the slope of the $H_A(T)$ curve for the latter compound around room temperature is much steeper than that of the former. In fact the $H_A(T)$ curve in $Nd_2Co_{14}B$ tends to approach the horizontal axis at a temperature of about 540K which is still far below the corresponding Curie temperature ($T_c = 1007K$). One of the reasons for this behaviour is the anisotropy contribution of the 3d sublattice, the corresponding anisotropy constant K_1 changing from positive to negative when going from $x = 0$ to $x = 1$. As a second reason one might mention the R-3d exchange interaction which in $R_2Co_{14}B$ is lower than in $R_2Fe_{14}B$.

Considerable effort was devoted to investigations dealing with the partial replacement of the rare earth component in $R_2Fe_{14}B$ by a different R element or by Th. Since there are two crystallographically nonequivalent rare earth sites in $R_2Fe_{14}B$, there exists the possibility that the substituted element is not distributed at random, but shows a preference for one of the two sites. Experimental evidence for such a site preference was obtained by Yelon et al. [16], using neutron diffraction data of the compound NdDyFe_{14}B to show that the larger Nd atoms tend to prefer to occupy the 4g sites, the smaller Dy atoms having a tendency to occupy the 4f sites. The purpose of investigating the magnetic properties of compounds of the type $(R_{1-x}R'_x)_2Fe_{14}B$ has been two-fold. One type of investigation focussed on optimizing the hard magnetic properties of these materials, whereas another type dealt primarily with fundamental aspects. Suitable substitutions of the R component can be expected to lead to changes in the easy magnetization direction, the spin reorientation temperature and the Curie temperature, from which conclusive information would be obtained regarding the exchange interactions and the crystal field interaction in this class of compounds.

With regard to improving the hard magnetic properties of $Nd_2Fe_{14}B$, interesting results were obtained by substitution of either Tb or Dy into the Nd sites. Results for $Nd_{2-x}Tb_xFe_{14}B$ are shown in Fig. 6 [17]. The anisotropic field shows initially a considerable increase with Tb concentration, while the decrease in magnetization is relatively modest. This offers the possibility of improving the temperature coefficient of the intrinsic coercivity of permanent magnets without too much lowering of the energy product. A similar increase of H_A is found in $Nd_{2-x}Dy_xFe_{14}B$. Plots of the intrinsic coercive fields ($_jH_C$) versus the corresponding anisotropy fields (H_A) of two sintered magnet bodies can be compared in Fig. 7 [18]. The lower curve shows the results for a normal NdFeB magnet. The upper curve was obtained from measurements on a Dy-doped NdFeB magnet. In both cases the temperature is taken as an internal parameter. In these plots extended regions can be identified where the experimental values of $_jH_C$ are linearly dependent on the experimental values of H_A. For the undoped NdFeB magnet, this linear behaviour ceases at temperatures only slightly higher

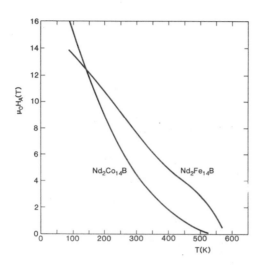

Figure 5.
Temperature dependence of the anisotropy field in $Nd_2Co_{14}B$ and $Nd_2Fe_{14}B$ [15].

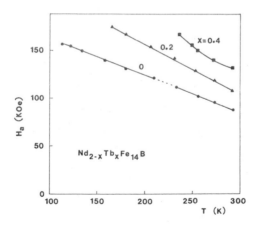

Figure 6.
Temperature dependence of the anisotropy field in various compounds of the type $Nd_{2-x}Tb_xFe_{14}B$ [17].

8

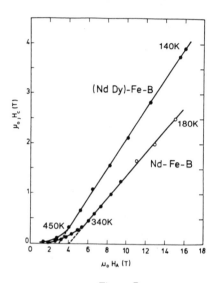

Figure 7.
Intrinsic coercive force versus experimentally determined values of the anisotropy field
in two sintered magnet bodies [18].

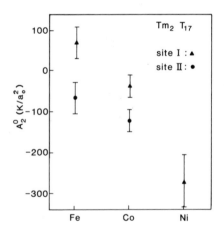

Figure 8.
Second order crystal field parameter A_2^0 derived by means of ^{169}Tm Mössbauer spec-
troscopy in Tm_2T_{17} compounds (T = Fe, Co or Ni). After Ref. [23].

than room temperature, but it extends to about 450K in the case of the Dy-doped magnet. The linear relationship between H_A and $_JH_C$ can be written as

$$_JH_C = s \cdot H_A + H_{eff}$$

where H_{eff} represents the intercept with the horizontal axis and s the slope of the linear part of the curve. The field H_{eff} may be considered as an effective demagnetizing field which depends on the magnetization. For this reason H_{eff} decreases strongly at higher temperatures. This means that the linear behaviour extends to much higher temperatures when $_JH_C$ is plotted versus $sH_A + H_{eff}$ and H_{eff} is taken to be proportional to the magnetization. On the other hand, the deviations from linear behaviour at high temperatures observed for both magnets may also be associated with a change in the prevailing coercivity mechanism from nucleation hardening (low temperatures) to pinning hardening (high temperature) [16,19].

A third type of substitution presents itself in the form of the B site in $Nd_2Fe_{14}B$. Successful attempts to replace B by other elements have been reported only for C. The substitution of C for B in $R_2Fe_{14}B$ presents, however, considerable difficulties from the metallurgical point of view. Extensive studies of the possibility of this type of substitution were reported recently by Liu et al. [20]. The room temperature anisotropy field in $Nd_2Fe_{14}B_{1-x}C_x$ increases with x, but it is accompanied by a slight decrease with x of both the Curie temperature and saturation magnetization [21,22].

TERNARY SYSTEMS

The simplest approach to finding new ternary phases is to start from binary intermetallic compounds in which at least one of the constituent elements has two or more crystallographically nonequivalent sites. Preferential occupation of one of these sites by a third element X may then lead to a ternary compound. From the viewpoint of permanent magnet applications the interest is primarily in compounds of relatively high Fe concentration which have crystal structures of sufficiently low symmetry to make uniaxial anisotropy possible. Another approach is to look for 3d-rich ternary rare earth compounds existing for Ni, Co or Mn and to replace all or at least a substantial fraction of the 3d element by Fe.

Guided by the above consideration, the choice is rather limited and remains virtually confined to four classes of compounds with one of the following structure types: Th_2Ni_{17} or Th_2Zn_{17}, $BaCd_{11}$, $ThMn_{12}$ or $SrNi_{12}B_6$.

$R_2(T,X)_{17}$. In the recent past many investigations were devoted to studying the magnetic properties of the Sm_2Co_{17} compound and its derivatives. In fact, it is this compound which is the main constituent of permanent magnet materials of the type $Sm(Co,X)_7$ where X is a mixture of Fe,Cu and Zr. The manufacture of the $Sm(Co,X)_7$ type of magnet is far from simple. Also the Co concentration is fairly high which makes this type of magnet material rather expensive. Attempts were, therefore, made by several authors to increase the Fe concentration or to start from Sm_2Fe_{17} and improve the properties of this material by suitable substitution. Here we recall that Sm_2Fe_{17} has the disadvantage of a low Curie temperature and an easy magnetization direction which, unlike Sm_2Co_{17}, is perpendicular to the c-direction. Several investigations have shown that it is possible to increase the Curie temperature by suitable substitution to a sufficiently high value. However, in view of the strong anisotropy found in R_2Co_{17} and R_2Ni_{17}, it is somewhat surprising that the positive Sm sublattice anisotropy was not strong enough in all these cases to overcompensate the negative Fe sublattice anisotropy to a sufficiently high level. Recent experimental studies of the crystal field induced R sublattice anisotropy made by means of Tm Mössbauer spectroscopy in Tm_2T_{17} (T = Ni, Co, Fe) showed that the crystal field interaction changes markedly when going from Ni to Fe [23]. It can be derived from the results shown in Fig. 8 that the value of the second order crystal field parameter A for both rare earth sites is strongly negative only in R_2Ni_{17}, favouring an easy magnetization direction parallel to the c-axis for R element with $\alpha_J > 0$. However, this property is considerably reduced in R_2Co_{17}, while in R_2Fe_{17} the effect of both rare earth sites tends to cancel each other out. For this reason one obtains the impression that efforts to obtain suitable permanent magnet materials based on R_2Fe_{17} are less likely to be successful.

$R(T,X)_{11}$. Little investigation has been carried out so far on ternary compounds with a crystal structure based on the tetragonal $BaCd_{11}$ type. Chevalier et al. studied the crystallographic and magnetic properties of silicides of the type $Nd(Co_{1-x}Fe_x)_9Si_2$ and found that single phase materials of the $NdCo_2Si_2$ structure can be obtained for Fe concentrations within the range $O \leq x \leq 0.55$ [24]. Curie temperatures up to 560 were reported, but the room temperature anisotropy fields were found to be comparatively small (15 kOe). In view of the fact that the concentration of the (expensive) Co is still fairly high in these materials, they do not look very promising for permanent magnet applications.

$R(T,X)_{12}$. Compounds of the type RMn_{12} have an almost vanishing Mn sublattice magnetization owing to an antiparallel coupling of the Mn moments. Attempts to break this antiferromagnetic coupling by substitutions have not been successful in so far as the overall sublattice magnetization remained relatively small. More promising are Fe- base compounds of the type $RFe_{10}V_2$. These are ternary compounds of the $YNi_{10}Si_2$ structure type. As can be seen from the results shown in Fig. 9, their Curie temperatures are comparatively high and the same can be said of the saturation magnetization of $YFe_{10}V_2$ and those $RFe_{10}V_2$ compounds, in which R is a light rare earth element [25]. Because of the antiparallel coupling between the moments of the heavy rare earths and Fe, the magnetization in the corresponding compounds is much lower than in those mentioned above. The Fe sublattice magnetization contributes favourably to the anisotropy, since it prefers a magnetization direction parallel to the c-axis. There are indications that the R sublattice anisotropy is fairly small in these ternaries. In the near future an investigation will be made of the degree to which the occurrence of $RFe_{10}X_2$ compounds extends to X elements which are different from V.

$RT_{12}B_6$. This structure type is not found in binary compounds. Up to now only ternary compounds in which T represents Co or Ni were reported in the literature. Substitution of Fe for Co in $RCo_{12}B_6$ is only possible in a limited concentration range. In the series $Nd(Co_{1-x}Fe_x)_{12}B$ the hexagonal $SrNi_{12}B_6$ structure was observed only for $0 \leq x \leq 0.6$. Because of low values of the saturation magnetization and the fact that the Curie temperatures were below room temperature, these materials are found to be unsuitable for permanent magnet purposes.

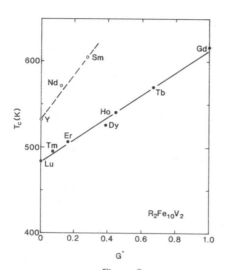

Figure 9.
Curie temperature versus the corresponding reduced de Gennes factor in tetragonal compound of the type $R_2Fe_{10}V_2$.

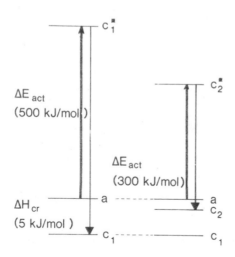

Figure 10.
Energy level diagram showing the relative energies of the amorphous state (*a*), the stable crystalline state (*c_1*) and the metastable crystalline state (*c_2*)

METASTABLE MATERIALS

Metastable intermetallic compounds can be obtained from amorphous alloys after partial crystallization. Two exotherms are usually found when the corresponding alloys are heated in a DSC unit. The first exotherm originates from the crystallization of the metastable compounds, the second exotherm corresponds to the transformation into the stable equilibrium phases. The metastable compounds which we found were obtained by heating to temperatures close to the first exotherm and then cooling to room temperature.

The underlying principle is the following. In most ternary systems a large number of widely different ternary compounds may be visualized to exist with a free energy which falls slightly above the plane, tangent to the free energy surfaces associated with three stable compounds or elements in the corresponding ternary system. Because of their higher free energy, these compounds are not equilibrium compounds. They will, therefore, not be observed experimentally in alloys cast normally. However, it is possible in some cases to reach the metastable energy states associated with these latter compounds by starting from amorphous alloys for which the free energy is even higher. It may be derived from the schematic representation of level schemes shown in Fig. 10 that the prerequisites for the formation of a metastable compound are (i) a free energy which lies in between that of the amorphous alloy and the stable crystalline phase, and (ii) an activation energy for crystallization that is smaller than the activation energy for crystallization into the stable crystalline materials. Only in the latter case may one expect that the transformation rate into the metastable phase predominates. If melt-spinning is used to obtain the amorphous alloys, one is faced with a third requirement, namely (iii) that the concentration considered should fall into the easy glass-forming region. The latter regions are generally located in those parts of the ternary diagrams where the temperature of the liquidus surface is relatively low. In view of the three restrictions mentioned, one may wonder whether there is much probability of finding novel ternary compounds in this way. Preliminary results obtained recently are quite encouraging in this respect. As may be seen from Table II we found three types of novel R-Fe-B ternaries [26,27]. Two of these have a novel ternary structure type. The compound $NdFe_{12}B_6$ has the same structure type as reported in the literature for $SrNi_{12}B_6$. Unfortunately, neither of the compounds found so far qualifies as starting material for permanent magnets. The two cubic compounds are unsuitable with regard to the development of uniaxial anistropy, while the hexagonal compound has a Curie temperature that is too low for practical applications. Nevertheless, the magnetic properties of these materials may be helpful in understanding the exchange interaction in related materials.

Table II

Metastable ternary R-Fe-B compounds.

Compound	Sym.	lattice constants (nm)		Curie temp. (K)
$Pr_2Fe_{23}B_3$	cub.	a = 1.418		$T_c = 644$
$Nd_2Fe_{23}B_3$	cub.	a = 1.419		$T_c = 655$
$Sm_2Fe_{23}B_3$	cub.	a = 1.406		$T_c = 663$
$Gd_2Fe_{23}B_3$	cub.	a = 1.411		$T_c = 689$
$NdFe_{12}B_6$	hex.	a = 0.9605	c = 0.7545	$T_c = 230$
$Y_3Fe_{62}B_{14}$	cub.	a = 1.235		$T_c = 510$

The magnetic properties of the two novel ternary compounds $Nd_2Fe_{23}B_3$ and $NdFe_{12}B_6$ can be compared with those of the other ternaries in the Nd-Fe-B system in Table III, where the various compounds have been listed in order of decreasing iron concentration. Decreasing iron concentration is seen to lower the saturation magnetization. The iron atoms do not carry a magnetic moment in the two compounds of lowest iron content.

Table III

Survey of magnetic properties of ternary compounds in the Nd-Fe-B system.

Compound	at% Fe	$T_c(K)$	$\sigma_s(Am^2/kg)$	$M_s(\mu_B/F \cdot U)$
$Nd_2Fe_{14}B$	82.3	588	194	37.6
$Nd_2Fe_{23}B_3$	82.1	655	171	49.1
$NdFe_{12}B_6$	63.1	230	125	19.7
$Nd_{1.1}Fe_4B_4$	43.0	10	30	2.2
Nd_2FeB_3	16.7	64	42	2.8

It can be seen from the data shown in Fig. 11 that the Curie temperatures behave generally in accordance with the iron concentration (the compound Nd_2FeB_3 was not taken into consideration, since here the value of T_c reflects the Nd-Nd interaction). Included in Fig. 11 is the value of T_c for Nd_2Fe_{17} (triangle) which is seen to be relatively low. The presence of iron sites, for which the nearest neighbour Fe-Fe separation is relatively short, is frequently considered as responsible for this too-low value. Such iron sites favour an antiparallel moment coupling which reduces the value of the Curie temperature. Iron sites of this kind are, however, present in all three ternary compounds in which iron is magnetic and determine the value of T_c. In table IV we have listed the types of sites (Wyckoff notation) that have iron neighbours at relatively small Fe-Fe

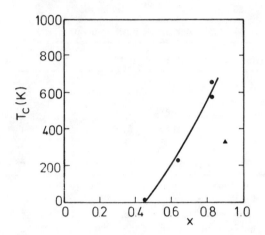

Figure 11.
Curie temperature versus Fe concentration in various Nd-Fe-B compounds (●) *and*
Nd₂Fe₁₇ (▲).

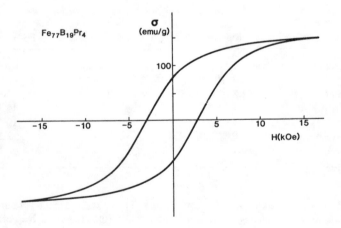

Figure 12.
Hysteresis loop observed for partially crystalline alloys of Fe₃B doped with a few
percent of Pr.

distances, together with the number of these neighbours and their distances. It follows from these data that the fraction of iron sites having an unfavourable nearest neighbour iron distance is lowest in Nd_2Fe_{17} (12%) and highest in $NdFe_{12}B_6$ (50%). In general, the correlation between low T_c values and the presence of short Fe-Fe distances does not appear to be very convincing.

Finally, it may be noted that the preparation of metastable compounds via the amorphous state or via sufficiently rapid quenching of cast alloys is not restricted to ternary systems. In fact, several authors have investigated the possibility of obtaining improved *soft* magnetic materials based on Fe_3B, this compound is also metastable at room temperature. This same phase can also be obtained after heat treatment of amorphous alloys consisting of the approximate composition Fe_3B, but containing a few percent of rare earths as an additive. Surprisingly, the result is a far from weak magnetic material. As seen from fig. 12, a considerable hysteresis has developed. The relatively high remanence and Curie temperature in conjunction with its extremely low R content make this material a possible candidate for applications in bonded isotropic magnets.

Table IV

Iron sites in three ternary Nd-Fe-B compounds that have one or two nearest neighbour iron atoms at relatively short distances.

Compound	Site	Number of nearest neighbour iron atoms	Nearest neighbour distances (pm)	Fraction of nearest neighbour iron atoms		T_c(K)
$NdFe_{12}B_6$	16(h)	2	238	0.50	0.50	230
$Nd_2Fe_{14}B$	16(k$_2$)	1	240	0.29 ⎫	0.43	588
	8(j$_1$)	1	240	0.14 ⎬		
$Nd_2Fe_{23}B_3$	24(d)	2	240	0.13 ⎫	0.39	655
	48(e)	1	240	0.26 ⎬		
Nd_2Fe_{17}	6(c)	1	242	0.12	0.12	327

References

[1[M. Sagawa, S. Fujimura, M. Togawa, H. Yamamoto and Y. Matsura, J. Appl. Phys. **55**, 2083 (1984).

[2] R.W. Lee, E.G. Brewer and N.A. Schafel, Proc. Intermag. Conf. St. Paul Minnesota, April 1985.

[3] K.H.J. Buschow, Mat. Sci. Repts. **1**, (1986) 3.

[4] H.M. Van Noort and K.H.J. Buschow, J. Less-Common Met. **113**, (1986) L9.

[5] J.F. Herbst and W.B. Yellon, J. Appl. Phys. **60**, (1986) 4224.

[6] M.Q. Huang, E.B. Boltich, W.E. Wallace and E. Oswald, J. Magn. Magn. Mater. **60**, (1986) 270.

16

[7] A.T. Pedziwiatr, S.Y. Jiang and W.E. Wallace, J. Magn. Magn. Mater. **62**, (1980) 29.

[8] F. Bolzoni, J.M.D. Coey, J. Gavigan, D. Givord, O. Moze, L. Pareti and T. Viadieu, J. Magn. Magn. Mater. **65**, (1987) 123.

[9] R. Grössinger, R. Krewenka, H.R. Kirchmayr, S. Sinnema, Yang-Fu-Ming, Huang Ying-kai, F.R. de de Boer and K.H.J. Buschow, J. Less-Common Met. **132** (1987).

[10 Zong-Quan Gu and W.Y. Ching, J. Appl. Phys. (at the press).

[11] H.H.A. Smit, R.C. Thiel and K.H.J. Buschow, Physica (1987).

[12] M. Boge, G. Czjzek, D. Givord, C. Jeandey, H.S. Li and J.L. Oddou, J. Phys. F. **16**, (1986) L67.

[13] C. Abache and H. Oesterreicher, J. Appl. Phys. **60**, (1986) 1114.

[14] S. Hirosawa, K. Tokuhara, H. Yamamoto, S. Fujimura, M. Sagawa and H. Yamamuchi, J. Appl. Phys. **59**, (1986) 873.

[15 R. Grössinger, R. Krewenka, X.K. Sun, R. Eibler, H.R. Kirchmayr and K.H.J. Buschow, J. Less-Comm. Met. **124**, (1986) 165.

[16] W.B. Yelon, B. Foley, C. Abache and H. Oesterreicher, J. Appl. Phys. **60**, (1986) 2982.

[17] L. Pareti, F. Bolzoni, M. Solzi and K.H.J. Buschow, J. Less-Comm. Met. **132** (1987) L5.

[18] R. Grössinger, R. Krewenka, H.R. Kirchmayr, P. Naastepad and K.H.J. Buschow J. Less-Common Met. (to be published).

[19] K.D. Durst and H. Kronmüller, J. Magn. Magn. Mat. (at the press).

[20] N.C. Liu, H.H. Stadelmaier and G. Schneider, (to be published).

[21] N.C. Liu and H.H. Stadelmaier, Mat. Lett. (at the press).

[22] P. Bolzoni, F. Leccabue, L. Pareti and J.L. Sanchez, J. Phys. (Paris) **46**, (1985) C6-305.

[23] P.C.M. Gubbens, A.M. van der Kraan, J.J. van Loef and K.H.J. Buschow, J. Magn. Magn. Mat. (at the press).

[24] B. Chevalier, G. Gurov, L. Fournes and J. Etourneau, J. Chem. Res. (to be published).

[25] F.R. de Boer, Huang Ying-kai, D.B. de Mooij and K.H.J. Buschow, J. Less Comm. Met. (to be published).

[26] D.B. de Mooij, J.L.C. Daams and K.H.J. Buschow, Philips J. Research (at the press).

[27] K.H.J. Buschow, D.B. de Mooij and H.M. van Noort, J. Less-Common Met. **125**, (1986) 135.

STUDIES OF SUBSTITUTED $R_2T_{14}B$ AND R_2Co_{17} SYSTEMS
(T = Fe or Co)

W. E. WALLACE, A. T. PEDZIWIATR, E. B. BOLTICH, H. KEVIN SMITH, S. Y. JIANG, S. G. SANKAR AND E. OSWALD
Mellon Institute and Magnetics Technology Center, Carnegie-Mellon University, Pittsburgh, PA 15213

ABSTRACT

Certain R_2Co_{17} and $R_2T_{14}B$ systems (R = a rare earth, T = Fe or Co) are of significance for use in the fabrication of high energy magnets. Improvements in these systems as high energy magnet materials are effected by partially replacing Co or T by appropriate d transition metals. A description is given of the magnetic behavior of a number of systems based on R_2Co_{17} and $R_2T_{14}B$. Also, an account is given of recent work dealing with the systems RCo_4B and $PrCo_{4-x}Fe_xB$. The latter materials have large magnetizations and high Curie temperatures. However, they are not useful permanent magnet materials because they exhibit planar anisotropy.

Results of some Auger spectroscopy studies of $R_2Fe_{14}B$ systems (R = Pr,Nd or Er) are presented. This work shows the intergranular region to be rich in B, O and R. This region is undoubtedly non-magnetic, a feature which is most probably of significance in regard to the observed coercivity of $R_2Fe_{14}B$ magnets.

INTRODUCTION

The subject matter of this paper is divided into 4 parts: (1) the magnetic behavior of $Nd_2Fe_{14}B$-based alloys, (2) the magnetic behavior of some R_2Co_{17}-based alloys, (3) the magnetic behavior of RCo_4B and $PrCo_{4-x}Fe_xB$ alloys and (4) a preliminary study of the surface features of $R_2Fe_{14}B$ systems by Auger spectroscopy.

$Nd_2Fe_{14}B$ is now recognized as an important material for fabricating high energy permanent magnets. It is well known that its utility is limited by its low Curie temperature, $\sim 315°C$. A number of alloy studies have been carried out in the author's laboratory and elsewhere to improve the performance of $Nd_2Fe_{14}B$ magnets at elevated temperatures. Some of this work is described in the present paper. A description is also given of the magnetic behavior of alloys based on R_2Co_{17}. Some of these systems are of potential significance as permanent magnet materials, except for their unfavorable anisotropies. Favorable anisotropy can be achieved in some instances if the Co is replaced by Fe or Mn or a combination of these. Alloys of composition $RCo_{4-x}Fe_xB$ have been studied because of their general similarity to the important $R_2Fe_{14}B$ magnetic phases and their potential as high energy magnet materials. Auger studies have been carried out on a number of $R_2Fe_{14}B$ systems to obtain information as to the chemistry of the boundary phases in sintered magnets made from these materials, hopefully leading to an improved understanding of the mechanism of coercivity in these materials.

$R_2Fe_{14}B$- AND $R_2Co_{14}B$-BASED ALLOY SYSTEMS

Many studies have been carried out recently in which substitutions have been made on the various sublattices in $R_2Fe_{14}B$ and $R_2Co_{14}B$ systems. These can be grouped into three categories as follows:

1. Substitutions in the iron sublattice:

    a) $Nd_2Fe_{14-x}M_xB$,   M = Ru,Ga,Si,Co,Mn
    b) $R_2Fe_{14-x}Si_xB$,   R = Y,Nd,Pr,Er
    c) $R_2Fe_{14-x}Mn_xB$,   R = Y,Nd,Pr,Gd
    d) $R_2Fe_{14-x}Co_xB$,   R = Pr,Nd,Dy,Er
    e) $R_2Co_{14}B$,   R = Pr,Nd,Tb,Sm,Gd,Y
    f) $Nd_2Fe_{14-x-y}Co_yAl_xB$

2. Substitutions in the rare earth sublattice:

    a) $Er_{2-x}R_xFe_{14}B$,   R = Y,Th,Ce,Pr,Gd,Sm
    b) $R_{2-x}Th_xFe_{14}B$,   R = Y,Dy,Er

3. Substitution for the boron:

    a) $R_2Fe_{14}C$,   R = Dy and Er.

This work has been briefly described recently in a review paper published in connection with the 3rd International Conference on Physics of Magnetic
Materials held in Szczyrk-Bila, Poland [1]. In the present paper mention will
be made of the two investigations in the list cited above and brief descrip-
tions will be given of (1) the $R_2Co_{14-x}Si_xB$ system and (2) the effect of
hydrogen on the magnetic behavior of $R_2Fe_{14}B$ alloys.

$R_2Fe_{14-x}Si_xB$ and $R_2Co_{14-x}Si_xB$ Alloys

These systems have been studied with R = Pr,Nd or Er for the Fe-containing
system [2,3]. The maximum Si content is about x = 2 for the two systems, resulting in a shrinkage of about 0.5% in the unit cell volume. The composition de-
pendencies of magnetization, T_c, and anisotropy fields for the $Nd_2(Fe,Si)_{14}B$
alloys are shown in Figs. 1-3. The most striking feature of this work is that
T_c increases with increasing Si content. Thus one has the unusual result that
replacing a magnetic constituent, i.e., Fe, with a non-magnetic substituent,
i.e., Si, causes a rise in T_c. This undoubtedly arises because Si replaces Fe
at positions in the lattice where the Fe-Fe distances are abnormally short,
reducing the negative exchange originating in these regions. As expected, insertion of Si for Fe results in a decrease of magnetization (Fig. 1). H_A is
not radically modified when Si replaces Fe in $Nd_2Fe_{14}B$. At low Si concentrations the magnetization decreases according to simple dilution. At higher
concentrations the decrease is more rapid. The overall behavior is rather
similar to that observed in Fe-Si binary alloys [4].

The exchange field, obtained by a molecular field analysis of the
magnetization-temperature behavior of $Er_2Fe_{14}B$, acting within the Er sublattice is $2 \mu_B H \approx 250$ K. This is close to the value for Gd in $GdFe_2$,
$2 \mu_B H \approx 300$ K [5,6]. This indicates that the interaction mechanism is very
similar for the ternary $R_2Fe_{14}B$ and binary RFe_2 intermetallic compounds.
The exchange field acting on Fe is $2 \mu_B H_{exch} \approx 1.3 \times 10^3$ K. Consequently,
the Fe magnetization decreases more slowly than that of the rare earth.

Since the molecular field acting on the rare earth is comparable with
the overall crystal field splitting, it appears that the moment of the rare
earth is not quenched by the action of the crystal field.

Results obtained for the $R_2Co_{14-x}Si_xB$ systems are largely summarized in
Table I and in Figs. 4-6. Magnetization, T_c and anisotropy fields all decline
with increasing Si content. Results for $Y_2(Co,Si)_{14}B$ (see Table II) show that
the Co moment decreases as the Si content of the alloy increases. This is
frequently observed in R-Co systems in which Co is replaced by a non-magnetic
substituent [7]. The decline in Co moment, together with the dilution of Co
is undoubtedly responsible for the very sharp decline in T_c for the quaternary
system.

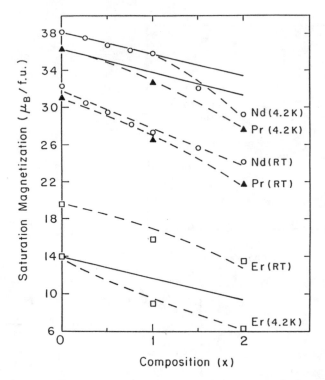

Fig. 1. The composition dependence of magnetization for
$R_2Fe_{14-x}Si_xB$ alloys at 4.2 K and room temperature.
The solid lines indicate simple dilution
behavior.

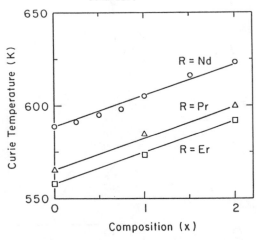

Fig. 2.

The composition dependence of Curie
temperature for $R_2Fe_{14-x}Si_xB$ alloys.

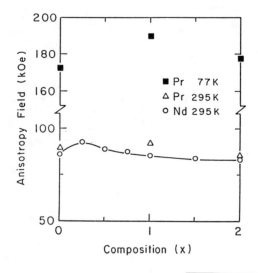

Fig. 3. The composition dependence of anisotropy fields at room temperature and 77 K for $R_2Fe_{14-x}Si_xB$ alloys, R = Pr or Nd.

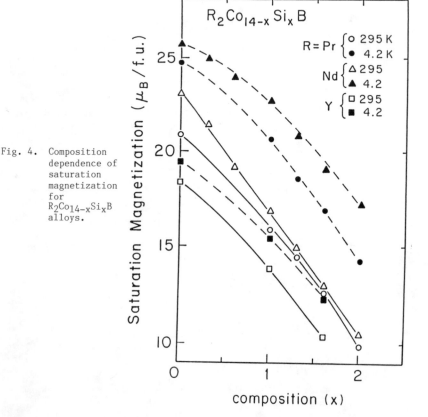

Fig. 4. Composition dependence of saturation magnetization for $R_2Co_{14-x}Si_xB$ alloys.

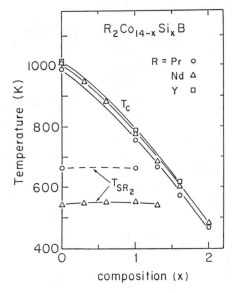

Fig. 5. The composition dependence of the Curie
         temperature, T_c, and spin-reorientation
         temperature, T_{SR_2}.

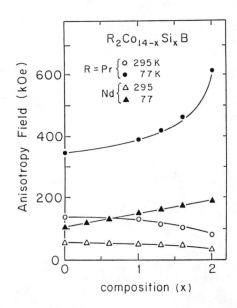

Fig. 6. The composition dependence of anisotropy
         fields.

TABLE I

Crystallographic and Magnetic Data Obtained for $R_2Co_{14-x}Si_xB$ Systems
(R = Pr,Nd,Y)

Composition	a (Å)	c (Å)	T_C (K)	T_{SR_1} (K)	T_{SR_2} (K)	M_s (μ_B/f.u.)		H_A (kOe)	
						295 K	4.2 K	295 K	77 K
R = Pr									
x = 0.0	8.630	11.870	986	–	664	20.9	24.8	137	348
1.0	8.629	11.790	754	–	662	15.9	20.7	131	390
1.3	8.626	11.773	666	–	–	14.5	18.7	113	420
1.6	8.620	11.751	574	–	–	12.6	17.0	102	465
2.0	8.618	11.744	470	–	–	9.9	14.3	82	618
R = Nd									
x = 0.0	8.651	11.871	1004	34	546	23.2	25.8	55	105
0.3	8.646	11.831	950	25	548	21.5	25.0	54	120
0.6	8.639	11.811	278	–	550	19.2	23.9	53	132
1.0	8.630	11.772	773	–	552	16.9	22.7	52	150
1.3	8.628	11.761	682	–	543	15.0	20.9	49	159
1.6	8.624	11.747	601	–	–	13.0	19.1	46	171
2.0	8.621	11.732	487	–	–	10.5	17.3	38	179
R = Y									
x = 0.0	8.581	11.711	1015	–	–	18.4	19.5	–	–
1.0	8.579	11.629	777	–	–	13.9	15.5	–	–
1.6	8.576	11.623	610	–	–	10.3	12.4	–	–

TABLE II

Co Moments in $Y_2Co_{14-x}Si_xB$

x	Average Co Moment (μ_B/Co)
0	1.39
1.0	1.19
1.6	1.00

It is well known that $Nd_2Fe_{14}B$ undergoes a magnetic transformation (often
called a spin reorientation) at about 135 K. Above this, the crystal is uniaxial. Below this, the magnetic moments lie on a cone. $Nd_2Co_{14}B$ undergoes a
similar transformation (cone-to-axis) at about 34 K [8]. In $Nd_2(Co,Si)_{14}B$
alloys this transformation is suppressed, i.e., it occurs at a reduced temperature, as Si content is increased and disappears entirely when x ≥ 1.6. This
happens because Si entry weakens exchange. It is exchange which is responsible
for the occurrence of the transformation.

The transformation at T ≤ 34 K is due to competition between exchange and
the crystal field interaction. $Nd_2Co_{14}B$ exhibits a second magnetic transformation at about 546 K. This is an axis-to-plane transformation with rising temperature. This high temperature transformation is a consequence of competing

rare earth and Co anisotropies. The latter, which is planar, becomes dominant at high temperatures. Because the Co anisotropy is weakened as Co is replaced by Si, the high temperature transformation is expected to occur at high temperatures, and this is indeed observed. For $x > 1.3$ the $Nd_2Co_{14-x}Si_xB$ crystals remain axial up to T_c.

These same physical principles are reflected in the dependence of H_A on x. For both the Pr and Nd systems H_A decreases with Si content at 295 K and increases at 77 K. The rise in H_A at 77 K follows since the rare earth anisotropy is dominant at low temperature. Entry of Si into the lattice weakens the Co anisotropy and makes the overall uniaxial anisotropy stronger.

It is of interest to note that $Y_2Co_{14-x}Si_xB$ exhibits planar anisotropy from 4.2 K to T_c.

Hydrogenated $R_2Fe_{14}B$ and $Pr_2(Fe,Co)_{14}B$ Systems

Rare earth-iron intermetallics are known [9] to be prolific absorbers of hydrogen. The $R_2Fe_{14}B$ systems with R = Y,Ce,Pr,Nd and Sm were examined in regard to their capacity for hydrogen and for the effect of hydrogenation on their magnetic behavior. They were found to absorb 4.4 to 6 atoms of hydrogen per formula unit. The effects of hydrogenation on magnetization and anisotropy for $R_2Fe_{14}B$ systems are indicated in Table III.

In regard to magnetic behavior of the $R_2Fe_{14}B$ systems, hydrogenation has two effects - (1) it increases magnetization and (2) it favors planar anisotropy. For R = Y,Pr and Sm the hydride has planar anisotropy. For R = Ce or Nd the anisotropy is still axial but the anisotropy field is reduced, compared to the parent material.

The temperature at which the magnetic transformation in $Nd_2Fe_{14}B$, alluded to above, occurs is shifted downward some 50 C. This is expected in view of the altered anisotropy of the Fe sublattices.

The effect of hydrogenation on $Pr_2Fe_{14-x}Co_xB$ alloys is illustrated in Table IV. The systems containing Co show a decrease in magnetization upon hydrogenation, which is in contrast with an increase which is noted for

TABLE III

Saturation Moments and Anisotropy Field for $R_2Fe_{14}B$ Systems
and Their Hydrides

Compound	$\mu_s(\mu_B/f.u.)$		H_A (kOe)		Direction of Magnetization	
	300 K	77 K	300 K	77 K	300 K	77 K
$Y_2Fe_{14}B$	27.5	30.4	27.3	21.0	axis	axis
$Y_2Fe_{14}BH_{4.5}$	29.7	32.3	--	--	plane	plane
$Ce_2Fe_{14}B$	24.0	29.4	37.1	36.6	axis	axis
$Ce_2Fe_{14}BH_{4.4}$	29.1	31.6	10.1	--	axis	plane
$Pr_2Fe_{14}B$	31.0	34.8	79.3	172	axis	axis
$Pr_2Fe_{14}BH_{5.0}$	33.9	36.6	--	--	plane	plane
$Nd_2Fe_{14}B$	32.2	36.4	70.7	--	axis	cone
$Nd_2Fe_{14}BH_{4.9}$	34.1	36.9	17.5	--	axis	cone
$Sm_2Fe_{14}B$	28.6	31.1	--	--	plane	plane
$Sm_2Fe_{14}BH_{\sim6}$	30.9	32.7	--	--	plane	plane

TABLE IV

Magnetization at 20 kOe and Anisotropy Field Data for
$Pr_2Fe_{14-x}Co_xB$ and Their Hydrides ($8 \leq x \leq 14$)

Compound	M (μ_B/f.u.)		H_A (kOe)		D.O.M.	
	R.T.	77 K	R.T.	77 K	R.T.	77 K
$Pr_2Fe_6Co_8B$	26.2	28.9	68	170	axis	axis
$Pr_2Fe_6Co_8BH_{3.6}$	25.1	26.4	--	--	plane	plane
$Pr_2Fe_4Co_{10}B$	24.5	27.2	67	187	axis	axis
$Pr_2Fe_4Co_{10}BH_{3.7}$	23.6	25.7	--	--	plane	plane
$Pr_2Fe_4Co_{10}B$	22.7	25.1	84	232	axis	axis
$Pr_2Fe_4Co_{10}BH_{3.3}$	21.0	19.3	--	53	plane	axis
$Pr_2Co_{14}B$	20.9	23.2	137	348	axis	axis
$Pr_2Co_{14}BH_4$	19.9	22.9	53	217	axis	axis

$R_2Fe_{14}B$. The origin of this difference is as yet unclear. It is generally
noted that the moment of Co in this and related systems is sensitive to the
environment, whereas there is no such sensitivity for Fe. For example, Burzo
et al. noted [7] many years ago that the Co moment is quite composition dependent in $GdNi_{2-x}Co_x$ ternaries, increasing with increasing x. As with $R_2Fe_{14}B$
systems, hydrogenation diminishes the tendency for axial anisotropy in $R_2Co_{14}B$
alloys, converting the systems to planar anisotropy in a number of instances.

$Gd_2Fe_{14}C$, $Dy_2Fe_{14}C$ and $Er_2Fe_{14}C$

Studies of the Dy and Er carbides show that they form in the well-known
$Nd_2Fe_{14}B$ tetragonal structure. Abache and Oesterreicher [10] found a similar
structure for $Gd_2Fe_{14}C$. The Curie temperature and magnetizations are lower
than is observed for the corresponding borides (see Figs. 7 and 8). Since the
lattice of the carbides is slightly expanded compared to the borides, the reduction in T_c seems likely not to be a consequence of geometrical factors. Instead, it is probably an electronic effect, the carbon contributing more electrons to the Fe d-band than boron, reducing the Fe magnetization and the
strength of the Fe-Fe exchange.

R_2Co_{17}-BASED ALLOYS

The R_2Co_{17} systems by themselves are not useful in the fabrication of high
energy magnets. They either have weak uniaxial anisotropy (e.g., Sm_2Co_{17}) or
planar anisotropy (e.g., Pr_2Co_{17} and Nd_2Co_{17}). However, the uniaxial anisotropy can be strengthened or the planar anisotropy can be converted to uniaxial
anisotropy by replacing Co in the R_2Co_{17} systems with an appropriate 3d or 4d
transition metal. This was originally observed by Schaller, Craig and Wallace
[11] and by Ray and Strnat [12]. A brief survey of some of the results obtained
follows. These involve partial replacement of Co in R_2Co_{17} systems (1) by Cr,
Mn or Fe, (2) by Ti,Zr or Hf and (3) by Cu and Ti,Zr or Hf.

$R_2Co_{17-x}M_x$ Systems (M = Cr,Mn or Fe)

Results obtained [13,14] for R = Ce or Pr show (see Table V) that replacement of Co by Cr,Mn and Fe leads to the following:

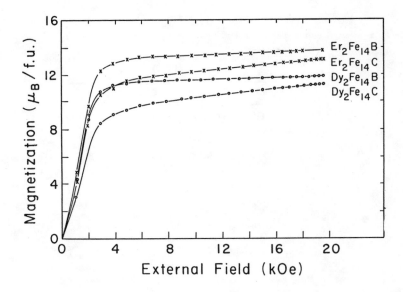

Fig. 7. The magnetization isotherms at 4.2 K for $R_2Fe_{14}B$ and $R_2Fe_{14}C$ compounds (R = Dy or Er).

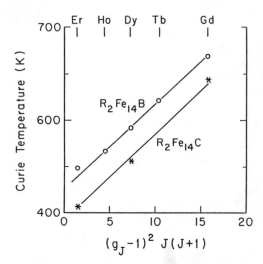

Fig. 8. The dependence of the Curie temperatures, as a function of the De Gennes factor for $R_2Fe_{14}C$ and $R_2Fe_{14}B$ compounds (R is a heavy rare earth). The T_c value for $Gd_2Fe_{14}C$ is also given.

TABLE V

Magnetic Data for $Pr_2Co_{17-x}T_x$ Alloys (T = Fe,Mn,Cr)

Alloy	T_c	μ_{sat} (μ_B/f.u.)		H_A (kOe)	
	(K)	293 K	77 K	293 K	77 K
Pr_2Co_{17}	1199	29.6	31.3	plane	plane
$Pr_2Co_{15}Fe_2$	1179	30.7	32.3	cone	plane
$Pr_2Co_{13}Fe_4$	1151	32.5	34.6	18	cone
$Pr_2Co_{11}Fe_6$	1099	34.0	36.6	31	30
$Pr_2Co_9Fe_8$	1033	34.7	39.8	32	68
$Pr_2Co_7Fe_{10}$	961	33.4	36.7	17	39
$Pr_2Co_5Fe_{12}$	818	32.7	35.0	cone	plane
$Pr_2Co_{16}Mn_1$	1041	30.0	29.9	cone	plane
$Pr_2Co_{15}Mn_2$	929	28.8	30.8	10.0	cone
$Pr_2Co_{14}Mn_3$	791	24.9	27.8	8.0	cone
$Pr_2Co_{13}Mn_4$	648	19.0	23.9	cone	plane
$Pr_2Co_{16}Cr_1$	996	24.1	23.0	plane	plane
$Pr_2Co_{15}Cr_2$	790	18.4	19.6	plane	plane
$Pr_2Co_{14}Cr_3$	514	11.2	15.7	cone	plane

(1) a decrease in T_c.
(2) a rise in magnetization for low concentrations of Fe followed by a decline at higher concentration.
(3) a decrease in magnetization for Cr and Mn as substituents.
(4) modification in anisotropy in the direction of uniaxial anisotropy.

Results are provided for Pr_2Co_{17}-based systems [13], but results for Ce_2Co_{17}- [14] and Sm_2Co_{17}-based [15] systems are generally similar.

The results in Table V indicate that while uniaxial anisotropy is favored by replacing Co with Cr,Mn or Fe, the uniaxial anisotropy actually achieved by alloying is weak. The largest room temperature H_A value observed for Ce_2Co_{17}- and Pr_2Co_{17}-based systems is $\sim$ 40 kOe, the value for $Pr_2Co_9Fe_8$.

$R_2Co_{17-x}M_x$ Systems (M = Ti,Zr,Hf or V)

It is well known that the so-called 2:17 magnets are substantially improved [16] by doping with Ti,Zr or Hf. Because of this effect experiments have been performed to ascertain the influence of doping by these elements on the magnetic behavior of R_2Co_{17} systems, R = Ce,Pr and Sm. Particular interest centered on the measured anisotropy fields, this being the factor that represents a deficiency for the use of R_2Co_{17} systems as a permanent magnet material. Anisotropy fields for $Ce_2Co_{17-x}Zr_x$ and $Sm_2Co_{17-x}Zr_x$ are shown [17] in Fig. 9. It is noted that H_A is incremented by nearly 50 kOe when 1 Co is replaced by Zr. This sharp increase is undoubtedly a band structure effect.

The effects of Ti,Zr,Hf and V on the anisotropy of Ce_2Co_{17} and Sm_2Co_{17} alloys are shown in Fig. 10. It is evident that all of these substituents raise H_A but of these, Zr is the most effective in incrementing H_A. The sequence of substituent effects in regard to H_A is Zr > V > Hf > Ti > Cr > Mn > Fe.

In regard to Curie temperature and magnetization, these are all decreased by replacement of Co with Ti,Zr,Hf or V.

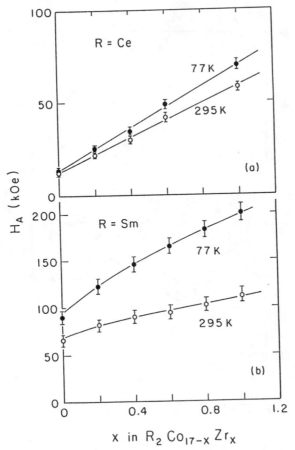

Fig. 9. Variation of anisotropy field as a function
of x at 77 K and 295 K in (a) $Ce_2Co_{17-x}Zr_x$
and (b) $Sm_2Co_{17-x}Zr_x$.

$Pr_2(Co,Cu)_{17-x}M_x$ Studies (M = Ti,Zr,Hf)

The 2:17 alloys used in permanent magnet fabrication contain Cu in partial
replacement of Co in the Sm_2Co_{17} phases. For this reason, experiments were
performed on R_2Co_{17} systems in which Co is partially replaced by Cu and by M,
where M = Ti,Zr or Hf. Results presented here are for the Pr alloy [18]. The
compositions studied can be represented by the formula $Pr_2(Co_{16}Cu_1)_{\frac{17-x}{17}}M_x$.

It is observed that all alloys studied ($0 \leq x \leq 1$) form in the rhombohedral
structure. The c/a ratio increases with x when M = Ti but decreases when
M = Zr or Hf, suggesting a different mode of substitution for Ti than for Zr
or Hf. It appears that Ti may substitute for one Co in the dumbbell sites,
but that Zr and Hf substitute for a pair of Co atoms.

T_c and magnetization decrease with increasing x over the composition range
studied. Anomalies in the magnetization versus temperature curves are observed

28

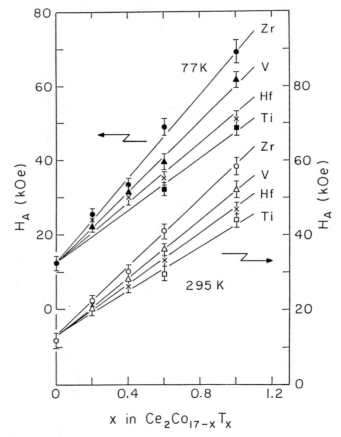

Fig. 10. Composition dependence of H_A for doped Ce_2Co_{17} alloys.

(see Figs. 11 and 12). This is attributed to the change from planar to uni-axial anisotropy as temperature is increased. This postulate was confirmed by x-ray studies of magnetically aligned powders.

The planar-to-axial anisotropy transition was observed in all the systems studied (see Fig. 13). It is to be noted that Zr is more effective than Ti or Hf in generating axial anisotropy. Some of the four-component alloys are rather strongly anisotropic (see Table VI).

RCo_4B AND $PrCo_{4-x}Fe_xB$

The RCo_4B structure is derived from the normal RCo_5 structure by ordered substitution of the 2c Co in alternate layers. Thus the c spacing of the RCo_4B systems is roughly double the c spacing of the RCo_5 compounds, whereas the a spacings of RCo_5 and RCo_4B are very similar. Magnetic studies were made

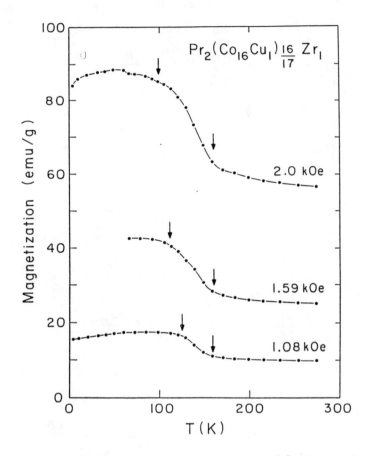

Fig. 11. Magnetization-temperature behavior of Pr_2Co_{17} doped with Zr and Cu.

TABLE VI

Magnetic Anisotropy Field of $Pr_2(Co_{16}Cu_1)_{\frac{17-x}{17}}M_x$ at 300 K

Compound	Anisotropy Field (kOe)
$Pr_2(Co_{16}Cu_1)_{16.5/17}Zr_{0.5}$	11
$Pr_2(Co_{16}Cu_1)_{16/17}Zr_1$	48
$Pr_2(Co_{16}Cu_1)_{16/17}Hf_1$	45

of the systems with R = Pr,Sm,Gd and Y in 1984 by Spada et al. [19] and, more recently, with R = Pr,Nd,Gd,Er and Y by Pedziwiatr et al. [20].

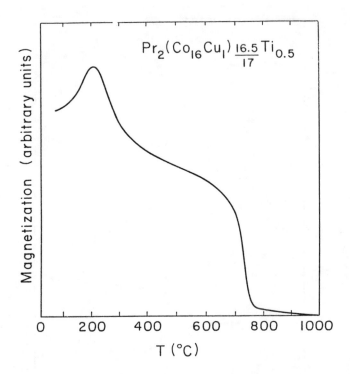

Fig. 12. Magnetization-temperature behavior of Pr_2Co_{17}
doped with Ti and Cu.

The RCo_4B systems exhibit reasonably high Curie temperatures (380 to
505°K) and the usual coupling systematics, i.e., parallel and antiparallel
coupling for light and heavy rare earths, respectively. Results obtained by
Pedziwiatr et al. are summarized in Table VII. The magnitude of the saturation magnetizations (see Table VII) and the temperature dependence of magnetization (Fig. 14) indicates that the Y,Pr or Nd systems are ferromagnetic.
The RCo_4B systems with R = Gd or Er are ferrimagnetic (Fig. 14).

The magnetizations given in Table VII permit one to evaluate the Co
moment in the several systems. The values computed are given in Table VIII.
In these, moments of 2.4, 3.0 and 9.0 μ_B/atom are assumed for R = Pr, Nd and
Er, respectively.

Anisotropy in YCo_4B and $GdCo_4B$ originates with the Co sublattice. It is
to be noted that the uniaxial anisotropy increases with increasing temperature
(see Table VII). This is similar to the behavior of $Y_2Fe_{14}B$ and $Gd_2Fe_{14}B$. It
appears that the two kinds of Co in the RCo_4B systems may contribute oppositely
to the anisotropy.

Data were obtained for $PrCo_4B$, $GdCo_4B$ and $ErCo_4B$ in the paramagnetic
region. Reciprocal susceptibilities are plotted versus temperature in Fig. 15.
The latter two systems in which there is ferrimagnetic coupling follow the
Néel hyperbolic law, whereas $PrCo_4B$ follows a modified Curie-Weiss law, the
modification involving a term added to account for the Pauli susceptibility
of the system. The equations used are

$$\chi^{-1} = \chi_o^{-1} + TC^{-1} - \sigma(T-\theta)^{-1}$$

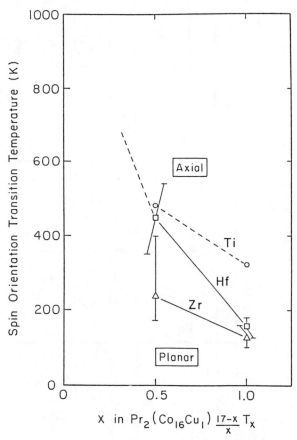

Fig. 13. Composition dependence of spin-reorientation
temperature for Pr_2Co_{17} doped with Cu plus
Ti, Zr, or Hf.

and

$$\chi = \chi_p + C(T-\theta_p)^{-1}.$$

The data are fit to these equations, and the values of the parameters are
listed in Table IX. In simple rare earth intermetallic compounds, e.g., RNi_2
or RAl_2, the rare earth exhibits the free ion value. Assuming this for RCo_4B
systems, the contribution of Co to the observed Curie constant and the effective Co moment can be readily computed. Results obtained are also given in
Table IX; they range from 2 to 2.5 μ_B/atom.

Molecular field coefficients were evaluated for the Er and Gd systems
in the paramagnetic and ordered regimes, using a two-sublattice model. Results obtained by this analysis are given in Table X. Although agreement is
not exact, the results are not in sharp disagreement, considering the simplified nature of the model used to interpret the data. The analysis reveals
the relative magnitudes of the three types of interactions and reveals the
antiferromagnetic nature of the R-Co interaction.

Table VII

The Lattice Parameters and the Magnetic Properties of RCo_4B Compounds

Compound		Y	Pr	Nd	Gd	Er
Lattice constants (Å)	a	5.020	5.114	5.079	5.034	4.979
	c	6.875	6.880	6.879	6.879	6.869
Curie temperature (K)		382	455	458	505	386
Saturation magnetization per formula unit at 4.2 K (μ_B)		2.65	5.10	5.80	3.25	5.10
Rare earth moment (μ_B)		–	2.4	3.0	7.0	9.0
Cobalt magnetization per formula unit (μ_B)		2.65	2.70	2.80	3.75	3.90
Anisotropy field (kOe)	77 K	15	–	122	32	78
	295 K	21.5	–	158	110	

PrCo$_4$B alloys were studied, in which Co was partially replaced by Fe. Single-phase PrCo$_{4-x}$Fe$_x$B systems were obtained for $0 \leq x \leq 1.5$. For larger concentrations of Fe, $\check{R}$(Co,Fe)$_4$B appeared along with R$_{1+\epsilon}$Fe$_4$B$_4$. These materials (i.e., $\check{R}$(Co,Fe)$_4$B) formed in the RCo$_4$B structure, the lattice being enlarged about 2% by volume as Fe replaces Co.

PrCo$_4$B has planar anisotropy at room temperature, and this remains unchanged when Co is replaced by Fe. Thus these materials (i.e., PrCo$_{4-x}$Fe$_x$B alloys) do not seem promising as a permanent magnet material. Otherwise,

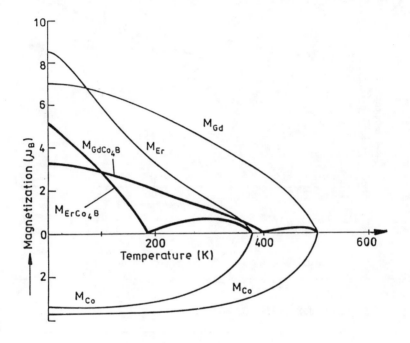

Fig. 14. Overall and sublattice magnetizations of ErCo$_4$B and GdCo$_4$B.

TABLE VIII

Average Co Moment (μ_B/Co atom) in RCo$_4$B Systems

R in RCo$_4$B	Co Moment
Y	0.66
Pr	.68
Nd	.70
Gd	.94
Er	.98

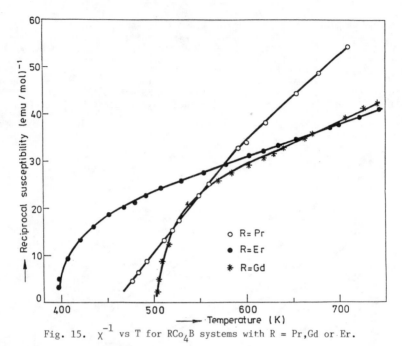

Fig. 15. χ^{-1} vs T for RCo_4B systems with R = Pr,Gd or Er.

TABLE IX

Data Obtained from Paramagnetic Measurements

Compound	C (emuK/f.u.)	χ_o^{-1} (emu/f.u.)$^{-1}$	σ [(emuK/f.u.)]$^{-1}$	$\theta(\theta_p)$ (K)	(emuK/f.u.)	μ_{eff} (μ_B/Co atom)
$PrCo_4B$	4.04	–	–	462	2.40	2.19
$GdCo_4B$	10.59	– 25	45	510	2.68	2.31
$ErCo_4B$	14.45	–105	89	402	2.92	2.42

TABLE X

Exchange Interaction Coefficients

		$GdCo_4B$			$ErCo_4B$		
		J_{CoCo}	J_{GdCo}	J_{GdGd}	J_{CoCo}	J_{ErCo}	J_{ErEr}
a)	From thermal variation of spontaneous magnetization	352	–34	20	320	–18	5
b)	From paramagnetic data	165	–26	43	120	–21	19

they are promising since they have a high Curie temperature and a substantial magnetization (see Fig. 16)..

The planar anisotropy of the $PrCo_{4-x}Fe_xB$ quaternary alloys is of interest when contrasted with the behavior of $NdCo_4B$. The latter material is uniaxial, which is somewhat surprising in view of the planar nature of $NdCo_5$. Dilution by boron of the Co sublattices which favors uniaxial anisotropy might be expected to lead to a planar material. This happens with $PrCo_4B$ and with the $Pr(Co,Fe)_4B$ alloys, and this is in accord with expectations. In the boride the planar tendency of Pr becomes dominant. Why this does not also happen with $NdCo_4B$ is unclear. In the $Pr(Co,Fe)_4B$ systems, both the Fe and B serve to reduce the uniaxial anisotropy which originates with Co leaving the rare earth as the dominant species. The planar nature of these quaternary alloys is in full accord with expectation.

Fig. 17 shows T_c as a function of the transition metal moment in $PrCo_{4-x}Fe_xB$. There is a close correlation between moment and T_c, strongly suggesting that T_c is essentially determined by the transition metal interactions.

STUDY OF $R_2Fe_{14}B$ SYSTEMS BY AUGER SPECTROSCOPY

The mechanism of coercive force in $R_2Fe_{14}B$ systems is currently an area of active inquiry. It is generally believed that the nature – structure and composition – of the boundary layer or intergranular region is significantly involved in regard to coercive force. However, details regarding the boundary layer chemistry are as yet sketchy. The present study [21] provides some information. Samples of $R_2Fe_{14}B$ with R = Pr, Nd and Er were fractured in air and inserted in the Auger spectrometer (10^{-10} torr vacuum) as rapidly as possible. Even so, there was exposure to air for 20 to 30 minutes. Because of the fracture in air and the possibility of oxidation, the present study is

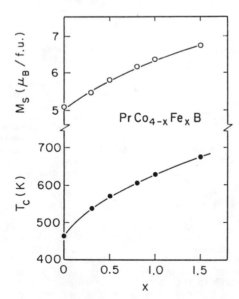

Fig. 16. The composition dependence of the Curie temperature and saturation magnetization of $PrCo_{4-x}Fe_xB$ alloys.

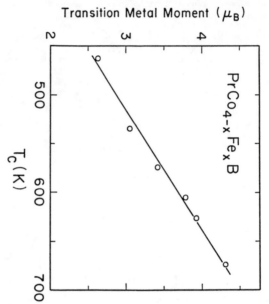

Fig. 17. Transition metal magnetization as a function of the Curie temperatures for PrCo$_{4-x}$Fe$_x$B alloys.

regarded as preliminary. At present, the work is being extended and improved by arranging for the samples to be fractured in the UHV chamber of the spectrometer [22].

Auger electron spectroscopy analysis of the fresh surface produced during fracture showed enrichment in rare earth and boron and a depletion in Fe. Representative results are given for Nd$_2$Fe$_{14}$B in Fig. 18. A substantial oxygen signal was in evidence. The R/Fe ratio was about 2 at the surface but, as shown in the depth profile obtained by Ar sputtering, the excess rare earth rapidly diminished with distance into the interior of the material.

Because of the chemical shift of about 8 eV in the Auger signal for oxidized and elemental boron, these two forms of boron can easily be distinguished in the Auger spectrum. The AES spectrum shows that the surface is rich in oxidized boron. It seems likely that the segregation of B and rare earth to the surface is chemically driven by the oxidation process.

A crude indication of the results obtained in the depth profiling is indicated in Table XI. These results and those shown in Fig. 18 make it appear that the oxygen present in BO$_x$ is only a minor part of the total oxygen. Undoubtedly the remainder is in chemical union with Nd. Hence the large Nd content present at the surface is probably the Nd in Nd$_2$O$_3$. Thus the composition of the outer layers is, on the basis of these experiments and with samples prepared as indicated, most probably a complex mixture of Nd and B oxides, the Nd and B having been drawn to the surface by the strong affinity of Nd and B for oxygen. It follows that the underlying areas have been significantly depleted in B and Nd. It is suggested that this underlying defected region serves to pin domain walls, and this is involved in determining the coercivity of Nd$_2$Fe$_{14}$B.

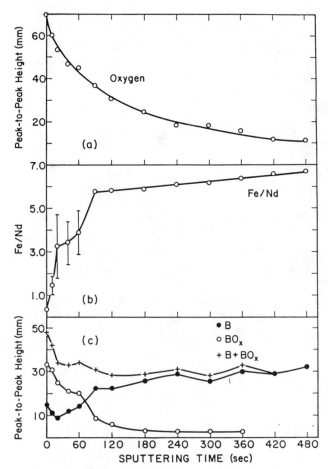

Fig. 18. Depth profiling of $Nd_2Fe_{14}B$. (a) Oxygen content
as a function of depth. (b) Fe/Nd ratio as a
function fo depth. The bar represents± one
standard deviation. (c) B and BO_x content as
a function of depth. With the sputtering rate
used, 1 min. corresponds to 20 to 30 Å.

It is well known that the coercive force of $Nd_2Fe_{14}B$ depends on the magnetizing field (H_m) to which it is exposed; the higher the magnetizing field,
the higher the coercive force. The domain wall moves readily through the
$Nd_2Fe_{14}B$ grains in $Nd_2Fe_{14}B$ magnets, but as it approaches the surface of the
grain it encounters the defected region alluded to in the preceding paragraph.
If H_m is small, the wall penetrates only a short distance into the defected
region. As H_m is increased, it is driven further ino this region. For small
H_m, the magnetization easily reverses, given low coercivity. For larger
H_m where the wall has been driven more deeply ino the defected region, magnetization is less easily reversed and a larger coercivity is expected. For very
large H_m the wall is driven completely out of the grain. When a reverse field

38

TABLE XI

Composition of $Nd_2Fe_{14}B$ Ingots Near Surface

Position	Nd/Fe	$\dfrac{BO_x}{BO_x+B}$	Total Oxygen (Peak-to-peak signal)
0 (surface)	2.00	0.7	70
− 25 Å	.30	.6	45
− 50 Å	.19	.2	28
−100 Å	.16	.1	20
− ∞ (bulk)	.14	0	7 (at −200 Å)

is applied, magnetization reversal must occur against the 70 kOe anisotropy field of $Nd_2Fe_{14}B$ or some fraction of this because of the defected structure near the surface.

The boundary phases of $Nd_2Fe_{14}B$ − BO_x and Nd_2O_3 − are non-magnetic and isolate one grain from another. Hence if magnetization reverses easily in one grain because of special features which give rise to a very small nucleation field, it will not spread to neighboring grains because of the insulating nature of the oxide boundary phases. The special interplay as noted between the matrix, the defected matrix layer and boundary phase plays the crucial role in determining both the pinning of domain walls and in the nucleation of reverse domains and in this way establishes the coercivity of the $R_2Fe_{14}B$ magnets.

ACKNOWLEDGEMENTS

Partial support of the work from the Magnetic Materials Research Group, through the Division of Materials Research, National Science Foundation, under Grant No. DMR-8613386, is gratefully acknowledged. Partial support was also provided by the U.S. Army Research Office.

REFERENCES

1. W. E. Wallace, A. T. Pedziwiatr, E. B. Boltich, F. Pourarian, E. Oswald and S. Y. Jiang, Acta Physica Polonica, to appear.
2. A. T. Pedziwiatr, W. E. Wallace and E. Burzo, IEEE Trans. Mag., accepted.
3. A. T. Pedziwiatr and W. E. Wallace, submitted.
4. R. M. Bozorth, Ferromagnetism, D. Van Nostrand Co., Inc. (1951).
5. E. Burzo and I. Ursu, J. Appl. Phys. 50, 1471 (1979).
6. E. Burzo, J. Phys. F. 10, 2025 (1980).
7. E. Burzo, D. P. Lazar and M. Ciorascu, Phys. Stat. Solidi B 65(2), K145 (1974).
8. M. Q. Huang, E. B. Boltich, W. E. Wallace and E. Oswald, J. Mag. Mag. Mat. 60, 270 (1986).
9. W. E. Wallace, R. S. Craig and V. U. S. Rao, Advances in Chemistry Series, No. 186, "Solid State Chemistry: A Contemporary Overview," Smith L. Holt, Joseph B. Milstein and Murray Robbins, eds. American Chemical Society, 1980, p. 207.
10. C. Abache and H. Oesterreicher, J. Appl. Phys. 57, 4112 (1985).
11. H. J. Schaller, R. S. Craig and W. E. Wallace, J. Appl. Phys. 43, 3161 (1972).
12. A. E. Ray and K. J. Strnat, IEEE Trans. Magn. MAG-8, 516 (1972).

13. M. V. Satyanarayana, H. Fujii and W. E. Wallace, J. Mag. Mag. Mat. 40, 241 (1984).
14. H. Fujii, M. V. Satyanarayana and W. E. Wallace, J. Appl. Phys. 2371 (1982).
15. W. E. Wallace and K. S. V. L. Narasimhan, in The Science and Technology of Rare Earth Materials, W. E. Wallace and E. C. Subbarao, eds. Academic Press, Inc., New York (1980), p. 329.
16. T. Ojima, S. Tomizawa, T. Yoneyama and T. Hori, IEEE Trans. Mag. MAG-13, 1317 (1977).
17. H. Fujii, M. V. Satyanarayana and W. E. Wallace, Solid State Commun. 41, 443 (1982).
18. E. Oswald, Ph.D. Thesis, University of Pittsburgh, 1986.
19. F. Spada, C. Abache and H. Oesterreicher, J. Less-Common Met. 99 L21 (1984).
20. A. T. Pedziwiatr, S. Y. Jiang, W. E. Wallace, E. Burzo and V. Pop, J. Mag. Mag. Mat., to appear.
21. H. K. Smith and W. E. Wallace, Lanthanides and Actinides, 1, 217 (1985).
22. Janean M. Elbicki and W. E. Wallace, to be published.

SYNTHESIS OF $R_2Fe_{14}B$ SINGLE CRYSTALS

B.N. DAS AND N.C. KOON
Naval Research Laboratory, Washington, DC 20375-5000

ABSTRACT

Single crystals of $R_2Fe_{14}B$ weighing 5 to 15 g were grown from a liquid melt by tri-arc and levitation Czochralski methods. The arc method was used for the growth of smaller size crystals from 5 to 7 g, and the levitation method was used for the growth of larger sizes, from 10 to 15 g. Crystals of $(R_1)_{2-x}(R_2)_xFe_{14}B$ could also be grown with isomorphous replacement of rare earth atoms. The starting alloy composition for the crystal growth process was chosen based on solidification microstructure. In this paper we discuss the solidification microstructure, isomorphous replacement, seeding and their interactions with the crystal growth processes and crystal quality.

INTRODUCTION

While investigating the magnetic properties of rapidly quenched rare earth-iron-boron amorphous alloys, we found that some of the crystallized alloys have interesting hard magnetic properties [1]. The magnetic properties of these alloys are related to a ternary compound in rare earth-iron-boron system [2]. Based on our experience with rare earth Laves phases [3], we decided to grow single crystals of this compound by tri arc Czochralski process in order to study the magnetic properties of this material [4].

The ternary compound was established to be $R_2Fe_{14}B$ by Herbst, Croat, Pinkerton and Yelon [5]. Sagawa *et al.*, [6] investigated the powder metallurgical process of preparing the compound. By varying the constituent elements of the starting material and measuring the resultant properties of the alloys, they [6] established the optimum starting composition, $R_{15}Fe_{77}B_8$, for the powder metallurgical process. Givord *et. al.*, [7] used this composition to grow large grains of $R_2Fe_{14}B$ from melt for magnetic experiments. Recently Swets [8] grew large single crystals of $Nd_2Fe_{14}B$ from Hukin crucible using RF power and a melt composition close to $Nd_{15}Fe_{77}B_8$ but with some variation.

In our crystal growth experiments, the optimum composition for crystal growth processes was obtained from a study of solidification microstructure. This paper reports the results of solidification microstructure of several rare earth alloys and shows how the optimum composition for the crystal growth process was established. The study of solidification microstructure was extended to establish the crystal growth region of quaternary substituted alloys. The nature of the solidification microstructures also provided information on crystal growth process.

EXPERIMENTAL PROCEDURES

To prepare the starting alloy specimens, Fe-B alloys were first prepared by arc melting proper mixture of iron and boron under a positive pressure of argon, (99.995% pure). The specimens were turned over on the copper hearth and remelted three times to achieve complete alloying between Fe and B. The binary Fe-B alloys were prepared first to avoid the formation of high melting rare earth borides in the ternary alloys. A proper amount of rare earth was added to the iron-boron alloy and the resultant mixture, about 30 to 40 g, was melted at least four times by repeated turning over and remelting. The purity of the elements used for alloy preparation were: 99.99% for rare earths, 99.95% for Fe and 99.9% for B. The prepared alloy was cleaned of surface scum and thin impurity film by surface grinding on a clean alumina wheel. The cleaned alloy was ready for use in the crystal growth experiment.

To obtain a single crystal of $R_2Fe_{14}B$, a portion of the prepared alloy, 12 g to 15 g, was arc melted in a water cooled rotating copper hearth (see Figure 1). The hearth was rotated at 12 rpm during melting and crystal growth processes. When the melt reached the proper temperature, as indicated by the liquid fluidity (the melt is fluid like water near proper temperature) on mechanical tapping the frame, the seed was lowered slowly into the melt. The seed melt interface also is a good indicator for proper crystal growth temperature. A low melt temperature results in rapid freezing of melt on seed surface, proper melt temperature indicates uniform wetting of the seed and high melt temperature rapidly dissolves the seed end in contact with the melt. Because of fogging on the inner glass walls, the precise melt temperature measurement by optical pyrometry was not possible. In order to get an idea of the actual temperature, a melt temperature of 1340°C was measured by optical pyrometry in the beginning of the crystal growth process. The proper melt temperature was maintained by control of power to the arcs.

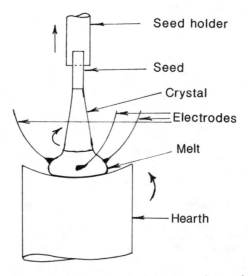

Figure 1. Schematic of tri arc Czochralski crystal growth apparatus.

In the beginning of the crystal growth process, a rod of small diameter, about 2 mm, was grown at a high temperature gradient and a slow pulling rate, 3 to 5 mm an hour. A faster growth rate, about 8 to 10 mm an hour, was used for the study of solidification microstructure. The temperature gradient was controlled by positioning the electrodes and by power imput to the arcs. For improved temperature uniformity at the seed-melt interface, the seed was rotated at approximately 15 rpm opposite to the rotation of the hearth. When the length of the growing small diameter rod reached approximately 3 to 5 times its diameter, the electrode positions were carefully changed (i.e., placed away from the growth interface) so as to increase the crystal rod diameter. The decrease of power to the arcs also increased the growing crystal rod diameter. The crystal growth process was monitored by a video camera and a monitor. The camera and monitor combination magnified the liquid-solid interface, i.e., crystal growth interface, 25 times for better control of the growth process. At the end of the crystal growth process, the crystal was lifted from the melt and the power to the arcs was turned off.

For levitation Czochralski crystal growth, a Hukin crucible, powered by a 15 KW induction generator, was used. The melt was in the range of 50 g. The melt stirring was provided by induction coupling and the seed rotation was maintained around 15 rpm. In the beginning of the crystal growth process, the liquid melt temperature was 1350°C and the proper temperature was maintained by control of power to the crucible.

Samples of the crystal and melt left either on the hearth or on the crucible were examined by optical microscopy, X-ray diffraction, transmission electron microscopy (TEM), and differential thermal analysis (DTA). For optical microscopy, the samples were prepared by standard metallographic technique. To minimize cracking and chipping, the specimens, mounted in epoxy, were ground in SiC abrasive papers with grit sizes 400, 600 and 1200. The ground specimens were then polished on wheels with microcloth and alumina particles of 5 μm and 0.5μm sizes. Water was used as a coolant during the grinding and polishing processes.

The X-ray diffraction methods were used to identify the presence of phases and to evaluate the crystal quality. Because of its advantages of using a small amount of sample, which can be obtained from selected regions, the Debye-Scherrer powder method was used for phase identification. Information on crystal quality and orientation were obtained from back reflection Laue photographs. The TEM and DTA experimental details were the same as mentioned earlier [2].

EXPERIMENTAL RESULTS

Alloy preparation

The alloys shatter on rapid heating and cooling in normal arc melting process because of poor thermal shock resistance. Therefore, it was necessary both to heat and cool slowly with a soft arc in the beginning and at the end of the melting process. This method prevented shattering and thereby avoided changes in alloy composition due to loss of material from the hearth. After the final remelting, the alloy was cooled rapidly to minimize segregation. The repeated turning over and remelting of the alloy button on the hearth slope for better liquid mixing reduced the segregation considerably.

The microstructure of cast starting alloys show the solidification microstructure of an arc melted samples, i.e., a microstructure ranging from frozen-in region near the hearth to equiaxed grain at the center changing to columnar dendrites at the top. No evidence of any other large scale alloy segregation is seen under the microscope.

Solidification microstructure and crystal growth

Information on solidification microstructure was obtained from an optical microscopic study of the grown crystal boules and residual melt. Because of the superior magnetic properties of the crystallized ribbons [1] prepared from an alloy of composition $R_{15}Fe_{78.3}B_{6.7}$, we decided to use this alloy composition for our beginning crystal growth experiments. A crystal boule grown at a pulling rate of 8 mm an hour from this alloy composition is shown in Figure 2. The solidification microstructure, Figure 3, shows the presence of a primary, matrix and impurity phases. The primary phase was confirmed by X-ray diffraction to be α-Fe. About this time, the matrix

44

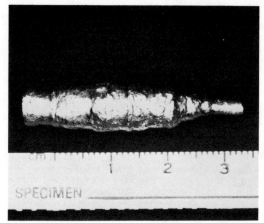

Figure 2. Crystal boule grown at 8 mm an hour from a melt composition of
$R_{15}Fe_{78.3}B_{6.7}$, here R_{15} corresponds to $Pr_{14.48}La_{0.52}$.

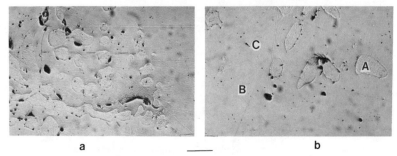

a   ———   b

Figure 3 a, b. Solidification microstructures from $R_{15}Fe_{78.3}B_{6.7}$ alloy
(R=$Tb_{7.5}La_{7.5}$) show the primary (A), matrix (B) and impurity
(C) phases. The amount of primary α-Fe phase decreases from
near edge (a) to center (b) of the boule. Marker 44μm.

ternary phase was established to be $R_2Fe_{14}B$, i.e., $R_{11.8}Fe_{82.3}B_{5.9}$ by
Herbst, Croat et. al. [5].

In our next series of crystal growth experiments, growth rate 8 mm an
hour and with alloy composition at $R_2Fe_{14}B$, the solidification
microstructure, Figure 4, shows an increase in the primary α-Fe content.
The microstructure does not show the presence of any other phases except the
presence of small amount of impurity phases as seen in Figure 3. The
positions of these two alloys are shown in the ternary diagram of Figure 5.
In Figure 5, alloy position 1 corresponds to composition $R_2Fe_{14}B$ (i.e.
$R_{11.8}Fe_{82.3}B_{5.9}$) and alloy position 2 corresponds to composition
$R_{15}Fe_{78.3}B_{6.7}$.

Our crystal growth experiments from these two series of alloy
compositions show that crystal boules can be grown at these compositions.
But the primary α-Fe formation is an inherent nature of the solidification
process in this region of the alloy composition. The results also suggest
that the α-Fe amount decreases as the alloy composition moves away from Fe
corner in the ternary diagram. From these results and based on the amount
of α-Fe in the alloys 1 and 2 in Figure 5, a starting alloy composition for

the growth of iron free crystals was estimated. The estimated composition is shown in position 3 in Figure 5 and it corresponds to $R_{16}Fe_{77}B_7$.

The solidification microstructure of the starting alloy, position 3 in Figure 5, is presented in Figure 6. The α-Fe does not crystallize at this starting alloy composition but minor amounts of two other phases are present at the grain boundaries of large primary $R_2Fe_{14}B$ grains. One of the phases is stained easily during the polishing process and the surface appears corroded. The other phase appears highly polished and smooth. The stained

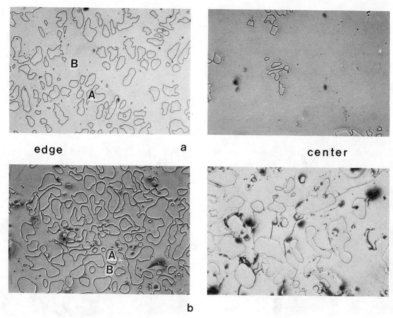

edge   a   center

b

Figure 4a, b. Solidification microstructure from $R_2Fe_{14}B$ alloys, Nd(a) and TbLa(b), show the primary (A) and matrix (B) phases. The amount of primary α-Fe is smaller in Nd compound (a) compared to TbLa compound (b). Marker 100μm.

corroded phase is probably primary rare earth phase with some Fe and B in it depending on the solubility limit of these two elements. The highly polished phase is a ternary boride phase. Stadelmaier et. al. [9] have reported the presence of these two types of phases in their ternary alloys. According to them, the non-reacted polished ternary boride phase is RFe_4B_4.

X-ray diffraction

The X-ray diffraction experiments confirmed the matrix phase to be $R_2Fe_{14}B$. Information on crystal growth direction and crystal quality are obtained from back reflection Laue photographs. The back reflection Laue photographs taken on three crystallographic directions, i.e., <001>, <100> and <110>, of the tetragonal $R_2Fe_{14}B$ crystal (space group $P4_2/mnm$, point group 4/mmm, structure no. 136) are presented in Figure 7 a, b and c. The Laue photograph taken in the <001> direction, (i.e., c axis direction, Figure 7a) shows the four fold symmetry. A high density of spots are present in the zone between {001} and {110}. Each of the photographs taken in the <100> and <110> directions shows the presence of two fold rotational

46

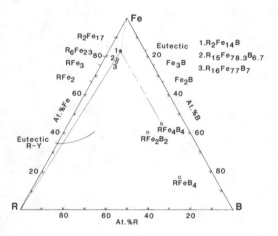

Figure 5. Alloy positions 1 ($R_2Fe_{14}B$), 2($R_{15}Fe_{78.3}B_{6.7}$) and 3($R_{16}Fe_{77}B_7$) in ternary system of R-Fe-B. The diagram also shows the positions of other ternary and binary compounds.

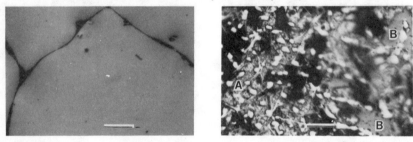

a   b

Figure 6a, b. Solidification microstructure from $R_{16}Fe_{77}B$ alloy shows a) the large primary grains of $R_2Fe_{14}B$ and b) the presence of rare earth (A) and a ternary boride (B) phase in the three phase intergranular region. The bright reflecting regions are $R_2Fe_{14}B$. Marker a) 50 μm, b) 10μm.

symmetries with two perpendicular mirror planes in agreement with point group 4/mmm.

The nature of the Laue spots gave information on crystal quality. The well formed spots shows higher quality of the crystal. The diffuse spots indicate strained crystal. The splitting of spots and the presence of extra zones, respectively, indicate the presence of low angle boundaries and twinning. A high crystal quality, as indicated by well formed Laue spots, was observed in the crystals grown at a slow growth rate, 3 to 5 mm an hour. However, the low angle boundaries were still present in some crystals. The twinning and straining were observed in many crystals grown at a rate of 8 to 10 mm an hour.

The crystal growth axis varied with the change in rare earth atoms in the crystal. The growth axis was close to <001> for crystals containing Y atom and the growth axis moved 30 to 60° away from <001> direction for crystals containing Nd atom. The mixed rare earth crystals, up to 10% substitution of Y atom, grew close to <001> direction. These growth directions are for naturally seeded crystals, i.e., for single crystal seeds

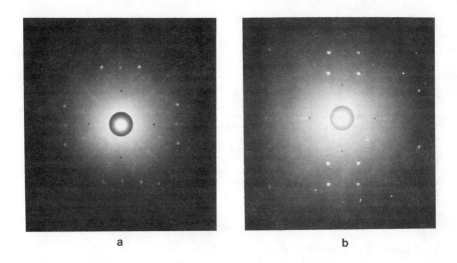

a   b

c

Figure 7a, b, and c. Back reflection Laue photographs of a tetragonal
    $R_2Fe_{14}B$ crystal. The X-ray beam is parallel to: a) <001>,
    b) <100> and c) <110>. Sample to film distance: 3cm.

obtained by growing a crystal from melt with a polycrystalline seed. Our crystals were grown in the direction of growth of the seed crystals.

DISCUSSION

As-prepared alloys

The nature and quality of the alloys were established by optical microscopic examination, X-ray diffraction and fluorescence analyses. A homogeneous alloy melt of $R_{16}Fe_{77}B_7$ composition is appreciably fluid under an arc power of 18 volts and 170 amps. During cooling in the hearth the homogeneous alloys normally break into two or more pieces. When an alloy shows major deviation in its physical characteristic, e.g., fluidity, fracture behavior, the alloy is remelted. When in doubt, the alloys are examined metallographically. The nature of the distribution of rare earth and iron atoms are checked by X-ray fluorescence analysis. In some special cases, the presence of the phases was examined by X-ray diffraction analysis.

Solidification microstructure

A two phase field of Fe and $R_2Fe_{14}B$ exists between the Fe corner and $R_2Fe_{14}B$ compound in the rare earth-iron-boron system according to the solidification microstructures, Figures 3 and 4. The microstructures also indicate the formation of $R_2Fe_{14}B$ compound by a peritectic reaction. This nature of formation by a peritectic reaction, Figure 8, explains the enrichment of α-Fe in alloy position 1, $R_{11.8}Fe_{82.3}B_{5.9}$, compared to alloy position 2, $R_{15}Fe_{78.3}B_{6.7}$.

A lesser amount of primary α-Fe phase is present near the center compared to the regions near the surface of the crystal boule, Figure 4. This phenomenon is related to partial homogenization in the comparatively warmer central region of the crystal boule during growth, Figure 9. In the solidification of $R_2Fe_{14}B$ from liquid, we have observed a smaller amount of primary α-Fe for R=Nd compound compared to R=Y, Tb ones. This is due to the change in liquid composition at the peritectic temperature for these alloy systems.

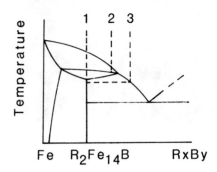

Figure 8. A vertical section of the ternary R-Fe-B diagram through $R_2Fe_{14}B$ and Fe corner, shows the schematic of peritectic formation of $R_2Fe_{14}B$ and the mode of solidification of alloys 1, 2 and 3.

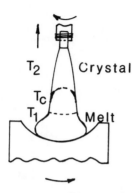

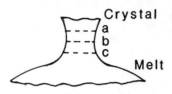

Figure 10. Crystal-melt interface
positions a, b and c
corresponding to various
melt temperatures T_1, T_2
and T_3 respectively, where
$T_1 > T_2 > T_3$.

Figure 9. Constant temperature
profile, T_c, in a
growing crystal; here
$T_1 \gg T_2$.

The solidification microstructure, Figure 6, of the starting alloy, position 3 in Figures 5 and 8, shows the absence of crystallization of α-Fe. For magnetic measurement, the absence of α-Fe in the samples is desirable. A minor amount of two other phases, rare earth and a ternary boride, are present at the grain boundaries of large $R_2Fe_{14}B$ grains. The $R_2Fe_{14}B$ phase is crystallizing as primary phase for the starting alloy composition.

Crystal growth from starting alloy composition

The solidification microstructure, Figure 6, indicate that our starting alloy composition is in the three phase field of $R_2Fe_{14}B$, a ternary boride (probably RFe_4B_4) and terminal rare earth phase. One can estimate from Figure 5 that 70% of the charge may crystallize into $R_2Fe_{14}B$ but, in reality, we could grow about 40% of the charge material into $R_2Fe_{14}B$. The changing of the liquid composition towards the eutectic with the removal of $R_2Fe_{14}B$ and longer diffusion distance complicate the growth process.

We have seen that proper melt temperature uniformly wets the seed without melting it (seed) and separating the seed and melt. A small change in the temperature near the proper melt temperature causes a significant change in the position of the crystal melt-interface, Figure 10. A small increase in temperature moves the interface up, position a in Figure 10, and brings a reduction in crystal diameter. On the other hand, a small decrease in temperature lowers the interface down, position c in Figure 10, and the crystal diameter increases with the growth of the crystal. A crystal of constant diameter grows when the melt temperature is controlled to keep the interface at b.

A rapid freezing of the crystal-melt interface has sometimes been observed during the increase of the crystal diameter, i.e., the interface position near but below c. The lowering of power, i.e. temperature probably triggers the dendritic freezing, Figure 11, due to constitutional supercooling because of the decrease in the ratio of temperature gradient, G, over crystal growth rate, R, i.e., G/R. The constitutional supercooling arises [10] in stirred melt when,

$$G/R< \left[\frac{mC(1-K)}{D(k+(1-k)\exp(-RS/D)} \right] \tag{1}$$

Here, G = temperature gradient in liquid at freezing interface, °c/cm
R = crystal growth rate, cm/sec.
m = slope of liquidus, °c - cm^3/mole.
C = solute ($R_2Fe_{14}B$) concentration in bulk melt, moles/cm^3.
k = equilibrium distribution coefficient, Cs/C.
D = diffusion coefficient of solute, $R_2Fe_{14}B$, in melt, cm^2/sec.
S = stagnant boundary layer thickness, cm.

A high temperature gradient and slow growth rate are required to avoid constitutional supercooling, i.e. to make G/R larger than the quantity on r.h.s. of equation 1. For this reason, the crystal growth rate was reduced

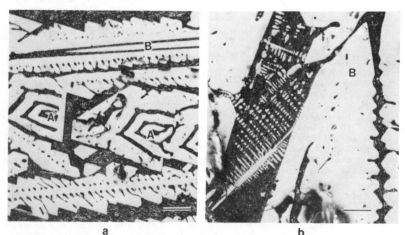

a   b

Figure 11 a,b. Dendritic freezing in the melt below the crystal-melt interface. The dendrites are growing normal (A) and parallel (B) to
the plane of the photograph. The dark regions are interdentric
spaces. Marker a) 40μm, b) 20μm.

to the range of 3 to 5 mm an hour for the growth of higher quality crystals. Even then we see some low angle boundaries in our crystals. These low angle boundaries may be formed by dislocation rearrangement (polygonization). The effect of seeding direction on polygonization and crystal quality are being investigated.

Extension of the crystal growth process to quaternary alloys

In order to explore magnetic properties it is desirable to substitute the nonmagnetic Y atom in $R_2Fe_{14}B$ crystals by other magnetic rare earth atoms. Many similar properties of the rare earth metals and compounds and the fact that almost all the rare earths form $R_2Fe_{14}B$[11] suggest the possibility of substituted crystal formation. In the rare earth-iron binary systems, Table I, similar compounds and close temperatures of formation are observed for both light and heavy rare earth atoms. We have also observed in our study [2] of rare earth-iron-boron amorphous ribbons that the rare earth atoms can be substituted isomorphously in the amorphous state. These behaviors suggest that one can grow substituted crystals from mixed rare earth alloy melt.

Table I. Rare earth-iron compound, the nature and temperature of formation

System	Compound	Nature of formation*	Temperature of formation °C	Reference
Y-Fe	Y_2Fe_{17}	C	1400±25	
	Y_6Fe_{23}	C	1300	Elliott[12]
	YFe_3	C	1400	
	YFe_2	P	1125±25	
Ce-Fe	$CeFe_5$	P	1060	Shunk[13]
	$CeFe_2$	P	925	
Pr-Fe	Pr_2Fe_{17}	P	1105±8	Moffatt[14]
	$PrFe_2$	P	670±7	
Nd-Fe	Nd_2Fe_{17}	P	1185	Moffatt[14]
	$NdFe_2$	P	1130	
Sm-Fe	$SmFe_2$	P	850	Shunk[13]
	$SmFe_5$	P		Nassau [15] et al.
Gd-Fe	Gd_2Fe_{17}	P	1355	Moffatt[14]
	Gd_6Fe_{23}	P	1280	
	$GdFe_3$	P	1180	
	$GdFe_2$	P	1080	
Tb-Fe	Tb_2Fe_{17}	P	1312	Moffatt[14]
	Tb_6Fe_{23}	P	1276	
	$TbFe_3$	P	1212	
	$TbFe_2$	P	1187	
Dy-Fe	Dy_2Fe_{17}	C	1375	Moffatt[14]
	Dy_6Fe_{23}	P	1285	
	$DyFe_3$	C	1305	
	$DyFe_2$	P	1270	
Ho-Fe	Ho_2Fe_{17}	C	1343	Moffatt[14]
	Ho_6Fe_{23}	P	1332	
	$HoFe_3$	P	1293	
	$HoFe_2$	P	1288	
Er-Fe	Er_2Fe_{17}	P	1295	Moffatt[14]
	Er_6Fe_{23}	P	1261	
	$ErFe_3$	P	1275	
	$ErFe_2$	C	1300	

---

*C-Congruent, P-Peritectic

In our experiments, we have grown substituted crystals from mixed rare earth alloy liquid melts. The substituted single crystals, $(R_1)_{1.8}(R_2)_{0.2}Fe_{14}B$, have been grown from 10% substituted $(R_1)_{14.4}(R_2)_{1.6}Fe_{77}B_7$ alloy melt. The X-ray fluorescence analysis indicated 10% substitution of rare earth atoms in the crystal.

SUMMARY AND CONCLUSIONS

The study of solidification microstructure established the peritectic solidification of $R_2Fe_{14}B$ from liquid melt and confirmed the existence of a two phase field between Fe and $R_2Fe_{14}B$ in the ternary rare earth-iron-boron system. The starting crystal growth alloy composition, $R_{16}Fe_{77}B_7$, was also established from solidification microstructures. Based on the isomorphous replacement of rare earth atoms and solidification microstructures, the mixed rare earth alloy melt for the growth of substituted crystals was $(R_1)_{16-16x}(R_2)_{16x}Fe_{77}B_7$, where X is the fraction of R_2 substituting R_1.

Information on crystal growth conditions was obtained from solidification microstructures, e.g., dendritic solidification and enrichment of eutectic mixture in the liquid melt. The dendritic solidification was prevented by maintaining a high temperature gradient and

52

by growing the crystals at a slow growth rate, 3 to 5 mm an hour. The enrichment of eutectic mixture in the liquid melt was avoided by stopping the crystal growth process when approximately 40% of the charge by weight has been pulled.

The seed crystals were obtained by growing a crystal from proper melt, $R_{16}Fe_{77}B_7$, and using either a polycrystalline $R_{16}Fe_{77}B_7$ or Fe rod for pulling. This seed crystal was then used to grow larger crystals in the direction of growth of seed. The natural growth directions changed with the change of rare earth atoms.

Using all the information and conditions mentioned above, single crystals of $R_2Fe_{14}B$, Figure 12, and $(R_1)_{2-x}(R_2)_xFe_{14}B$, up to 10% substitution, have been grown by tri arc Czochralski and levitation Czochralski processes. The water cooled copper hearth and levitation process eliminated the reaction between crucible material and reactive rare earth containing melt.

A high crystal quality, as indicated by well formed Laue spots, was observed in the crystals grown at a slow growth rate, 3 to 5 mm an hour. However, the low angle boundaries were still present in some crystals. The twinning, straining and dendritic growth were observed in many crystals grown at a rate of 8 mm an hour.

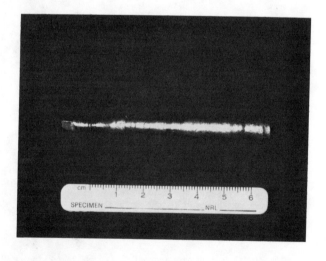

Figure 12. $Nd_2Fe_{14}B$ single crystal boule grown with proper conditions.

REFERENCES

1. N.C. Koon and B.N. Das, J. Appl. Phys. 55, 2063 (1984).

2. B.N. Das, P. D'Antonio and N.C. Koon (unpublished - manuscript prepared for publication).

3. J.B. Milstein, N.C. Koon, L.R. Johnson and C.M. Williams, Mat. Res. Bull. 9, 1617 (1974).

4. N.C. Koon, B.N. Das, M. Rubinstein and J. Tyson, J. Appl. Phys. 57, 4091 (1985).

5. J.F. Herbst, J.J. Croat, F.E. Pinkerton and W.B. Yelon, Phys. Rev. B229, 4176 (1984).

6. M. Sagawa, S. Fujimura, N. Togawa, H. Yamamoto and Y. Matsuura, J. Appl. Phys. 55, 2083 (1984).

7. D. Givord, H.S. Li and R. perrier de la Bathie, Solid State Commun. 51, 857 (1984).

8. D.E. Swets, J. Crystal Growth, 75, 277 (1986).

9. H.H. Stadelmaier, N.A. El-Masry and S. Cheng, Materials Letters, 2, 169 (1983).

10. G.A. Chadwick, in Fractional Solidification, Volume I, edited by M. Zief and W.R. Wilcox (Marcel Dekker, Inc., Publishers, New York, 1967), p. 113.

11. M. Sagawa, S. Fujimura, H. Yamamoto, Y. Matsuura and K. Hiraga, IEEE Trans. Magn., MAG-20, 1584 (1984).

12. R.P. Elliott, in Constitution of Binary Alloys, First Supplement, (McGraw-Hill, New York, NY 1965), p. 442.

13. F.A. Shunk, in Constitution of Binary Alloys, Second Supplement, (McGraw-Hill, New York, NY, 1969), p. 231, 349.

14. W.G. Moffatt, in Binary Phase Diagrams Handbook, (General Electric Company, Schenectady, NY, 12305, 1976).

15. K. Nassau, L.V. Cherry and W.E. Wallace, J. Phys. Chem. Solids, 16, 123 (1960).

THIN FILM FABRICATION OF $R_2Fe_{14}B$ COMPOUNDS

J.F. ZASADZINSKI, C.U. SEGRE, E.D. RIPPERT, J. CHRZAS AND P. RADUSEWICZ

Department of Physics, Illinois Institute of Technology, Chicago, IL 60616

ABSTRACT

Thin films ($\simeq$4000 Å) of $R_2Fe_{14}B$ compounds have been synthesized by d.c. triode sputtering for R=Nd, Sm and Er. Deposition onto single-crystal substrates heated to 600°C and with rates of 1.3 Å/s results in nearly epitaxial film growth such that the c-axis of the tetragonal structure is perpendicular to the film plane. As a consequence, intrinsic anisotropy effects are observed in the magnetic properties of the as-made films. Although sapphire and quartz substrates give the best results, similar directed growth is observed on recrystallized Nb foils. Magneto-optical measurements on the Nd-based compounds give a remnant polar Kerr angle θ_r=0.22° and coercive field H_c=5.5 kOe. The easy axis is in the film plane for R=Sm and Er and a remnant magnetization greater than 80% of saturation is observed for H applied parallel to the film plane. Electrical resistivity measurements on $Nd_2Fe_{14}B$ films indicate that spin-disorder scattering from the rare earth site is important and features from both the Curie and spin reorientation temperatures are observed.

INTRODUCTION

The discovery of Fe-based permanent magnet materials with energy products greater than 45 MGOe for $Nd_2Fe_{14}B$ and related alloys [1,2] has spurred a tremendous amount of research into the fabrication of these compounds in bulk and thin film form. Thin film fabrication is a powerful method of materials synthesis which offers a unique way of investigating both the basic and applied aspects of the magnetic compounds, $R_2Fe_{14}B$ (R=rare earth). Micro-stepper motors [3] and electronic circuits [4] are but a few of the applications in which thin film techniques may be used. The strong anisotropies found in the $R_2Fe_{14}B$ compounds are a source of fundamental interest and similarly provide some unique magnetic structures. It is well-known that in order to obtain the best magnetic properties, it is important to have small grain size and multi-phase material as well as preferential orientation. Nevertheless, the ability to grow single crystal films epitaxially is of fundamental importance for being able to measure the intrinsic properties of these materials and to better understand the origin of the magnetic interactions which make these materials promising. It is also the first step in fabricating superlattices on this class of compounds.

We report that single-phase thin films of $R_2Fe_{14}B$ compounds can be grown with a high degree of epitaxy such that essentially all of the crystallites are oriented with their c-axis perpendicular to the film plane. This permits investigation of intrinsic properties such as the magnetic anisotropy, the electron scattering and the magnon spectrum. In addition, it is found that, to a reasonable degree, the orientation is maintained for a variety of substrates including sapphire, quartz and niobium. Thin films of the Sm- and Er-based compounds can be formed with the magnetic easy axis lying entirely in the film plane whereas $Nd_2Fe_{14}B$ aligns perpendicular to the film plane. In this configuration, $Nd_2Fe_{14}B$ has potential as a perpendicular recording medium and we report measurements of the polar Kerr rotation.

EXPERIMENT

The films were prepared by d.c. sputtering using a triode source which allowed independent control of target potential and deposition rate. The target was a small button ($\simeq 0.5$ inch diameter) formed by melting the elements in an Argon arc furnace. A cryopumped, bakeable, stainless steel chamber provided a typical vacuum of 1.5×10^{-7} Torr prior to sputtering. Those sputtering conditions which remained fixed for most of the depositions were: Argon pressure (3.5 mTorr), target potential (700 V) and deposition rate (1.3 A/s). With these conditions the film structure and texture were most sensitive to the substrate temperature, which was varied from 400°C to 700°C. Substrates were single crystal sapphire and quartz mounted to a molybdenum block heater capable of producing temperatures of 1000°C at the substrate surface. In addition, we used Nb and Ta foils as substrates. The Nb foils were recrystallized by ultra-high vacuum annealing near the melting point which resulted in large, single-crystal grains typically 1.0 mm diameter or larger. The Ta foils were ordinary, untreated, cold-rolled samples.

Measurements included X-ray diffraction, SEM micrographs, low and high temperature electrical resistivity and Kerr rotation. Magnetization measurements were performed with a SQUID susceptometer capable of reaching temperatures as low as 5.0 K and fields up to 50 kOe.

RESULTS AND DISCUSSION

Structural Properties

Both the structure and texture of the $R_2Fe_{14}B$ films were quite sensitive to substrate temperature. Films deposited at temperatures of 500°C and below were amorphous as determined by X-ray diffraction. At a temperature of 600°C an unusually high degree of crystalline orientation is observed as shown in Fig. 1, where a <u>wide-spectrum</u> diffraction pattern is presented for $Nd_2Fe_{14}B$. The dominant peaks are (004) and (006) which indicate that the c-axis of the tetragonal structure is perpendicular to the film plane. The identification of the (006) plane has been firmly established due to the sharpness of all of the X-ray peaks and the reproducible thin-film geometry which easily gives a precision of 0.05°. Lattice parameters calculated from the indexed peaks (a=8.80 Å, c=12.23 Å) agree well with those reported for bulk samples [5]. The ability to distinguish between an (006) plane (2θ=44.65°) and the nearby (331) plane (2θ=44.34°) is straightforward. We can also rule out the presence of α-Fe which has a (110) reflection at 2θ=44.65°. This will be evident when we discuss the X-ray patterns of $Sm_2Fe_{14}B$ and $Er_2Fe_{14}B$. Note that the other peaks in Fig. 1 also correspond to planes with c-axes close to the film normal. To obtain a more quantitative measure of film texture we have compared the integrated intensities of all the visible peaks to their

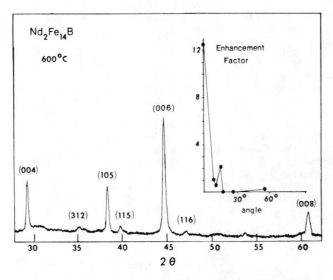

Fig. 1 X-ray diffraction pattern for $Nd_2Fe_{14}B$ deposited at 600°C substrate temperature. The inset shows the diffraction peak intensity enhancement, relative to a randomly oriented powder, for an (hkl) plane plotted versus the angle that the normal to the plane makes with the c-axis of the tetragonal structure.

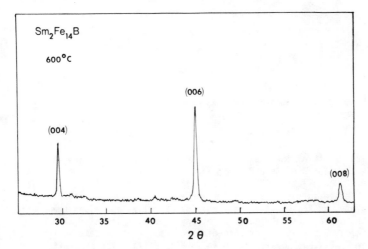

Fig. 2 X-ray diffraction pattern for $Sm_2Fe_{14}B$ deposited at 600°C substrate temperature.

corresponding intensities in a powder pattern, giving an enhancement factor. By plotting the enhancement factor of each reflection versus the angle the c-axis (of any crystallite giving rise to that diffraction line)

makes with the film normal (inset of Fig. 1), the high degree of
perpendicular orientation becomes apparent.

The X-ray pattern for $Er_2Fe_{14}B$ films are quite similar to Fig. 1 when
the substrate temperature is 600°C [6]. A more startling result was
obtained for $Sm_2Fe_{14}B$ films deposited at 600°C as shown in Fig. 2. Here we
observe virtually no reflections other than those which correspond to the
c-axis growing along the film normal (c=12.11 Å from these peaks). As in
Fig. 1, the sharp line widths are instrumentally broadened, which give a
lower bound estimate of the grain size as 3000-5000 Å. The absence of α-
Fe is easily observed in the Sm- and Er-based films. SEM micrographs as
well as etched films viewed under an optical microscope show grains with
lateral dimensions of a few microns. We are not aware of any other ternary
compound fabricated by sputter beam deposition which shows such a high
degree of epitaxial growth. Single crystal films might be possible by
optimizing sputtering conditions.

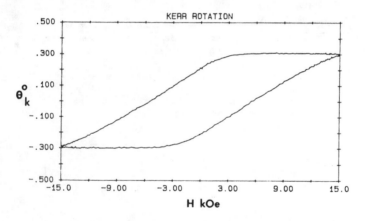

Fig. 3 Polar Kerr hysteresis loop for frontside of $Nd_2Fe_{14}B$ film on quartz. Backside
measurement gives a higher saturation value θ_k=0.45° indicating there is some surface
degradation.

The growth of the c-axis perpendicular to the film plane was
reproducible for R=Nd,Sm and Er as long as the substrate temperature was
near 600°C. Considering this, we anticipated magnetic anisotropy effects
to be observed in the as-made films and such was the case. At room
temperature, $Nd_2Fe_{14}B$ has an easy axis of magnetization along the c-axis
and, considering its intrinsic anisotropy energy, this material has
potential as a perpendicular recording medium. A polar Kerr hysteresis
loop for $Nd_2Fe_{14}B$ on quartz is shown in Fig. 3 [7]. The saturation Kerr
rotation angle is 0.3° and the remanence is 0.22° while H_c=5.5 kOe. No
protective layer was deposited in situ and there is evidence of surface
degradation. This was determined by measuring the Kerr angle on the back
side of the film through the transparent quartz substrate. This
measurement gave a saturation value of 0.45°, a value which is probably
indicative of the intrinsic magneto-optical properties of this material.
Curiously, the backside Kerr hysteresis loop showed much smaller remanence

and coercive field than the front side. It should be remembered that properties such as the remnant magnetization and coercive field in $R_2Fe_{14}B$ compounds are most strongly coupled to the microstructure such as grain size and grain boundary composition. Single crystals often show no hysteresis at all [8] whereas bulk [9] and thin film [10] samples in which the grain size has been reduced to less than 300 A, display large coercivities. Considering the large grains in our films we expect modest coercivities, nevertheless H_C=5.5 kOe is a relatively large value compared to typical magneto-optic thin films.

The film texture of $Nd_2Fe_{14}B$ was identical for single-crystal substrates of sapphire and quartz and there was no evidence of interdiffusion of film and substrate at 600°C. It is of interest to investigate whether oriented $R_2Fe_{14}B$ films can be grown on other substrate materials, particularly transition metals. We have successfully grown films on foils of Ta and Nb also heated to 600°C. The Ta was ordinary, cold-rolled sheet metal and although the $Nd_2Fe_{14}B$ structure was observed in the X-ray diffraction pattern, the reflections indicated a random orientation. The Nb foils were pre-treated by ultra-high vacuum annealing for several hours near the melting point. This resulted in

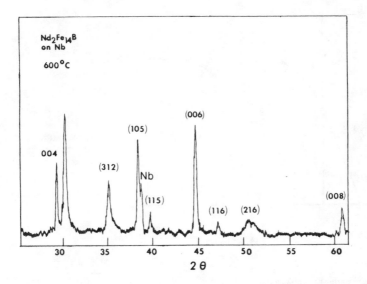

Fig. 4 X-ray diffraction pattern for $Nd_2Fe_{14}B$ on recrystallized Nb foil. Preferred orientation is maintained. Unidentified peak at 30° is possibly from Nd_2Fe_{17}.

significant recrystallization with large single-crystal grains of order 1.0 mm^2 up to 500 mm^2. The resulting X-ray diffraction pattern for $Nd_2Fe_{14}B$ grown on such foils heated to 600°C is shown in Fig. 4. Again, directed grain growth is observed with the dominant peaks of $Nd_2Fe_{14}B$ structure coming from the (006) and (105) planes. The large peak at 2θ=30.35° has not been identified but certainly does not correspond to any allowed reflection of $Nd_2Fe_{14}B$. We suspect that substrate/film interdiffusion has possibly taken place. It is well-known that loss of even small amounts of Boron leads to the Nd_2Fe_{17} structure and the (113) reflection of that phase is at $2\theta=29.95°$. Perhaps an alloy of this phase is forming at the Nb surface but this is conjecture at this point. Despite the impurity phase, it is encouraging to observe that oriented growth of the $Nd_2Fe_{14}B$ structure is possible on metallic substrates. It appears that the epitaxy of $R_2Fe_{14}B$ films does not depend on registry with the substrate's atomic structure. Rather, what is necessary is an atomically smooth and ordered surface provided by large single crystals.

Magnetization and Resistivity

Magnetization curves measured perpendicular to the film plane of $Nd_2Fe_{14}B$ on Nb are shown in Fig. 5. In the perpendicular direction, the internal field in the thin film geometry $H=H_a-4\pi M$ where H_a is the applied field. Subtracting $4\pi M$ from the virgin curve of Fig. 5 reveals that saturation occurs for $H\simeq 7$ kOe, similar to the c-axis direction for single crystals. The return curve is characterized by a remnant field which is 63% of saturation and a coercive field of 4.5 kOe. The reduced saturation magnetization compared to bulk values is possibly due to the impurity phase.

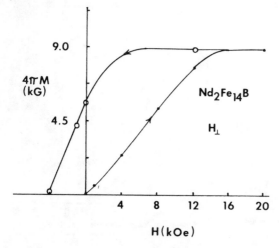

Fig. 5 Magnetization versus field applied perpendicular to film plane for $Nd_2Fe_{14}B$ on Nb. Closed circles are the virgin curves; open circles are the return curve.

The electrical resistivity of $R_2Fe_{14}B$ compounds can be an important probe of magnetic phase transitions and can give insight into the RKKY interaction which mediates the exchange coupling between the local rare earth 4f moments. This interaction can be quite strong as indicated in recent work by Fuerst et al [11] on $Er_2Fe_{14-x}Mn_xB$ pseudoternaries. At high Mn concentrations, the Fe sublattice becomes diluted and the Curie temperature, T_c, drops to a value below what was previously the spin reorientation temperature, T_s. At this point the new Curie temperature is T_s and it is determined solely by the Er-Er interaction. These results indicate that spin reorientation is really a result of the rare earth sublattice magnetization. One expects therefore to observe a freezing-out of the spin disorder scattering of conduction electrons below the reorientation temperatures in $Nd_2Fe_{14}B$ and $Er_2Fe_{14}B$.

We observe such an effect in the electrical resistivity of $Nd_2Fe_{14}B$ as shown in Fig. 6. Just below 200 K, the resistivity drops more rapidly than the extrapolated linear dependence observed between 200 K and 580 K. From 200 K to 100 K the a-axis component of magnetization develops and this must correspond to a final ordering of the rare earth sublattice. The size of the resistivity change at T_s is a strong indication that a principal scattering mechanism for conduction electrons in this compound is the magnetic scattering off the local 4f moments. Similarly large decreases in resistivity are observed in metallic rare earth compounds such as NdB_6 [12] just below the magnetic ordering temperature. We notice that the

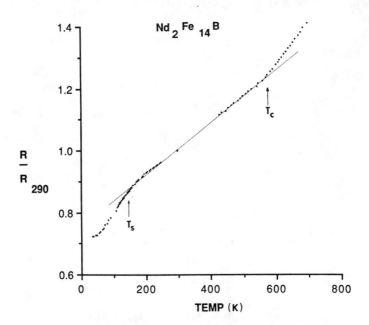

Fig. 6 Resistivity versus temperature for $Nd_2Fe_{14}B$ film normalized to the room-temperature value ($\simeq 170\ \mu\Omega$-cm). Changes in spin disorder scattering are evident at both the Curie temperature, T_c, and spin-reorientation temperature, T_s.

resistivity again deviates from the linear dependence for temperatures above 590 K. Since this temperature is quite close to the Curie temperature, T_c, of $Nd_2Fe_{14}B$, it suggests that magnetic scattering is playing a role. However, the increased slope of R vs. T above T_c is opposite to that found in transition metal elements such as Fe or Ni where intraband d-d scattering dominates [13]. The high temperature behavior may be a result of a less-interesting effect such as decomposition of the film, but the proximity to T_c would then be mere coincidence. Considering the low temperature results, we believe that the resistivity increase above T_c

is a consequence of spin disorder scattering from both the Fe and rare earth sublattices.

The magnetization behavior of the Er- and Sm-based compounds is similar as both compounds have an easy axis in the basal plane of the tetragonal structure and therefore parallel to the film plane. Our earlier work on $Er_2Fe_{14}B$ [7] showed strong anisotropy between parallel and perpendicular directions which was consistent with the X-ray diffraction analysis. Similarly, the $Sm_2Fe_{14}B$ films display an easy axis in the parallel direction as shown in Fig. 7 but a significantly higher saturation magnetization than the Er-based compound as a result of the parallel alignment of Sm and Fe moments. The remnant field is 80% of saturation or close to 9.0 kG but the H_c=1.0 kG is lower than in $Er_2Fe_{14}B$. This is consistent with the presence of large crystallites in these films, approaching the single-crystal behavior of no hysteresis [8].

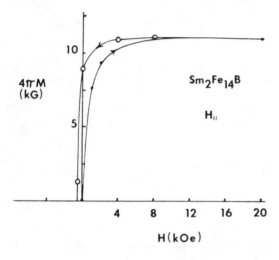

Fig. 7 Magnetization versus field applied parallel to film plane for $Sm_2Fe_{14}B$. Closed circles represent the virgin curve which is similar to the easy axis results on single crystals. Return curve (open circles) shows high remanence.

SUMMARY

Under proper conditions, thin films of $R_2Fe_{14}B$ compounds can be grown which have the following properties: single-phase structure; crystallites with lateral dimensions of a few microns; and an unusually high degree of preferential orientation. The most favorable growth direction on 600° substrates is such that the c-axis is perpendicular to the film plane. The best samples have bulk magnetization values and display anisotropies consistent with the easy and hard axis results found in single crystals of these compounds. Indeed, single crystal films might be possible. The most important result is that our films display properties which are intrinsic to $R_2Fe_{14}B$ compounds. This makes it possible to quickly synthesize, by co-sputtering, and study wide ranges of pseudoternary alloys which are important in understanding the magnetic properties of these materials.

REFERENCES

[1]   J.J. Croat, J.F. Herbst, R.W. Lee and F.E. Pinkerton, J. Appl. Phys. 55, 2078 (1984).

[2]   H. Sagawa, S. Fujimura, N. Togawa, H. Yamamoto and Y. Matsura, J. Appl. Phys. 55, 2083 (1984).

[3]   B.E. Carlisle, Machine Design, July 1985, p.73.

[4]   D.D. Stancil, J. Appl. Phys. 61, 4111 (1987).

[5]   S. Sinnema, R.J. Radwanski, J.J.M. Franse, D.B. de Mooij and K.H.J. Buschow, J. Magn. Magn. Mat. 44, 333 (1984).

[6]   Kerr measurements provided by S.C. Shin of Kodak Research Laboratories, Rochester, NY, using a 632.8 nm He-Ne laser.

[7]   J.F. Zasadzinski, C.U. Segre and E.D. Rippert, J. Appl. Phys. 61, 4278 (1987).

[8]   H. Hiroyski, H. Yamauchi, Y. Yamaguchi, H. Yamamoto, Y. Nakagawa and H. Sagawa, Solid State Commun. 54, 41 (1985).

[9]   F.E. Pinkerton, J. Magn. Magn. Mat. 54-57, 579 (1986).

[10] F.J. Cadieu, T.D. Cheung, L. Wickramasekara and N. Kamprath, IEEE Trans. Mag. 22, 752 (1986).

[11] C.D. Fuerst, G.P. Meisner and F.E. Pinkerton, J. Appl. Phys. 61, 2314 (1987).

[12] Z. Fisk and G.W. Webb in Treatise on Materials Science and Technology Vol. 21, (Academic, New York, 1981).

[13] R.H. White and T.H. Geballe, Long Range Order in Solids (Academic, New York, 1979).

INITIAL MAGNETIZATION BEHAVIOR OF RAPIDLY QUENCHED
NEODYMIUM–IRON–BORON MAGNETS

F. E. Pinkerton

Physics Department, General Motors Research Laboratories, Warren, MI 48090-9055

ABSTRACT

Initial magnetization and demagnetization data are reported for three forms of rapidly solidified Nd–Fe–B permanent magnet materials: melt-spun ribbons, hot pressed magnets, and die upset magnets. In all three materials the results are consistent with domain wall pinning at grain boundary phases as the coercivity mechanism. Optimally quenched ribbons are comprised of randomly oriented single domain $Nd_2Fe_{14}B$ grains, and both initial magnetization and demagnetization are controlled by strong domain wall pinning at grain boundaries. Maximum coercivity is accompanied by a low initial permeability. Coercivity is reduced in overquenched ribbons by partial retention of a magnetically soft amorphous or very finely crystalline microstructure. Coercivity decreases in underquenched ribbons because wall pinning weakens as the grain size increases above optimum. Correlation of magnetization and demagnetization behaviors suggests that maximum coercivity in ribbons is largely determined by the resistance to domain wall formation in grains smaller than the single domain particle limit. Grain size is much less important in the aligned die upset magnets. Domain walls are initially free to move until they become strongly pinned at grain edges, and complete magnetization requires an applied field greater than the coercive field. Hot pressed magnets show a mixture of ribbon and die upset behavior.

1. INTRODUCTION

The recent development of permanent magnet materials based on the ternary compound $Nd_2Fe_{14}B$[1] has sparked considerable scientific and technological interest.[2,3] Excellent permanent magnet properties have been achieved by both rapid solidification[4] and powder metallurgy[5] techniques. Rapid solidification by melt-spinning produces magnetically isotropic ribbon with a remanence B_r and intrinsic coercivity H_{ci} of 8 kG and 15 kOe, respectively. Ribbons can be consolidated into fully dense bulk magnets by hot pressing[6] to obtain a nearly isotropic magnet with an energy product $(BH)_{max} \sim 15$ MGOe. Subsequent hot deformation, or die upsetting, induces alignment into a previously hot pressed magnet, yielding energy products in excess of 40 MGOe. This paper reviews the initial magnetization and demagnetization process in the three stages of rapid solidification-based magnet materials.[7,8] The strong links between magnetization behavior and microstructure are discussed, together with the implications for the coercivity mechanism in each stage.

Much of the emphasis in this paper will be on the properties of melt-spun Nd–Fe–B ribbons, which have a unique finely microcrystalline microstructure with a grain size smaller than the estimated single domain particle limit.[9] While this microstructure is suggestive of a material in which the coercivity mechanism is of the single-domain particle type, it is known that ribbons do not strictly adhere to this description. Coercivity appears to be controlled primarily by domain wall pinning at an intergranular second phase, rather than arising from coherent rotation.[10] Nevertheless, there is ample evidence that the magnetization strongly resembles single-domain behavior. Transmission electron microscopy studies[9] have shown that individual grains do consist of a single magnetic domain in the thermally demagnetized state. Consequently initial magnetization and demagnetization are controlled by the same domain wall pinning mechanism,[8] but with quantitative differences which may be ascribed to particle interaction effects.[11] The quench rate dependence of the magnetization suggests that the small grain size is itself an important contribution to the pinning energy.[7] This will be a recurring theme in the description of ribbon magnetization: Single-domain behavior provides the starting framework for describing the magnetization, but it must be modified to include domain wall motion and magnetic interactions between grains.

In die upset magnets grain size plays a much less important role in the coercivity. The unconventional microstructure consisting of aligned platelet-like grains[12] again leads to strong domain wall pinning at grain boundaries and second phases. The die upset magnets have free domain walls in the thermally demagnetized state, and magnetizing fields in excess of the coercive field are required in order to develop full remanence and coercivity.[8]

2. EXPERIMENT

Rapidly solidified Nd–Fe–B ribbons were prepared by melt-spinning, in which molten alloy is quenched on the surface of a rapidly spinning wheel. The details of the process have been given elsewhere.[4] The magnetic properties of the ribbons are sensitive to the substrate velocity v_s, or quench rate, of the melt-spinner, and the best combination of intrinsic coercivity and remanence was obtained for ribbons quenched at substrate velocities of v_s= 20-21 m/s. Slower substrate velocities produced lower quench rate (underquenched) ribbons, while ribbons spun faster than optimum were overquenched.

Initial magnetization curves of coarsely crushed ribbon samples were generated using a vibrating sample magnetometer (VSM) and a superconducting magnet with a maximum applied field of 90 kOe. Demagnetization curves were obtained by subsequent recoil from 90 kOe. Magnetization behavior at intermediate fields was explored using a VSM in conjunction with an 18 kOe maximum field electromagnet. Initially unmagnetized samples were first partially magnetized by a positive field H_m and then the recoil magnetization from H_m was measured. The remanent moment $B_r(H_m)$ is a measure of moment reversal during the magnetization process. The development of coercivity and remanence was traced by repeating the measurement for a series of values of H_m, using a fresh sample from a single homogenized batch of crushed ribbon for each value of H_m. Demagnetization behavior was similarly traced by first fully magnetizing a sample with a 110 kOe pulsed field, and then measuring the remanent moment $B_d(H_m)$ after demagnetizing in a reverse field of magnitude H_m, for a series of increasing values of H_m. For comparison the first quadrant remagnetization behavior after dc field demagnetization was also examined.

Hot pressed and die upset magnets[6] were prepared from overquenched ribbons by pressing for ~2 min at ~20 kpsi and ~725 C. Magnetic measurements were made on cubes cut from the finished piece with one face normal to the press direction (which for hot-deformed magnets corresponds to the direction of preferential alignment of the magnetically easy c-axis). The family of demagnetization curves along the press direction after magnetization to a field H_m were generated in the manner described above for ribbons. The cubes were thermally demagnetized between each measurement by heating to 625 K. As with the ribbons, the initial magnetization curve was measured up to an applied field of 90 kOe. Remagnetization curves after dc field demagnetization were also measured on the die upset sample. Demagnetizing fields have been taken into account on all measurements.

3. OPTIMALLY QUENCHED RIBBONS

3.1 Microstructure and Domain Structure

High coercivity in ribbons arises from a favorable combination of several microstructural features. Transmission electron microscope studies have shown that optimum ribbons are comprised of small spheroidal $Nd_2Fe_{14}B$ crystallites, each of which is surrounded by a thin (~3 nm) shell of a non-crystalline Nd-rich phase.[4,9] The bulk isotropy reflects the random orientation of the uniaxial crystallites. Because the quench rate varies across the thickness of the ribbon, the grain size tends to increase in going from the quench surface to the free surface of the ribbon.[4] The grain diameters (20-80 nm) are less than the critical diameter below which an isolated spherical particle will have a single magnetic domain (estimated to be 150 and 300 nm from Lorentz microscopy[9] and optical Kerr effect observations,[13] respectively).

Lorentz microscopy on thermally demagnetized ribbons has indeed confirmed that individual $Nd_2Fe_{14}B$ grains consist of a single magnetic domain, with the domain walls pinned at the grain boundaries.[9] (These are not classical 180° domain walls, but rather reflect the more subtle and variable magnetization changes between the randomly oriented grains.) There is ample evidence, however, that coercivity does not simply originate from a classical single domain particle mechanism. In particular, the observed coercivities are less than half of the value of $H_{ci} = 37$ kOe predicted by Stoner and Wohlfarth for the coherent rotation model.[10] Magnetization reversal in ribbons is characterized by a field considerably smaller than the coherent rotation field and presumably involves the propagation of a domain wall through each grain. The grains are nevertheless "single domain particles" in that the formation of domain walls within the very small grains is energetically unfavorable in the absence of an applied field. The walls are expelled to the grain boundaries, where they are pinned at the amorphous Nd-rich grain boundary phase.[9] This domain structure suggests that the energy barrier to the formation of domain walls within the very small grains is an important contribution to the wall pinning energy. The behavior of underquenched ribbons provides further indirect evidence supporting this hypothesis, as will be discussed in Section 5. This particle size effect is intimately tied to the random orientation: We will see in Section 7 that aligned grains, such as those in die upset magnets, can form extended domain structures across many grains, so that the size of a single grain is no longer a controlling feature in the magnetization behavior.

Lorentz microscopy also shows evidence of a more complex cooperative behavior among the grains. Superimposed on the ubiquitous local domain wall structure which arises from the random orientation, there are extended zig-zag domain walls following grain boundaries.[9] These boundaries seem to demarcate regions within which the magnetizations of the grains are correlated, giving rise to small clusters having a net magnetization. Such correlations have observable effects on the magnetic behavior, as will be seen in Section 3.3.

3.2 Initial magnetization

The initial magnetization curve of optimally quenched ribbons having a coercivity of $H_{ci} = 12.2$ kOe is shown in Fig. 1. The characteristic "s" shape, marked by a low initial permeability, divides the curve into two regimes. At small fields, the gradual magnetization increase is due largely to moment rotation away from the easy axis in misaligned $Nd_2Fe_{14}B$ grains. The behavior for rotation only is indicated by the dashed curve; the departure from it is due to the presence of a small volume fraction of lower coercivity material in the sample. Fields near H_{ci} are required to overcome the wall pinning energy and reverse the $Nd_2Fe_{14}B$ grains, at which point the magnetization increases rapidly. The demagnetization curves after partial magnetization to field values H_m along the initial magnetization curve show that in this regime the sample develops the bulk of its remanence and coercivity. This is summarized in Fig. 2, where the remanence $B_r(H_m)$ and coercivity $H_{ci}(H_m)$ are plotted as a function of H_m. These results agree with other measurements on rapidly solidified ribbons.[14,15] While this type of magnetization behavior is frequently associated with bulk domain wall pinning,[14] it is clear from Lorentz microscopy that in this case the pinning occurs in the grain boundaries, and not within the $Nd_2Fe_{14}B$ grains themselves.

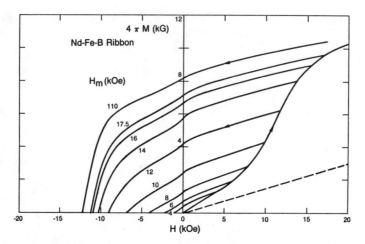

Figure 1. Initial magnetization and demagnetization curves for melt-spun Nd–Fe–B ribbons. The number on each curve is the maximum internal field H_m to which the sample was exposed prior to demagnetization. The dashed curve represents the magnetization change due to moment rotation away from the easy axes in a randomly oriented collection of uniaxial particles, calculated according to Reference 10.

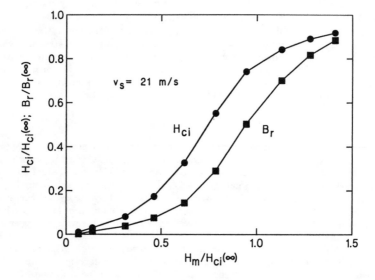

Figure 2. Remanence B_r (normalized by the remanence $B_r(\infty)$ of a fully magnetized sample) and intrinsic coercivity H_{ci} (normalized to $H_{ci}(\infty)$) as a function of the internal magnetizing field H_m (normalized to $H_{ci}(\infty)$) for optimally quenched Nd–Fe–B ribbons. The lines are guides to the eye.

The family of demagnetization curves in Fig. 1 provides another important clue to the magnetization behavior. The recoil curves in the first quadrant are parallel to the recoil curve for the fully magnetized sample, which in turn is described very well by the Stoner-Wohlfarth model of random uniaxial particles.[10] The magnetization change as the field is reduced from H_m occurs only by moment rotation back toward the easy axis; there is no significant domain wall relaxation. Domain walls are pinned at grain boundaries at all stages of sample magnetization, from which we surmise that each grain reverses as a unit, and that magnetization proceeds by reversing one grain (or at most a small group of grains) at a time.

The correspondence between the onset of reversal during magnetization and the coercivity H_{ci} observed during demagnetization follows as a consequence of the single domain structure, in which domain walls are pinned at grain boundaries in both the thermally demagnetized and fully magnetized states. The initial magnetization and demagnetization processes occur, at least qualitatively, by the same process of domain wall motion across small $Nd_2Fe_{14}B$ grains, and are controlled by the same coercivity mechanism. This is illustrated also by the close resemblance of the B_r values shown in Fig. 2 to the quantity $\frac{1}{2}(1 - B_d/B_r(\infty))$ plotted in Fig. 3, which measures the fraction of reversed magnetization during demagnetization. Here B_d is the remanence after demagnetizing a fully magnetized sample in a reverse field $-H_m$.

The initial magnetization behavior of Nd–Fe–B ribbons is characteristic of all high coercivity melt-spun R-Fe-B materials we have studied (R= Pr, Dy, and Tb). The initial magnetization of Pr-Fe-B (with v_s chosen to give optimum magnetics) is virtually identical to that of Nd–Fe–B. The giant room temperature coercivities of melt-spun Dy-Fe-B and Tb-Fe-B (60 kOe and greater than 90 kOe, respectively)[16] are reflected in their initial magnetization,[8] which has an extended regime of moment rotation, with a dramatic increase in magnetization as moment reversal occurs at fields near $H_{ci}(\infty)$.

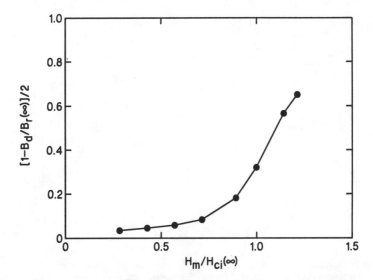

Figure 3. Fraction of moment reversed during demagnetization, calculated from the remanence B_d, as a function of the internal reverse field H_m for optimally quenched Nd–Fe–B ribbons. The line is a guide to the eye.

3.3 Evidence of Particle Interaction Effects

A quantitative comparison of the initial magnetization and demagnetization behaviors shows that the simple picture outlined above is only an approximate description. The quantities $B_r(H_m)$, which measures the degree of magnetization reversal at H_m on the initial magnetization curve, and $B_d(H_m)$, which gives similar information along the demagnetization curve, provide a direct test for the symmetry of the initial magnetization and the demagnetization processes. For a noninteracting assembly of uniaxial single domain particles, these two quantities obey the relation:[17]

$$B_d(H_m) = B_r(\infty) - 2B_r(H_m). \tag{1}$$

Although the coercivity mechanism is not of the coherent rotation variety, Gaunt et al. have noted that the above relation will still hold provided that domain walls interact with the same density and distribution of pinning sites during both initial magnetization and demagnetization.[18] A plot of B_d as a function of B_r in Fig. 4 shows a clear departure from the straight line representing Eqn. 1. We ascribe the relative difficulty in demagnetizing the sample at low B_r (small H_m) to magnetic interactions between

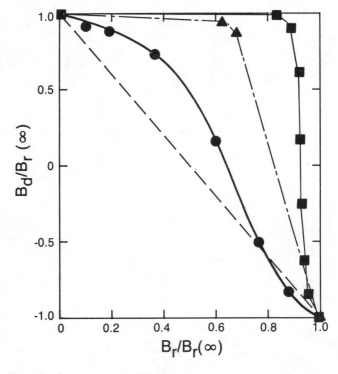

Figure 4. Normalized remanence B_d after demagnetization to a reverse field H_m vs. normalized remanence B_r after magnetization to H_m for melt-spun Nd–Fe–B ribbons (circles), hot pressed magnets (triangles), and die upset magnets (squares). Non-interacting particles would fall on the dashed line. The lines are guides to the eye.

neighboring grains, by which we mean that the wall pinning energy depends not only on the orientation of a given grain and the applied field, but also on the magnetizations of neighboring grains. For small H_m, it is then easier to initiate the reversal of any given grain during initial magnetization, where the local pinning energy may be reduced by neighboring grains whose magnetizations are already along the direction of the applied field, than during demagnetization, where all surrounding grains have magnetizations tending to oppose reversal. Reversal may occur in individual grains whose neighborhood happens to particularly favor reversal, or by the growth of aligned clusters at the expense of misaligned ones, or both. As more grains reverse, it becomes easier to reverse the remaining grains, and the bulk magnetization increases faster than it would for isolated particles.[11] Once more than half the grains have reversed, so that the moment distributions on magnetization and demagnetization are similar, Eqn. 1 is obeyed, as in the right-hand portion of Fig. 4.

The magnetic interaction between neighboring particles appears to be strong enough to favor alignment of their magnetizations to the extent allowed by the misorientation of their easy axes. Although such strong interactions might ordinarily be expected to give rise to extended domain structures in a crystallographically aligned material, such as those observed in die upset magnets,[12] in ribbons such extended structures are frustrated by the random orientation. Cooperative behavior is limited to correlations between the magnetizations of neighboring particles, which are organized into small clusters having a net magnetization. Along grain boundaries demarcating adjoining clusters, the domain walls coalesce into the extended zig-zag walls observed in Lorentz microscopy.[9]

In the fully magnetized state, the extended domain walls have been swept out (in effect forming a single cluster), but the local walls due to random orientation remain at all fields less than the anisotropy field. Upon reversing the field, there are initially no grains with magnetizations along the field direction, hence there is no tendency to lower the wall energy, and a somewhat higher field is required to begin reversing grains. It is primarily this delayed onset which appears as an asymmetry in plots of B_d vs. B_r.[7,8] Once started, however, reversal will be accelerated by the interactions, and because of the higher onset field will actually progress more rapidly than it did for initial magnetization, in effect "catching up."

Evidence for particle interactions is also found in the family of remagnetization curves for dc field demagnetized samples shown in Fig. 5. Each curve was generated from a fully magnetized sample by partially demagnetizing the sample in a dc reverse field and then following the subsequent remagnetization into the first quadrant. Remagnetization occurs along rotation lines, showing that, as in the case of initial magnetization, the domain walls are pinned at all stages of demagnetization. The dashed curve shows the initial magnetization from the thermally demagnetized state from Fig. 1. Magnetization reversal near $H_{ci}(\infty)$ is more abrupt in the field demagnetized sample than in the thermally demagnetized sample, which we attribute to a difference in the morphology of clusters in the field demagnetized state compared to that in the thermally demagnetized state. The alignment of clusters in thermally demagnetized material is random. By virtue of its magnetic history, however, the clusters in field demagnetized material will tend to be aligned either parallel or antiparallel to the direction in which the field was applied. This more ordered moment distribution encourages moment reversal, once initiated, to proceed more rapidly.

4. OVERQUENCHED RIBBONS

Coercivity is reduced in overquenched ribbons because formation of the $Nd_2Fe_{14}B$ phase is incomplete at very high quench rates. The ribbon quenched at the highest rate $(v_s = 32 \text{ m/s})$ is almost entirely amorphous,[9] and its soft magnetization behavior, shown in Fig. 6, indicates that there is little resistance to domain wall motion. The filled squares

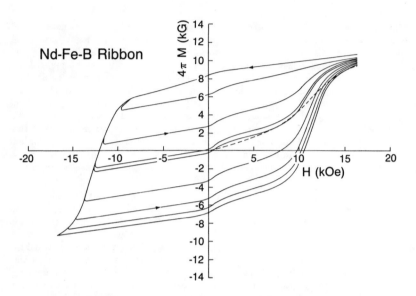

Figure 5. Remagnetization curves for a dc field demagnetized melt-spun Nd–Fe–B ribbon sample. The dashed curve indicates the initial magnetization from the thermally demagnetized state as shown in Figure 1.

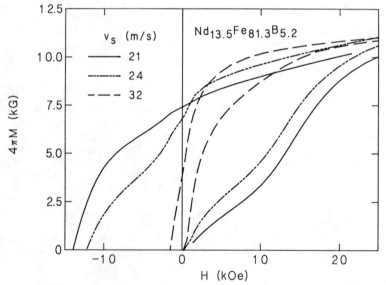

Figure 6. Initial magnetization and demagnetization curves for overquenched and optimally quenched ribbons. Parts of the first quadrant demagnetization curves have been suppressed for clarity.

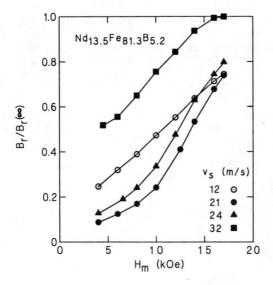

Figure 7. Remanence B_r (normalized by the remanence $B_r(\infty)$ of a fully magnetized sample) as a function of the internal magnetizing field H_m for $Nd_{13.5}Fe_{81.3}B_{5.2}$ ribbons melt-spun at various substrate velocities. The lines are guides to the eye.

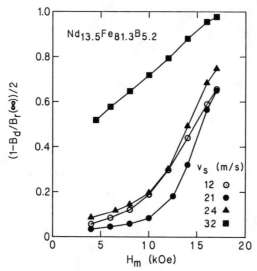

Figure 8. Fraction of moment reversed during demagnetization, calculated from the remanence B_d, as a function of the internal reverse field H_m for $Nd_{13.5}Fe_{81.3}B_{5.2}$ ribbons melt-spun at various substrate velocities. The lines are guides to the eye.

74

in Fig. 7 represent the remanent moment B_r obtained after partial magnetization to a field H_m. Most of what little remanence there is develops rapidly at low fields. Similarly, Fig. 8 shows that reversal during demagnetization occurs readily at low reverse fields. In contrast to the optimally quenched ribbon, the plot of B_d vs. B_r in Fig. 9 does fall on the dashed line representing Eqn. 1. This is consistent with the suggestion of Mishra[9] that coercivity in highly overquenched ribbons is due to weak domain wall pinning at isolated $Nd_2Fe_{14}B$ nuclei which act as independent pinning sites.

While Figs. 7 and 8 indicate that most of the moment reversal occurs at low fields, the much higher fields required to obtain full remanence imply the presence of a high coercivity component. The distribution of coercive fields responsible for magnetic hardness can be roughly visualized by examining the rate of moment reversal, $-dB_d/dH_m$, as a function of H_m. The moment reversal distribution, shown as the filled squares in Fig. 10, is dominated by very high reversal rates at low fields, but it has a secondary maximum near the field where the corresponding curve for optimum ribbons is sharply peaked. This small volume fraction of the ribbon which more closely resembles optimally quenched material most likely results from slower quenching at inhomogeneities in the ribbon.

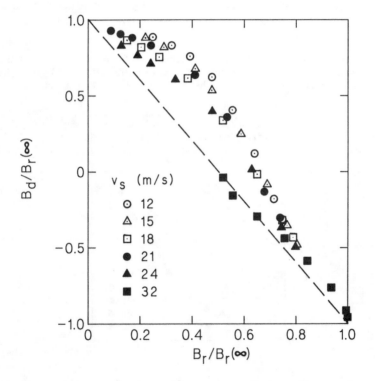

Figure 9. Normalized remanence B_d after demagnetization to a reverse field H_m vs. normalized remanence B_r after magnetization to H_m for $Nd_{13.5}Fe_{81.3}B_{5.2}$ ribbons melt-spun at various substrate velocities. Non-interacting particles would fall on the dashed line.

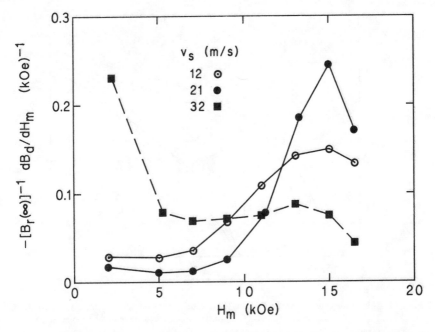

Figure 10. Rate of moment reversal on demagnetization as a function of the internal reverse field for $Nd_{13.5}Fe_{81.3}B_{5.2}$ ribbons melt-spun at various substrate velocities. The lines are guides to the eye.

When spun at $v_s = 24$ m/s, just above optimum, the ribbon contains a much larger fraction of the high coercivity $Nd_2Fe_{14}B$ component, and the initial magnetization and demagnetization behaviors shown in Fig. 6 approach those of the optimally quenched ribbon. For the magnetically hard portion, the development of B_r (Fig. 7) and moment reversal during demagnetization (Fig. 8) occur by a mechanism nearly identical to that of the optimally quenched ribbon, as can been seen in Fig. 9. The coercivity and remanence remain depressed, however, because the ribbon still contains a substantial amount of magnetically soft material. This material corresponds to the substrate surface of the ribbon, where the quench rate is still fast enough to produce amorphous or very finely crystalline material.[4]

5. UNDERQUENCHED RIBBONS

The loss of coercivity as the quench rate is reduced below optimum is associated with the growth of some $Nd_2Fe_{14}B$ grains larger than the optimum particle size,[9] especially large multi-domain grains on the outer surface of the ribbon.[4] Fig. 10 illustrates the shift of the coercivity distribution toward lower fields. At the same time, the degradation in the second quadrant loop shape is accompanied by an increase in the initial permeability from the thermally demagnetized state, as shown in Fig. 11. Moment reversal occurs at lower fields both during initial magnetization and demagnetization, as seen in Fig. 7 and Fig. 8. Results for $v_s = 15$ and 18 m/s lie between those for 12 and 21 m/s, and have been omitted for clarity.

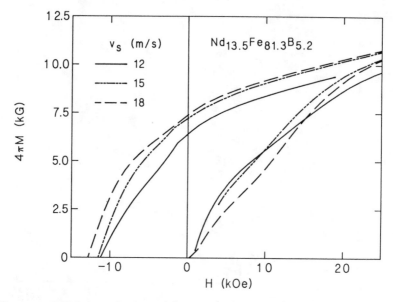

Figure 11. Initial magnetization and demagnetization curves for underquenched ribbons. Parts of the first quadrant demagnetization curves have been suppressed for clarity.

While domain wall pinning weakens as the grain size increases with underquenching, Fig. 9 shows that grain size has little effect on the relationship between B_r and B_d. Initial magnetization and demagnetization behavior are strongly coupled regardless of the grain size or coercivity distribution. Although a substantial fraction of the ribbon has grains large enough to contain domain walls in the thermally demagnetized state,[9] there is little tendency to adopt a behavior characteristic of die upset magnets,[8] in which B_r develops rapidly at low fields as domain walls are swept easily to the grain boundaries, but strong pinning there requires much higher fields for demagnetization. Apparently grains large enough to have easy initial domain wall motion also have reduced coercivity, with walls that are no more effectively pinned on demagnetization than on magnetization. This gives indirect evidence that the strength of the grain boundary pinning, and hence the maximum coercivity, may be largely determined by the resistance to domain wall formation within $Nd_2Fe_{14}B$ grains smaller than the single domain particle limit.

6. HOT PRESSED MAGNETS

When overquenched ribbons are consolidated to full density by hot pressing, the resulting microstructure is similar to that of the optimally quenched ribbons. The grains remain spheroidal, with an average diameter of about 100 nm.[19] The Nd-rich intergranular phase remains non-crystalline in some parts of the sample, but elsewhere in the sample it is crystalline. The grain orientations are still nearly isotropic, although a small degree of alignment is induced during densification as ribbons are partially deformed to fill in the voids between them.[6]

Fig. 12 shows the initial magnetization and demagnetization behavior of a hot pressed sample parallel to the press direction. (This sample was fabricated from overquenched ribbons of a somewhat different composition than those in Section 3, so a direct comparison of the coercivity to that of the ribbons is not relevant.) Although the demagnetization curve closely resembles that of the ribbons, the initial magnetization behavior is substantially different. The magnetization rises sharply at low fields, and the remanent magnetization develops rapidly, as indicated in Figs. 13 and 14. As a consequence, the comparison of $B_d(H_m)$ with $B_r(H_m)$ in Fig. 4 shows the extreme behavior characteristic of easy domain wall motion during initial magnetization and strong pinning during demagnetization. The recoil curves from low values of H_m no longer occur along rotation lines, but show considerable domain wall relaxation.

At higher fields the initial magnetization once again takes on an "s" shaped ribbon-like character, with recoil behavior parallel to rotation only lines. Part of the material in the hot pressed magnet has thus been transformed to a die-upset-like behavior, the details of which will be discussed in the next section, while part of the material retains its ribbon-like characteristics.

7. DIE UPSET MAGNETS

Grain growth during hot deformation produces a significant change in microstructure from that of ribbons.[12] The $Nd_2Fe_{14}B$ grains in the die upset magnet are platelets, roughly 60 nm thick and 300 nm in diameter, with the c-axis oriented normal to the plate and along the press direction. A key difference from ribbons is the grain alignment, which allows extended domain structures to form. Lorentz microscopy[12] shows extended domain walls running through the grains parallel to the c-axis. The walls are unaffected by crossing grain boundaries perpendicular to the wall, but are pinned at the parallel edges of the grains. Thus the coercivity of die upset magnets is dominated by domain wall pinning, and, in contrast to ribbons, grain size does not play an important role in the magnetization behavior.

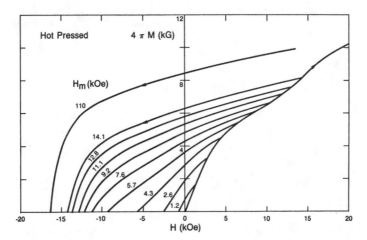

Figure 12. Initial magnetization and demagnetization curves for a hot pressed Nd–Fe–B magnet, with the field applied parallel to the press direction. The number on each curve is the maximum internal field H_m to which the sample was exposed prior to demagnetization.

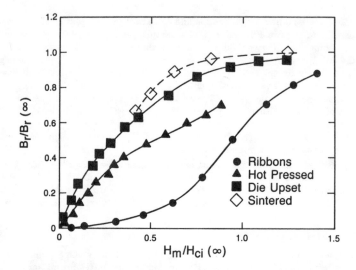

Figure 13. Remanence B_r (normalized by the remanence $B_r(\infty)$ of a fully magnetized sample) as a function of the internal magnetizing field H_m (normalized to $H_{ci}(\infty)$) for various Nd–Fe–B permanent magnet materials. Solid circles–melt-spun ribbons; Solid triangles–hot pressed magnet; Solid squares–die upset magnet; Open diamonds–sintered magnet (reference 5).

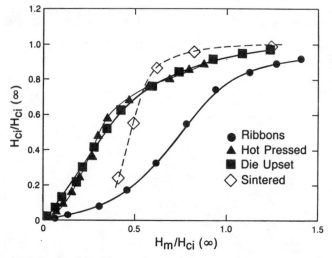

Figure 14. Intrinsic coercivity H_{ci} as a function of the internal magnetizing field H_m for various Nd–Fe–B permanent magnet materials. Both axes are normalized by the coercivity $H_{ci}(\infty)$ of a fully magnetized sample. Solid circles–melt-spun ribbons; Solid triangles–hot pressed magnet; Solid squares–die upset magnet; Open diamonds–sintered magnet (reference 5).

Initial magnetization and demagnetization data are shown for a die upset magnet in Fig. 15. The large initial permeability along the aligned direction indicates that, unlike optimum ribbons, the domain walls are initially free to move, and can be swept through the sample at low fields. Remanence and coercivity develop rapidly at low fields (Figs. 13 and 14) as domains are swept out and the walls become pinned at an intergranular second phase.[12] Complete magnetization becomes difficult, however, as the pinning of domain walls makes further wall movement difficult. A magnetizing field in excess of $H_{ci}(\infty)$ is required to achieve maximum B_r and H_{ci}. The recoil lines are steep and curved, indicating substantial domain wall relaxation to local pinning sites as the applied field is reduced.

The remagnetization curves shown in Fig. 16 provide strong evidence that coercivity is strongly influenced by domain wall pinning. The curves remain nearly flat well into the forward field regime, with little or no domain wall relaxation, indicating that in the field demagnetized state the walls remain strongly pinned. While the initial stages of demagnetization in this aligned, fully magnetized material must take place by reverse domain nucleation, the remagnetization behavior demonstrates that, once nucleated, the subsequent propagation of domain walls is controlled by wall pinning. This situation is in marked contrast to the initially free walls in the thermally demagnetized sample, indicated by the dashed line in Fig. 16. It also differs considerably from the remagnetization behavior observed in Nd–Fe–B sintered magnets,[20,21] which show substantial domain wall motion at both reverse and forward fields. The development of H_{ci} and B_r in hot deformed magnets is also distinct from that observed in sintered magnets,[5] shown for comparison in Figs. 13 and 14. Sintered magnets acquire coercivity and remanence rather abruptly and completely when the magnetizing field is roughly one-half of the coercive field. The coercivity mechanism of sintered magnets has not yet been firmly established, but it is likely that reverse domain nucleation plays an important role.[2]

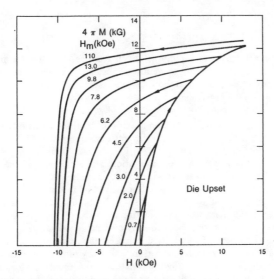

Figure 15. Initial magnetization and demagnetization curves for a die upset Nd–Fe–B magnet, with the field applied parallel to the press direction. The number on each curve is the maximum internal field H_m to which the sample was exposed prior to demagnetization.

80

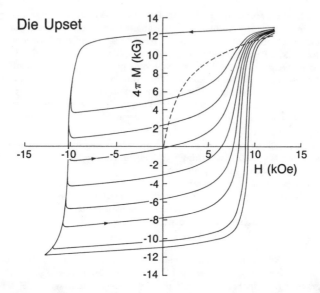

Figure 16. Remagnetization curves for a dc field demagnetized die upset Nd–Fe–B magnet, with the field applied in the press direction. The dashed line indicates the initial magnetization from the thermally demagnetized state, as in Figure 15.

8. SUMMARY

The coercivity mechanisms of rapidly solidified ribbons, hot pressed magnets, and die upset magnets are all governed by domain wall pinning at intergranular phases. The unique microstructure of each material, however, dictates how the wall pinning is manifested in the magnetization curves, and each material has a distinct initial magnetization, recoil, and remagnetization behavior.

Maximum coercivity in melt-spun ribbon is achieved at the quench rate which simultaneously minimizes the amount of magnetically soft amorphous material and maximizes the effectiveness of domain wall pinning at $Nd_2Fe_{14}B$ grain boundaries. The latter is closely tied to the grain size and random orientation, and strongest pinning is obtained in grains with diameters below the estimated single domain size. In the absence of domain walls within grains in the thermally demagnetized state, the initial magnetization as well as demagnetization is controlled by strong domain wall pinning at grain boundaries, and maximum coercivity is complemented by a minimum initial permeability. Over a broad range of quench rates encompassing from slightly overquenched (24 m/s) to heavily underquenched (12 m/s) ribbons, B_r and B_d develop by the same mechanism of moment reversal, suggesting that the strength of the grain boundary pinning, and hence the maximum coercivity, is largely determined by the resistance to domain wall formation within very small $Nd_2Fe_{14}B$ grains. Coercivity decreases in underquenched (slow quench rate) ribbons as the wall pinning weakens in large $Nd_2Fe_{14}B$ grains. Low coercivity in the overquenched (fast quench rate) condition is a consequence of incomplete formation of the $Nd_2Fe_{14}B$ phase. Some of this ribbon character is retained in the intermediate stage hot pressed magnets.

Grain size is much less important in the die upset magnets. Rather, it is the platelet morphology and alignment of the grains which leads to a strong pinning of extended domain walls at the grain edges. The ease of domain wall motion leads to a fast initial magnetization, but fields larger than $H_{ci}(\infty)$ are required to completely magnetize the sample. Coercivity and remanence develop in a manner different from that observed in sintered magnets.

9. ACKNOWLEDGEMENTS

The author wishes to thank J. F. Herbst, C. D. Fuerst, R. W. Lee, G. P. Meisner, and R. K. Mishra for many stimulating discussions. I am grateful to E. A. Alson, E. G. Brewer, C. B. Murphy, and T. H. VanSteenkiste for their technical support. D. J. VanWingerden was instrumental in performing the magnetic measurements.

10. REFERENCES

1. J. F. Herbst, J. J. Croat, F. E. Pinkerton, and W. B. Yelon, Phys. Rev. B **29**, 4176 (1984).

2. J. D. Livingston, in *Proceedings of the 8th International Workshop on Rare-Earth Magnets and Their Applications*, edited by K. J. Strnat (University of Dayton, Magnetics, KL-365, Dayton, Ohio 45469), pp. 423-440.

3. J. F. Herbst, R. W. Lee, and F. E. Pinkerton, Annu. Rev. Mater. Sci. **16**, 467 (1986).

4. J. J. Croat, J. F. Herbst, R. W. Lee, and F. E. Pinkerton, J. Appl. Phys. **55**, 2078 (1984).

5. M. Sagawa, S. Fujimura, N. Togawa, H. Yamamoto, and Y. Matsuura, J. Appl. Phys. **55**, 2083 (1984).

6. R. W. Lee, Appl. Phys. Lett. **46**, 790 (1985); R. W. Lee, E. G. Brewer, and N. A. Schaffel, IEEE Trans. Mag. **MAG-21**, 1958 (1985).

7. F. E. Pinkerton, IEEE Trans. Mag. **MAG-22**, 922 (1986).

8. F. E. Pinkerton and D. J. VanWingerden, J. Appl. Phys. **60**, 3685 (1986).

9. R. K. Mishra, J. Mag. Magn. Mater. **54-57**, 450 (1986).

10. E. C. Stoner and E. P. Wohlfarth, Philos. Trans. Roy. Soc. London **A240**, 599 (1948).

11. F. E. Pinkerton (unpublished).

12. R. K. Mishra and R. W. Lee, Appl. Phys. Lett. **48**, 733 (1986).

13. J. D. Livingston, J. Appl. Phys. **57**, 4137 (1985).

14. G. Hilscher, R. Grössinger, S. Heisz, H. Sassik, and G. Wiesinger, J. Mag. Magn. Mater. **54-57**, 577 (1986).

15. K.-D. Durst and H. Kronmüller, in *Proceedings of the 4th International Symposium on Magnetic Anisotropy and Coercivity in Rare Earth-Transition Metal Alloys*, edited by K. J. Strnat (University of Dayton, Magnetics, KL-365, Dayton, Ohio 45469, USA), pp. 725-735.

16. F. E. Pinkerton, J. Mag. Magn. Mater. **54-57**, 579 (1986).

17. E. P. Wohlfarth, J. Appl. Phys. **29**, 595 (1958).

18. P. Gaunt, G. Hadjipanayis, and D. Ng, J. Mag. Magn. Mater. **54-57**, 841 (1986).

19. R. K. Mishra, J. Appl. Phys. (to be published).

20. D. Li and K. J. Strnat, J. Appl. Phys. **57**, 4143 (1985).

21. U. Heinecke, A. Handstein, and J. Schneider, J. Mag. Magn. Mater. **53**, 236 (1985).

MICROSTRUCTURE-PROPERTY RELATIONSHIPS IN MAGNEQUENCH MAGNETS

RAJA K. MISHRA
Physics Department, General Motors Research Laboratories,
Warren, MI 48090-9055

ABSTRACT

Transmission electron microscopy has been used to characterize the microstructure of Nd-Fe-B magnets produced by melt-spinning and subsequent hot-pressing/die-upsetting. For a material of starting composition $Nd_{.135}Fe_{.815}B_{.05}$, the basic microstructure of melt-spun, hot-pressed and die-upset magnets consists of two phases. In the optimally processed melt-spun ribbons and hot-pressed samples, small and randomly oriented $Nd_2Fe_{14}B$ grains are surrounded by a thin noncrystalline Nd-rich phase. The die-upset material consists of closely stacked flat $Nd_2Fe_{14}B$ grains surrounded by a second phase of approximate composition Nd_7Fe_3. No $Nd_{1.1}Fe_4B_4$ phase is observed in these materials, but it can form if the chemical composition and/or processing parameters are varied. In all these materials, Lorentz microscopy reveals that magnetic domain walls are pinned by the second phase. The differences in the hard magnetic properties of the three kinds of MAGNEQUENCH magnets closely correlate with the differences in the distribution of $Nd_2Fe_{14}B$ crystallites and the pinning sites.

INTRODUCTION

Ever since the discovery of the hard magnetic Nd-Fe-B materials[1-4] there has been growing scientific and technological interest in this field. Hundreds of papers and articles have been published[5], and Nd-Fe-B permanent magnets have become commercially available. Today these magnets are principally produced by two different methods: one starting with melt-spun ribbons-[1] and the other via powder metallurgy[2]. MAGNEQUENCH is the trade name designating the group of magnets produced by General Motors. Three different grades of MAGNEQUENCH magnets can be made from the rapidly quenched Nd-Fe-B ribbons by (i) resin bonding, (ii) hot-pressing and (iii) die-upsetting[6]. In this paper we shall survey the available information on the microstructural features and their interactions with magnetic domains in the MAGNEQUENCH magnets. The microstructure of the sintered magnets will be referred to for comparison. The paper ends with a discussion on the still unanswered yet important questions about the microstructural variables requiring further research.

The main magnetic phase in all types of Nd-Fe-B magnets is the tetragonal $Nd_2Fe_{14}B$ phase[7]. However, depending on the grain size, crystallinity, grain orientations, phase distributions, etc., widely varying hard magnetic properties are possible. The oriented MAGNEQUENCH magnets produced by die-upsetting[8] and the sintered magnets (which are oriented in a magnetic field before sintering) exhibit similar energy product values of over 45 MGOe and intrinsic coercivity values of over 12 kOe[9]. The bonded and hot-pressed MAGNEQUENCH magnets are isotropic and have energy products and coercivities around 15 MGOe and 20 kOe respectively[6]. The microstructural features responsible for such widely varying properties are summarized below.

MICROSTRUCTURE

Transmission electron microscopy is a powerful technique to study the microstructure of materials if a suitable electron transparent specimen can be prepared. Modern electron microscopes are capable of providing information about the microstructural details at high resolution along with information on local crystal structure and chemical composition[10]. Magnetic domains and their interactions with the microstructure can also be studied in a transmission microscope in the Lorentz mode[11]. Although study of hard magnetic materials can pose operational problems due to the interaction of the magnetic field of the specimen with the magnetic field of the lenses of the microscope, these effects seem to be minimal for the $Nd_2Fe_{14}B$ magnets.

The hard magnetic properties[1] and the microstructure[12] of the melt-spun ribbons vary with the cooling rate of the ribbons during melt spinning. For a typical sample of $Nd_{135}Fe_{815}B_{05}$, the coercivity can rise from ~1 kOe to ~20kOe if the quench rate is reduced by varying the wheel speed from 35 m/s to 20 m/s. For still slower quench rates, the coercivity drops again; it is nearly zero for very slowly cooled ingots. Other hard magnetic parameters behave similarly, going through a maximum and falling off for faster or slower quench rates.

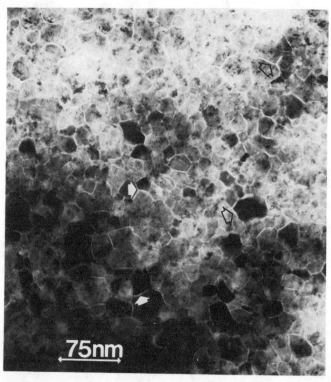

75nm

Fig. 1   Bright field micrograph of an optimally quenched ribbon showing randomly oriented grains surrounded by a thin intergranular phase (marked by arrows).

The optimally quenched ribbons have a two-phase microstructure as shown in figure 1. The main magnetic phase is $Nd_2Fe_{14}B$, and it comprises 95% of the volume. The second phase appears as an intergranular phase surrounding all the grains. The average diameter of the grains is about 30 nm, and the average width of the grain boundary phase is about 2 nm. The grains are randomly oriented and are polyhedral in shape with no well defined dihedral angles between boundaries. The spread of the grain sizes through the thickness of the ribbon is narrow. The intergranular phase is uniform in width and shows no sign of crystallinity in dark field images or lattice images. There is a difference in the Nd to Fe ratio of the grain boundary phase and the grains, the former being enriched in Nd. Both energy dispersive x-ray analysis and Auger spectroscopic analysis show that the Nd to Fe ratio in the boundaries is roughly twice that of the grains[12]. In addition, the Auger spectra in figure 2 show that B is nearly absent in the grain boundary phase.

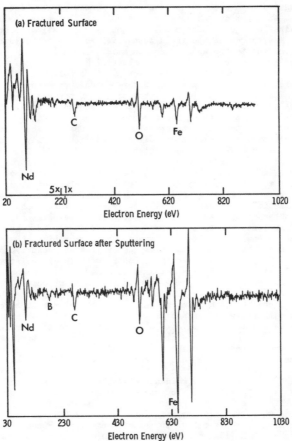

Fig. 2   Auger electron spectra from fractured surface of optimally quenched ribbon (a) immediately after fracture and (b) after removing the boundary layer. Note the absence of the B signal in (a) even when the spectrum is magnified 5 times.

The microstructure of the ribbons quenched very fast from the melt (~35 m/s) shows no $Nd_2Fe_{14}B$ grains. As the quench rate is reduced, small grains of $Nd_2Fe_{14}B$ are observed. The average size of the grains increases as the quench rate is reduced. Starting from a noncrystalline state in the over-quenched condition the microstructure changes with decreasing quench rate such that $Nd_2Fe_{14}B$ crystallites nucleate and then grow until they impinge at the boundaries, trapping the excess material from the nonstoichiometric melt. Based on the structural building blocks of the $Nd_2Fe_{14}B$ phase[7] it is proposed that nucleation of the crystalline grains is related to the trigonal prisms containing B. Growth of the grains is influenced by diffusion and rearrangement of Fe and Nd. Starting with a random distribution of the trigonal prisms in the noncrystalline state, $Nd_2Fe_{14}B$ crystallites nucleate homogeneously and grow till they impinge. Corresponding to the starting composition, all the B atoms properly coordinated in the trigonal prisms are consumed when 95% of the material is converted to the $Nd_2Fe_{14}B$ phase. The remaining material left at the boundaries then consists of the excess Fe and Nd, consistent with our observations. When the quench rate is so slow that the atoms are still mobile when the grains impinge, grain growth and boundary migration occur, and the second phase rearranges at the grain boundaries and the grain junctions. This is in fact observed in the under-quenched materials where grain sizes are large and the grain junctions contain pockets of Nd-rich material[12].

In the over-quenched state, the material has a nonequilibrium microstructure. The microstructure of under-quenched ribbons is close to equilibrium. The phases appearing in the pockets at the grain junctions in these ribbons are products of phase transitions in the terminal binary composition rather than the starting ternary composition.

Consistent with the microstructural observations and the phase distribution in all ribbons, including the underquenched ones, it may be argued that the microstructure of the ribbons evolves through a solid state reaction as opposed to a solidification process. A noncrystalline solid forms from the molten alloy as it touches the spinning wheel. As the ribbon cools further, if the cooling rate is not very fast, atoms and clusters of atoms (such as the trigonal prism containing B) diffuse and occupy crystallographic sites, and crystallization takes place. Similarly, when an overquenched ribbon is annealed at a temperature above the crystallization temperature, $Nd_2Fe_{14}B$ crystallites nucleate homogeneously and grow[12]. It can be argued that nuclei of the crystalline phase are formed during quenching, their number depending on the quench rate, and these nuclei grow when the ribbons are annealed back. Thus super-overquenched ribbons with fewer nuclei anneal to a microstructure with large grains compared to a moderately overquenched ribbon with many nuclei which more or less anneals to a microstructure comparable to that of the optimally quenched ribbons. Similarly, changing the chemical composition of the starting material can affect the microstructure depending on (i) whether the trigonal prisms are destabilized such that nucleation is not homogeneous any longer, (ii) the alloying atoms are more or less mobile affecting the growth of the crystallites, or (iii) how the terminal solid solution decomposes during cooling and heating.

The microstructure of the hot-pressed samples is in general similar to those of the optimally quenched or annealed ribbons as shown in figure 3. Since the ribbon fragments are maintained at the hot-pressing temperature for longer times (when densification occurs), some grain growth occurs inside the ribbon fragments, and the average grain size is larger. The microstructure at the interfaces where two ribbons have fused during hot-pressing is different. First, the grains near the ribbon interface are larger. Second, the interface region is richer in Nd[13]. The ribbons have an oxide layer on the surface before hot-pressing. This layer breaks up during heating in vacuum, leaving behind the Nd-rich surface layer. This layer is highly reactive and assists in bonding during hot-pressing. Grain boundary migration in the Nd-rich

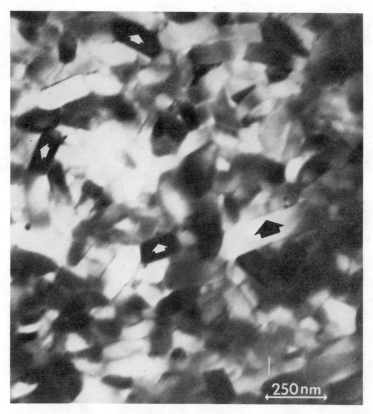

Fig. 3   Bright field micrograph of a hot-pressed sample. Note the shape of the grains marked by arrows.

surface region as well as local deformation of the grains near the surface give rise to the grain growth in the fusion zone. Some of the grains in the ribbon interior show evidence of deformation. These grains in figure 3 appear elongated in such a way that their flat faces are parallel to the (001) plane of the crystalline phase. Possibly, at the hot-pressing temperature and under pressure, mass flow occurs so as to have the low energy face of the grain as the grain boundary.

Die-upset magnets have a somewhat different microstructure. The main feature of the microstructure of these magnets as shown in figure 4 is that the majority of the $Nd_2Fe_{14}B$ grains are now platelets with their c-axes normal to the flat surface. These platelets are stacked against each other so as to have their c-axes parallel to the stress axis. There is a Nd-rich second phase around the grains. Chemical microanalysis of this phase using energy dispersive microanalysis gives a Nd:Fe ratio of 70:30 if one neglects the B content. Besides, we also find a third phase at some of the ribbon interfaces in the form of small particles of nearly free Nd[14]. The origin of these particles can be traced to the Nd rich layer of the fusion zone

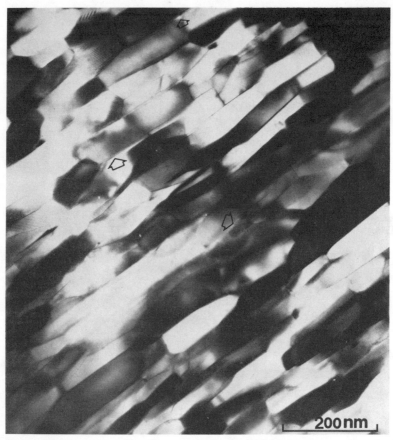

Fig. 4   Microstructure of a die-upset magnet showing a cross section of the
platelet shaped grains.The second phase is marked by arrows.

during hot-pressing. Large undeformed grains near the fusion zone and also
small undeformed grains between two fusion zones are observed.
        It is believed that deformation in this material begins with dislocation
motion in the grains that are favorably oriented to have the highest resolved
shear stress[15]. Accompanying this deformation, the grain boundary phase
also deforms. Since enough slip systems are not available for a generalized
deformation by dislocation motion alone[16], diffusion slip, boundary migration and grain boundary sliding must play decisive roles during the die-upset
process to produce the platelet microstructure. The grains rotate so as to
have their slip planes normal to the compressive stress axis[15]. In the
regions where the material does not experience the applied stress, grain
growth continues as if the material is being annealed, and no alignment
occurs. The same is true if the grains are so large that accommodation
processes cannot overcome the obstacle to general deformation due to lack of
enough slip systems.

The magnetic domains can be imaged in all three groups of materials. In the melt-spun and hot-pressed materials (figs. 5a and 5b), extended domain walls are seen to wrap around grains[12]. If the grain size is over 150 nm, domain walls also appear inside the grains. Grains with a diameter of about 80 nm tend to have a single domain whereas grains smaller than that show

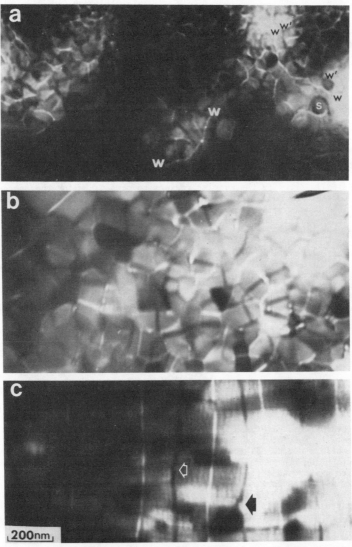

Fig. 5   Lorentz electron micrograph of (a) melt-spun, (b) hot-pressed and (c) die-upset magnets showing magnetic domain wall images (appearing as alternate black and white lines marked w-w or w'-w'). Note the kinks in the domain wall images in (c) as a wall passes near the edge of a grain.

several grains wrapped around by a domain wall. It must be pointed out that these grains are oriented randomly and have their easy axes of magnetization randomly distributed. The existence of extended domain walls suggests that on either side of the wall the components of the magnetization in the plane of the specimen are anti-parallel. The width of the domain wall measured from these images corresponds to the planar component of the magnetization and thus cannot be taken as a true measure of the wall width. In fact, different wall images show different widths.

The domain wall images in the die-upset magnets can be used for quantitative measurements of wall width. Since the walls tend to lie parallel to the c-axis as shown in figure 5c, using known values of specimen parameters (thickness, saturation magnetization, anisotropy constant, lattice parameters, orientation etc.) and microscope parameters (electron wavelength, objective lens defocuss, magnification etc.), actual domain wall widths can be computed from measured image widths[17]. Such measurements yield a value of ~3 nm for the domain wall width and 150 nm for the diameter of a single domain particle. Alternate estimates of the wall energy from domain width measurements in figure 5c give a value of 3.8 nm for the domain wall width[14].

The domain wall images in figure 5c can be seen to deviate from a straight line, indicating their affinity to pass near the grain boundary region so that maximum area overlap occurs between the two. From energetic considerations, such zigzags increase the wall area. There must be a decrease in the wall energy in the overlap regions to compensate for the increased area so that the total energy of the domain walls is reduced for this configuration. Domain wall pinning at the grain boundary is also verified through in situ observation of domain walls moving in an applied magnetic field from one pinned configuration to another pinned configuration[18].

Finally, if Lorentz images are taken from die-upset samples such that the c-axes do not lie in the plane of the specimen, interesting domain patterns, including "bubble domains", can be formed[18]. In such cases proper care must be taken to ascertain the component of the magnetization in the plane of the foil (which is responsible for the- Lorentz image) and the magnetic field of the microscope before analyzing the domain patterns or measuring domain wall widths, "bubble diameters", etc.

DISCUSSION

In this section we shall point out the similarities and dissimilarities between the microstructure of the MAGNEQUENCH magnets and the magnets prepared by powder metallurgy. First of all, the sintered magnets have large grains compared to those of the MAGNEQUENCH magnets[19,20]. Secondly, all sintered magnets have been reported to have at least four different phases[21]. The total amount of the magnetic $Nd_2Fe_{14}B$ phase in the sintered magnets is about 80-85% (by volume) compared to over 95% of the MAGNEQUENCH magnets being the $Nd_2Fe_{14}B$ phase. In the sintered magnets, in addition to the Nd-rich grain boundary phase, a B-rich $Nd_{1.1}Fe_4B_4$ phase and oxide phases appear[21]. This is a necessary consequence of the starting composition and the processing method used to prepare fully dense magnets. As we have discussed already, in MAGNEQUENCH magnets, these extra phases do not appear if we prepare them from the starting composition stated earlier and we follow the preparation methods discussed elsewhere[1,6]. If the starting composition or the processing scheme or both are changed, then other phases[22] can be formed and accompanying them there is a degradation of the magnetic properties.

The mechanism of magnetization reversal in both types of magnets has been discussed by many authors[23-25]. The importance of domain wall pinning in the MAGNEQUENCH magnets has been discussed in connection with the domain wall observations[12,18] and magnetization measurement studies[26]. Domain walls have to overcome pinning before they can move in the early stages of the magnetization of the melt-spun ribbons or the hot-pressed magnets. In the die-upset magnets, domain walls can move relatively easily in the early stages of the magnetization process, however the pinning field must be overcome before the material can be demagnetized. The area overlap between the domain wall and the grain boundary phase as well as the nature of the grain boundary phase determine the available coercivity in these materials.

As per the sintered magnets, experimental evidence of domain wall pinning at the grain boundary phase has been published[25,27,28]. In spite of the conflicting interpretation of the nature of the grain boundary phase in the sintered magnets, it is generally accepted that it is a Nd-rich layer and plays an important role in the magnetic hardening. Recent experimental evidence of the ease of reverse domain nucleation at the surface of the sintered magnets emphasizes the importance of the pinning properties of the intergranular phase for the overall coercivity of these magnets. In addition to the belief that post sintering annealing suppresses the nucleation sites[29] for reverse domains, it is also very likely that the annealing improves the pinning strength of the grain boundary phase through chemical change. It will be interesting to perform in situ experiments and examine reverse domain nucleation in magnets before and after annealing. In a material containing many phases, nucleation of reverse domains can occur easily, and these domain walls can propagate through a grain without any hindrance. High field magnetization studies[30] show that even after saturating the sintered magnets in a field of 100 kOe remanent domains are still present which can move easily during demagnetization.

When these questions are resolved , we may see that not only the magnetic phases in the sintered and melt-spun magnets are the same, but they have similar second phases that are decisive for the magnetic hardening and possibly also have the same mechanism for magnetization reversal. Conceivably, larger energy product magnets can be made from the melt-spun MAGNEQUENCH magnets since they have a higher percentage of the magnetically anisotropic $Nd_2Fe_{14}B$ phase.

ACKNOWLEDGMENTS

The author gratefully acknowledges the many helpful discussions with J. F. Herbst, R. W. Lee and F. Pinkerton.

REFERENCES

1. J. J. Croat, J. F. Herbst. R. W. Lee and F. E. Pinkerton, Journal of Applied Physics, 55, (1984) 2078.
2. M. Sagawa, S. Fujimura, N. Togawa, H. Yamamoto and Y. Matsuura, Journal of Applied Physics, 55 (1984) 2083.
3. G. C. Hadjipanayis, R. C. Hazelton and K. R. Lawless, Journal of Applied Physics, 55 (1984) 2073.
4. N. C. Koon and B. N. Das, Journal of Applied Physics, 55 (1984) 2063.
5. "Source Book on Nd-Fe-B Permanent Magnets", Rare Earth Information Center, Iowa State University, Ames, Iowa, IS-RIC-9, 1986.
6. R. W. Lee, N. Schaffel and E. G. Brewer, IEEE Trans. Magnetics, MAG-21 (1985) 1958.
7. J. F. Herbst, J. J. Croat, F. E. Pinkerton and W. B. Yelon, Physical Review B, 29 (1984) 4176.
8. R. W. Lee, Applied Physics Letters, 46, (1985) 790.

9. K. S. V. L. Narasimhan, Journal of Applied Physics, 57 (1985) 4081.
10. G. Thomas and M. J. Goringe, Transmission Electron Microscopy of Materials (John Wiley, New York, 1979).
11. J. N. Chapman, G. R. Morrison, J. P. Jakubovics and R. A. Taylor, Journal of Magnetism and Magnetic Materials 49 (1985) 277.
12. R. K. Mishra, Journal of Magnetism and Magnetic Materials 54-57 (1986) 450.
13. R. K. Mishra, Unpublished Research.
14. R. K. Mishra, "Microstructure of hot-pressed and Die-Upset Nd-Fe-B Magnets", in press, Journal of Applied Physics, 1987.
15. R. W. K. Honeycombe, The Plastic Deformation of Metals, Second Edition (Edward Arnold Ltd., London, 1984).
16. R. Von Mises, Z. Angew. Math. Mech. 8 (1928) 161.
17. S. Tsukahara and H. Kawakatsu, Journal of Magnetism and Magnetic Materials 36 (1983) 98.
18. R. K. Mishra and R. W. Lee, Applied Physics Letters, 48, (1986) 733.
19. M. Tokunaga, M. Tobise, N. Meguro and H. Harada, IEEE Transactions on Magnetics, MAG-22, (1986) 904.
20. M. Sagawa, S. Fujimori, H. Yamamoto, Y. Matsuura and S. Hirosawa, Journal of Applied Physics, 57, (1985) 4094.
21. J. Fidler and L. Yang, Proc. 4th Inter. Symp. on Magnetic Anisotropy and Coercivity in Rare Earth-Transition Metal Alloys, ed. K. Strnat, University of Dayton Press, Dayton, (1985) 647.
22. A. Hutten and P. Haasen, Scripta Metallurgica, 21, (1987) 407.
23. J. D. Livingston, Proceedings of the 8th International Workshop on Rare Earth Magnets and Their Applications, edited by K. Strnat, University of Dayton Press, Dayton (1985) 423.
24. K. -D. Durst and H. Kronmuller, ibid, pp. 725-735.
25. G. C. Hadjipanayis, Y. F. Tao and K. R. Lawless, IEEE Transactions on Magnetics, MAG- 22, (1986) 1845.
26. F. E. Pinkerton and D. van Wingerden, Journal of Applied Physics, 60, (1986) 3685.
27. R. K. Mishra, J. W. Chen and G. Thomas, Journal of Applied Physics, 59, (1986) 2244.
28. T. Suzuki and K. Hiraga, Journal of Magnetism and Magnetic Materials, 54-57, (1986) 527.
29. K. Hiraga, M. Hirabayashi, M. Sagawa and Y. Matsuura, Japanese Journal of Applied Physics, 24, (1985) 699.
30. Y. Otani, H. Miyajima, S. Chikazumi, S. Hirosawa and M. Sagawa, Journal of Magnetism and Magnetic Materials, 60, (1986) 168.

OPTIMIZATION OF LIQUID DYNAMIC COMPACTION FOR Fe-Nd-B MAGNET ALLOYS

Mauri K. Veistinen*, Y. Hara, E.J. Lavernia , R.C. O'Handley and N.J. Grant
Department of Materials Science and Engineering, Massachusetts Institute of Technology, Cambridge, MA 02139
*Permanent address, Outokumpu Inc., Toolonkatu 4, P.O.Box 280, 00100 Helsinki, Finland

ABSTRACT

Permanent magnets from the $Fe_{77}Nd_{15}B_8$ alloy were produced using liquid dynamic compaction (LDC). Process parameters such as gas and metal mass flow rates were carefully controlled to refine the microstructure. Accordingly, grain sizes between 0.1 and 5 microns were achieved in the LDC compact. In contrast to our earlier LDC results high intrinsic coercivity in excess of 7500 Oe was obtained for thick (> 7 mm) as-LDC'd samples with lower neodymium content.

INTRODUCTION

The liquid dynamic compaction (LDC) technique has been successfully used to produce isotropic permanent magnets of iron-neodymium-boron alloys directly from the melt in one processing step [1,2,3]. There are inherent advantages in the simplicity achieved by LDC processing particularly when compared to conventional powder metallurgy (PM) processes and melt-spinning which require several steps for magnet fabrication [4,5]. The LDC process can eliminate some of the steps such as fine-milling of master alloys, the handling of very reactive powders as well as the separate compacting and sintering stages.

So far, however, the magnetic properties reported for Nd-Fe-B based permanent magnets produced by LDC have been less than optimum when compared to those materials produced by conventional PM processes as well as rapid solidification techniques such as melt spinning. This has been attributed in part to the isotropic microstructure of LDC magnets. In addition, optimization of the processing parameters necessary for the spray deposition of Fe-Nd-B-magnets by LDC is yet to be completed. The best permanent magnetic properties recorded so far for LDC materials have been for an $Fe_{76}Nd_{16}B_8$ alloy, in a section of the compact having a thickness of approximately 2 mm [2]: $H_{ci} \cong 7.5$ kOe, $B_r \cong 7$ kG, $H \cong 5.5$ kOe and $(BH)_{max} \cong 9-10$ MGOe. However, when the thickness of the LDC compact was increased from 2mm to 10 mm the intrinsic coercivity was found to be 30-40% lower [2]. This substantial decrease in coercivity was attributed to the relatively larger grain sizes associated with thicker sections of the deposit. Also, higher coercivity values were associated with high concentrations of neodymium in this earlier work [2]. For example, the intrinsic coercive force for alloys containing 18 a/o Nd was measured to be about 10 kOe while that for alloys with 14 a/o Nd was only about 3 kOe.

Recently, isotropic $Fe_{76}Nd_{16}B_8$ permanent magnets have also been produced by plasma spraying [6], which in principle, is a process similar to gas atomization. Intrinsic coercivities between 5 kOe and 11 kOe have been reported for the plasma-sprayed deposits. However, in these plasma-sprayed samples the average magnet thickness was under 0.5 mm.

In the previous study [2] it was found that the refinement of the microstructure leads to better magnetic properties. One of the most important process parameters in LDC is the ratio of the gas to metal flow rates. The purpose of the present study is to examine the effect of this ratio on the magnetic properties of Fe-Nd-B permanent magnets produced by LDC. It has been shown that in gas atomization small average particle size and higher cooling rates can be achieved by increasing this ratio [7]. In this study the gas to metal mass flow rates were increased using a recently developed ultrasonic gas atomization die which allows for a six-fold increase in gas flow rates over those reported for earlier atomizer designs [1-3]. Furthermore, the position of the metal delivery tube was optimized for low metal flow rates. It has been shown that, depending on the position of the metal delivery tube relative to the atomizer, there may be either overpressure (pushing the metal back into the crucible) or underpressure (drawing the metal from the crucible) at the exit end of the metal pouring tube [8,9]. Hence, the metal flow rate can be controlled by changing the position of the metal delivery tube with respect to the gas jets. In the present study the position and the geometry of the metal delivery tube were optimized by measuring the pressure inside the tube in simulated gas atomization experiments [10].

Finally, the microstructure of the LDC compact was studied in order to correlate it with LDC process parameters and magnetic properties. The present results are also compared with those reported in earlier LDC studies [1-3].

EXPERIMENTAL

Premelted pieces of the master alloy (prepared by Rare Earth Products, England) having a nominal composition of $Fe_{77}Nd_{15}B_8$ (33.3 w/o Nd, 1.2 w/o B) were used in the present LDC study. To minimize oxidation and rare-earth element losses, the melting chamber was first evacuated to approximately 0.3 Torr and then backfilled with argon to a small positive pressure. About 1 kg of the master alloy was induction melted under an Ar atmosphere and superheated to about 1743 K before atomization. The molten alloy was atomized by argon using the ultrasonic gas atomization process (USGA) with a gauge pressure of 200 psig ($\cong$15 atm) atomization pressure. The atomization facilities and the LDC process have been described in detail elsewhere [1,8]. The gas mass flow rate J_{gas} was increased in the present work by increasing the gas exit area to 57 mm^2; in the previous LDC-Fe-Nd-B studies [1-3] the gas exit area was about 9 mm^2.

Further, the position of the metal delivery tube was adjusted in order to optimize the metal mass flow rate J_{met} [10]. The underpressures and overpressures at the tip of the pouring tube measured in this study are given in Figure 1 as a function of pouring tube position below (y<0) or above the atomizer. The zero-point position of the metal delivery tube was chosen so that the geometrical extensions of the gas nozzles intersect the edge of the metal delivery tube. Negative pressure difference (ΔP<0) corresponds to an underpressure which increases the metal flow rate and the positive pressure difference to an overpressure which decreases the flow rate. The position of the metal delivery tube used in this investigation was selected in order to avoid both pressurization and aspiration of the melt. This position is shown by an arrow in Figure 1. Despite careful adjustment of the delivery tube

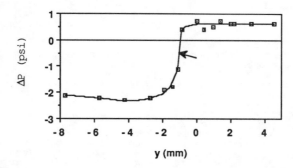

Figure 1. The variation of pressure at the tip of the pouring tube used in this study as a function of its position; the arrow indicates the position used.

position a small overpressure was observed during the experiment; this may be due to the erosion of the tip of the pouring tube by the Fe-Nd-B-melt. This erosion effectively made the melt delivery tube shorter and caused the observed overpressure (see Fig. 1). The ratio of the gas flow to metal flow rate, J_{gas}/J_{met}, was measured to be greater than 4 in this study, while it was found to be close to 1 in the previous LDC studies on magnetic alloys [1-3].

The copper substrate, which was not water-cooled, was located at a distance of 35 cm below the atomizer. This distance led to an optimum density of the LDC compact in the previous investigations [1]. Heat treatments were performed to modify the magnetic properties. The samples were placed inside quartz capsules which were then evacuated and backfilled with argon before sealing. The annealing cycle consisted of a rapid heat-up to an annealing temperature (450-650 °C), at which the samples were held for an hour. After annealing the capsules were water-quenched. The microstructures of the compact were investigated with optical microscopy and scanning electron microscopy (SEM) after metallographic polishing and etching in 1% Nital solution. For SEM the samples were also cooled to the liquid nitrogen temperature and then fractured by an impact hammer. Magnetic properties were measured at Department of Materials Science and Engineering with a Digital Measurement System`s model 880 A vibrating sample magnetometer (VSM) with the maximum field of 14 kOe and at the Francis Bitter National Magnet Laboratory using a conventional DC integrating fluxmeter in the fields up to 100 kOe. When the VSM was used, the samples were magnetized before measurements using a Bitter magnet. Specimens were cut from the LDC deposit to cylindrical shapes of 8 mm in diameter. The thickness of the samples ranged from 2 to 7 mm depending on the thickness of the LDC compact.

RESULTS AND DISCUSSION

The master alloy and the LDC compact were chemically analyzed to determine their composition, the neodymium loss and oxygen contamination during LDC. In this study about 1.4 w/o of neodymium was lost during the melting and deposition stage. This result is consistent with the earlier LDC studies [1,2]. The oxygen content increased from 570 ppm in the master alloy to 1850 ppm in the as deposited LDC material. This oxygen content of the LDC material is relatively high compared to those reported in earlier investigations where the oxygen contents of Fe-Nd-B-LDC alloys were found to vary between 350-1100 ppm [1,2]. The reported oxygen levels correspond to an increase of about 200 to 700 ppm over those of the master alloy. The unusually high oxygen content may be attributed to one or a combination of various factors which include poor control of the atmosphere prior to atomization, relatively high humidity content of the induction insulation, or a reaction between the molten alloy and the alumina insert as well as the metal delivery tube which was made of carbon steel. The latter would also explain the increase in carbon content observed in the present LDC compact.

Furthermore, the oxygen content was also determined as a function of the radius in the circular LDC compact. However, no variations in oxygen content were observed as a function of the radial distance from the center of the deposit. Also, no variations were found in oxygen content as a function of the cross-sectional thickness of the LDC compact.

Microstructural observations

In this study the LDC compact consisted of two different microstructural regions. The lower part of the deposit, about half of the thickness close to the copper substrate, contained mostly equiaxed grains. The average grain size varied in this region between 1 and 5 μm, being smallest closer to substrate surface. The relatively coarser grains found toward the top of this region were a result of the greater thermal exposure of this area due to the heat released by droplets that were subsequently splat-quenched, and to the relatively poor heat conductivity of the Fe-Nd-B-alloy. On the other hand, the secondary layer, which makes up the rest of the compact cross-section is dominated by 2-25 μm thick splats having well defined boundaries, Figure 2. These boundaries are probably oxides indicating the high oxygen level in the chamber. Interspersed in this structure, there are droplets which have solidified before impact. The diameter of these spherical particles was found to vary between 2 and 50 μm. The grain structure of these solid droplets was either equiaxed, columnar or dendritic with an average grain size between 1 and 5 μm. In the splats, microstructural details could not be determined with optical microscopy. SEM studies on polished and etched surfaces revealed that the grain size of the splats was under 1 μm. This was verified by samples fractured at liquid nitrogen temperature and studying the fracture surfaces using SEM. The average grain size was observed in many splats to be between 0.1-0.4 μm. The grain structure was mostly equiaxed; only occasionally were elongated grains observed, Figure 3.

In addition to the matrix phase (white in optical micrographs) the microstructure revealed at least two other phases. A secondary phase, located mainly at the grain boundaries, is seen as black in optical micrographs because it is easily removed by mechanical polishing and etching. Accordingly, this structure is seen as holes or valleys in SEM pictures, Figure 4. In addition to this dark phase there is another phase observed as bright spots in SEM photographs (see Fig. 4). This bright phase is seen at grain boundaries and triple points. Furthermore, there are also pores in the structure which are usually difficult to distinguish from secondary phases. In the present work porosity was clearly found to increase in the upper layer of the compact, while the lower part was almost free of pores. Although the concentration of porosity has been reported to decrease when nitrogen instead of argon is used as an atomization gas [11], it was not possible to use nitrogen in this study due to the nitride formation in Fe-Nd-B-alloys. Additionally, oxides decorating the splat boundaries can be found in the microstructure (see Fig. 2). The so-called "B-rich" type phases are not as clearly distinguished in the optical micrographs of the LDC-magnets as in those of sintered Fe-Nd-B-magnets. Instead, also in sintered magnets, Nd-rich phases are partially removed during sample preparation suggesting that two possible secondary phases observed in the present work may also be a single phase.

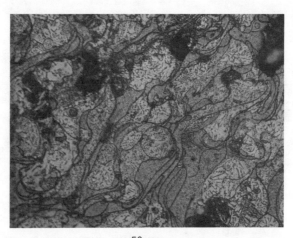

50 μm

Figure 2. Cross-section of the upper part of the compact; splats, entrapped solid particles and pores are seen in the microstructure.

Comparison of the microstructures obtained in the present study to those from the previous LDC studies is shown in Figure 5. While two different microstructural regions were found in this work there were

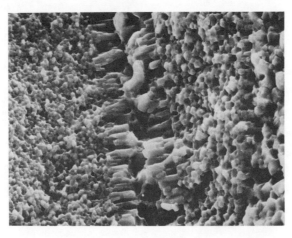

4 μm

Figure 3. SEM picture of the fractured splats in the upper part of the compact, where some elongated grains are observed (cross-section).

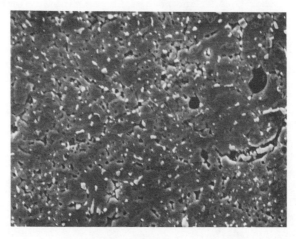

10 μm

Figure 4. SEM picture of the lower part of the compact cross-section, where secondary phases and pores are seen in addition to matrix phase.

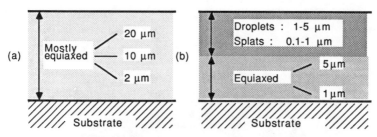

Figure 5. Schematic presentation of the microstructures observed in LDC; (a) previous LDC-magnet studies [1,2], (b) present work. Numbers refer to typical grain sizes in the various parts of the LDC compact.

mostly equiaxed structures reported in previous works. Only occasionally were these splats with coarse columnar grains seen in the earlier LDC deposits [1]. Futhermore, it is found that increasing the gas to metal mass flow ratio resulted in a significant refinement of the microstructure of the as-deposited LDC material. In earlier studies [1,2] the grain size was about 20 μm in the center of the compact whereas in the present study the largest average grain size was about 5 μm.

Compacting process during LDC

The various microstructural features observed throughout the as deposited LDC material are a result of the splatting conditions during spray deposition. At the beginning of the experiment when the spray density as well as the droplet size are largest, it is possible that complete solidification of splats may not occur before the arrival of the next partly solidified droplets [12,13]. Accordingly, a thin film of liquid will be retained on the surface. The mechanical action of the splatting will break off any oxide layer on the droplet surface as well as the thin dendrite arms in both the depositing droplets and in the liquid film. This results in a fine equiaxed microstructure with no splat boundaries in the lower part of the compact.

In contrast, in the upper part of the deposit the splatting process is different. This difference is mainly due to the decreasing metal flow rate (hence also decreasing droplet size) toward the end of the experiment. In the present work there was a small overpressure in the metal delivery tube and accordingly the metal flow rate decreased as the metallic head in the crucible decreased. Correspondingly, the ratio of gas to metal flow rate increased, resulting in finer droplet sizes in addition to lower spray densities toward the end of the run. If the splats solidify completely before the arrival of next droplets, the oxide on their surface reveals the splat boundaries. Because of the small droplet size and, accordingly, the high velocity of the droplets, the cooling rate achieved is high. Additionally, the violent splatting action against a rigid surface, the thinness of the splats and the undercooling of the droplets all contribute to the formation of a very fine-grained microstructure. Furthermore, as the average droplet size decreases during atomization, the proportion of solidified droplets arriving at the substrate increases and some of them retain their spherical morphology on impact (see Fig. 2). The smallest of these particles can appear as single crystals because of difficulties involved in nucleation within very fine droplets [13]. Because in this compaction process there are individually solidified splats there exist irregular interstices which can not be filled with liquid and, accordingly, porosity levels will increase relative to the lower part of the compact.

If instead of the slight pressurization of the melt (overpressure) used in this investigation there is a strong underpressure at the tip of the pouring tube, the metal flow rate would be increased dramatically by the aspiration effect, and would remain fairly constant throughout the experiment. Consequently, the compacting mechanism would not change significantly during the atomization run because the droplet size and spray density remain relatively constant. These have been the splatting conditions prevailing in the previous LDC studies [1-3], where the position of the metal delivery tubes corresponded to the maximum underpressure in the curve shown in Figure 1. The compacting mechanism of the earlier LDC studies corresponds approximately to that in the lower part of the compact in this work. However, due to the much lower gas to metal flow ratios used previously the droplet size as well as the spray density have been much larger and, correspondingly, the cooling rate has been much lower, resulting in a much coarser grain structure. For the same reason there could be seen in some compacts splats with coarse columnar grains [1], where the droplets remained completely liquid on impact [13]. The oxide revealing the splat boundaries was probably left on the splat surface due to the weak mechanical action of subsequently depositing droplets.

Magnetic properties

The intrinsic coercivity values (H_{ci}) of the as-LDC'd deposits are given in Figure 6 as a function of average sample thickness. Generally, the thickness of the compact for the parts where the samples were cut was about 2 mm thicker than corresponding specimens. Two samples were also cut from a 10 mm thick part of the compact; one sample from the upper part of the deposit (marked with U in Figure 6) and the other sample from the lower part of the compact (marked with L in Figure 6). The intrinsic coercive forces are found to be 6.1-7.8 kOe, which are regarded as relatively high values for the alloy composition and the process used in this study. Perhaps even more important is the fact that H_{ci} no longer depends on the sample thickness as it did in the previous LDC study [2]. Until now high LDC coercivities have been achieved only in 2 mm thick specimens; the intrinsic coercivity dropped 30-40% when the sample thickness increased from 2 mm to 10 mm. For the previous $Fe_{76}Nd_{16}B_8$ alloy the highest H_{ci} value for the 2 mm thick sample was about 7.5 kOe [2]. However, in the previous LDC study, the coercivity was also found to decrease when the Nd content was reduced so that for the 15 a/o Nd content used in this work the H_{ci} value was expected to be about 5 kOe for thin samples and only 3.5 kOe for thick samples. The decrease of the intrinsic coercive force with the compact thickness results probably from the coarser grain structure in the thicker sections of the deposit.

Figure 6 shows that the upper part of the compact gives slightly higher intrinsic coercivities when compared to the lower part. Both samples (U and L in Fig. 6) were magnetically isotropic. Typical demagnetization curves for the as-LDC'd deposit are shown in Figure 7 and the typical magnetic properties for the $Fe_{77}Nd_{15}B_8$ alloy used in this study are given in Table I. Also some heat treatment studies were done for this composition. However, H_{ci} values were not found to change

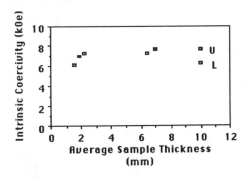

Figure 6. Intrinsic coercive force as a function of average sample thickness.

Table I.

Typical magnetic properties of as-LDC`d $Fe_{77}Nd_{15}B_8$ alloy

B_r (kG)	H_{ci} (kOe)	H_c (kOe)	$(BH)_{max}$ (MGOe)
6.5-7.2	7.5-8.0	4.2-5.0	7.0-8.0

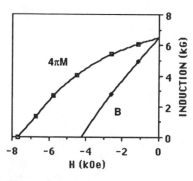

Figure 7. Typical demagnetization curves for as-LDC'd deposit.

significantly in the temperature range of 450-650°C. The same result was found in the earlier LDC study [2] for low neodymium contents.

When compared to the earlier LDC-magnet studies [1-3] the intrinsic coercivity was increased in this work compared to expectations for the equivalent Nd content and, especially, for thicker sections. Probably, this is partly due to the refinement of the microstructure. Obviously, for the same reason the coercivity for the upper part of the compact was higher than for the lower part. Also some other changes combined with the microstructural refinement can affect the coercivity. For example, the distribution and proportion of secondary phases as well as their compositions are probably changing when the cooling rate is increased with the higher ratio of the gas to metal flow rates.

SUMMARY AND CONCLUSIONS

Permanent magnets from the $Fe_{77}Nd_{15}B_8$ alloy were produced using liquid dynamic compaction (LDC). The ratio of the gas to metal mass flow rate was carefully controlled in order to refine the microstructure. The LDC compact consisted of two microstructural regions. The lower part of the deposit contained mainly equiaxed grain structrure, where the average grain size varied from 1 to 5 μm. The upper part of the compact contained mostly splats and entrapped solid particles. The grain size was typically between 0.1 and 0.4 μm in these splats while it varied from 1 to 5 μm in the entrapped particles. Based on these results, it is concluded that the droplet size and the spray density are more important in determining the microstructure than the heat transfer through the deposited material and the substrate. Both the lower and upper part of the compact were found to be magnetically isotropic. The typical magnetic properties of the as-LDC`d deposit were as follows: $H_{ci}≅7.8$ kOe, $H_c≅4.5$ kOe, $B_r≅6.7$ kG and $(BH)_{max}≅7.6$ MGOe. Heat treatments in the temperature range of 450-650°C were not found to change the magnetic properties significantly.

Higher intrinsic coercivity values were observed in this work for thicker (>7 mm) as-LDC`d samples and for lower neodymium content compared to our earlier LDC-magnet studies. In the previous studies high intrinsic coercive forces were achieved only in thin samples (≅2 mm) with neodymium content exceeding 16 a/o. The improved magnetic properties, probably, resulted from the refinement of the microstructure due to the high ratio of the gas flow to metal flow rate.

ACKNOWLEDGMENTS

The authors are grateful to TDK Corporation for the chemical analyses of the LDC compacts as well as Mr. A. Fukuno (TDK) and Mr. M. Tobise (Hitachi Metals) for many valuable discussions on Fe-Nd-B-magnets. This work was supported by the U.S Army Research Office Contract No. DAAG29-84-K-0171.

REFERENCES

1. T.S. Chin, Y. Hara, E.J. Lavernia, R.C. O`Handley, N.J. Grant, J.Appl Phys. 59, 1297 (1986).

2. S. Tanigawa, Y. Hara, E.J. Lavernia, T.S. Chin, R.C. O`Handley, N.J. Grant, IEEE Trans. Mag. MAG-22, 746 (1986).

3. T.S. Chin, D.S. Tsai, R.C. O`Handley, Y. Hara, N.J. Grant, S. Tanigawa, submitted to J. Mag. Mag. Mat.

4. J.J. Croat, in Soft and Hard Magnetic Materials with Applications, edited by J.H. Salssgiver, K.S.V.L. Narasimhan, P.K. Rastagi, H.R. Sheppard, C.M. Maucione (ASM,1986), p.81.

5. M. Sagawa, S. Fujimura, N. Togawa, H, Yamamoto, Y. Matsuura, J. Appl. Phys. 55, 2083 (1984).

6. R.A. Overfelt, C.D. Anderson, W.F. Flanagan, Appl. Phys. Lett. 49, 1799 (1986).

7. G. Rai, E.J. Lavernia, N.J. Grant, in Progress in Powder Metallurgy, Vol 41, 1985 Annual Powder Metallurgy Conference Proceedings, edited by H.I. Sanderow, W.L. Giebelhausen and K.M. Kulkarni (MPIF Press, Princeton, New Jersey, 1986), p. 55.

8. E. Ting and N.J. Grant, in Progress in Powder Metallurgy, Vol 41, 1985 Annual Powder Metallurgy Conference Proceedings, edited by H.I. Sanderow, W.L. Giebelhausen and K.M. Kulkarni (MPIF Press, Princeton, New Jersey, 1986), p. 67.

9. M.J. Couper and R.G. Singer, in Proc. 5th Intern`l Conf. on Rapidly Quenched Metals, edited by S. Steeb and H. Warlimont (North Holland, Amsterdam, 1985), p. 1737.

10. M.K. Veistinen, E.J. Lavernia, M. Abinante, N.J. Grant, (to be published).

11. U. Backmark, N. Backstrom, L. Arnberg, (private communication).

12. R.W. Evans, A.G. Leatham, R.G. Brooks, Powder Metall.,28, 13 (1985).

13. A.G. Leatham, R.G. Brooks, M. Yaman, in Modern Developments in Powder Metallurgy, Vol. 15-17, (MPIF Press, Princeton, New Jersey, 1986), p. 157.

MAGNETIC ANISOTROPY AND SPIN REORIENTATION IN $Nd_{2-x}Pr_xCo_{14}B$

F. POURARIAN, S. G. SANKAR, A. T. PEDZIWIATR[*], E. B. BOLTICH AND W. E. WALLACE
MEMS Department, Carnegie-Mellon University, and Mellon Institute, Pittsburgh, PA 15213.

ABSTRACT

$Nd_{2-x}Pr_xCo_{14}B$ intermetallics crystallize in tetragonal form isostructural to $Nd_2Fe_{14}B$. Structural and magnetic properties of the polycrystalline materials have been determined. Room temperature magnetic measurements indicate that these intermetallics exhibit uniaxial anisotropy throughout the entire composition range. Substitution of praseodymium in $Nd_{2-x}Pr_xCo_{14}B$ significantly enhances the anisotropy field both at 78 K and at room temperature. Magnetization versus temperature plots show spin reorientation T_{SR_2} at 546 K for $Nd_2Co_{14}B$ and at 668 K for $Pr_2Co_{14}B$. T_{SR_2} increases linearly with composition x. The 3d sublattice anisotropy which seeks the tetragonal plane appears to dominate above these temperatures. In addition to the above, low temperature magnetic measurements demonstrate the spin reorientation in $Nd_2Co_{14}B$ and $Nd_{1.9}Pr_{0.1}Co_{14}B$. Further additions of praseodymium strengthen the axial anisotropy at temperatures down to 4.2 K. Effect of competing crystal field interactions and the consequences of praseodymium substitution that result in considerable increase in anisotropy field in these quaternaries are discussed.

INTRODUCTION

It has been established that the outstanding permanent magnet characteristics of the Nd-Fe-B material originate from a large magnetic moment, a reasonable uniaxial anisotropy of the $Nd_2Fe_{14}B$ phase and a significant coercivity [1,2]. One of the disadvantages of these magnets is their relatively low Curie temperature [3]. Replacement of cobalt for iron leads to an increase of T_c [3-6].

Generally, the anisotropy in $R_2Fe_{14}B$ compounds arises from two contributions – the 3d sublattice and 4f sublattice anisotropies [7-9]. The 3d sublattice favors an easy direction of magnetization parallel to the c-axis, as in $Y_2Fe_{14}B$ and $Lu_2Fe_{14}B$ [2,10]. At room temperature the 4f anisotropy dominates over the 3d anisotropy in all cases where the R atoms are magnetic, but the easy direction of magnetization along the c-axis develops only in those compounds in which the R component has a negative second-order Stevens coefficient. In the case of $R_2Co_{14}B$ compounds, the 4f sublattice anisotropy shows qualitatively the same behavior as in $R_2Fe_{14}B$. However, the conribution of the 3d anisotropy is different. The 3d sublattice gives rise to a planar anisotropy, such as in the cases of $Y_2Co_{14}B$ and $La_2Co_{14}B$ [6].

In $Nd_2Fe_{14}B$-type structures, there are four formula units per unit cell, with six inequivalent Fe sites and two Nd sites in the space group $P4_2/mnm$ [11]. The inequivalent crystallographic Fe sites are: $16k_1$, $16k_2$, $8j_1$, $8j_2$, 4e and 4c. Neutron diffraction studies of powdered $Nd_2(Fe,Co)_{14}B$ revealed that the cobalt atoms show a preference for occupying 4e sites, while Fe atoms occupy preferentially the $8j_2$ sites [12,13].

An understanding of the basic magnetic properties of the Pr- and Nd-based 2:14-type compounds is important for fabricating permanent magnets. Therefore, an investigation of the magnetic properties of the partially substituted Nd by Pr and Fe by Co in $Nd_2Fe_{14}B$ was undertaken. The present investigation mainly includes measurements of magnetization, anisotropy fields and spin-reorientation temperatures caused by Pr substitution for Nd in $Nd_{2-x}Pr_xCo_{14}B$.

EXPERIMENTAL DETAILS

The purities of the starting metals are 99.9% for Pr and Nd, 99.95% for Co and 99.98% for B. The materials were prepared by induction melting of the constituent metals in a water-cooled copper boat under a flowing argon atmosphere. As-cast ingots were wrapped in a tantalum foil, sealed into quartz tubes filled with $\sim$ 1/3 atmosphere of argon gas, heat treated at $\sim$ 900 C for 14 days and then rapidly cooled to room temperature. X-ray diffraction, thermomagnetic analysis and optical microscopy were employed to ensure the homogeneity of the alloys. X-ray diffraction analysis was performed at room temperature on randomly oriented powders with the use of a Rigaku diffractometer and Cr-K$_\alpha$ radiation. Aligned powders (in a field of 20 kOe) were also studied by x-ray diffraction in order to establish their anisotropy. Lattice parameter refinement was done by a computer refinement based on Cohen's method. Thermomagnetic analysis was performed using a Faraday balance by recording magnetization versus temperature curves at low external field ($\sim$ 0.5 to 3 kOe) in the temperature range 4.2 K to $\sim$ 1100 K for a rough chunk of the sample. The spin-reorientation temperatures, T_{SR}, were determined from these M vs T curves. The Curie temperatures, T_c, were obtaind by plotting M^2 vs T and extrapolating the steep part of the curve to $M^2 = 0$. The magnetic measurements at 295 K and 77 K were carried out using a PAR vibrating sample magnetometer in external fields up to 20 kOe. Saturation magnetizations, M_s, were obtained from magnetization isotherms usin Honda (M α 1/H) plots. Anisotropy fields, H_A, were determined at 295 K and 77 K for powders with size < 37 μm which were aligned in wax, by measuring the M vs H curves in the easy and hard directions. H_A values were deduced from the extrapolated intersection of these curves.

RESULTS AND DISCUSSION

Structural and magnetic properties of all the alloys of $Nd_{2-x}Pr_xCo_{14}B$ (with x = 0 to 2) are presented in Tables I and II. They crystallize in a tetragonal $Nd_2Fe_{14}B$-type structure with the space group $P4_2/mnm$. The lattice parameter c increases non-linearly with increasing Pr content, while the parameter a shows almost no change across the $Nd_{2-x}Pr_xCo_{14}B$ series. These results may indicate some preferential site occupation of the Pr atom on R (4f or 4g) crystal sites. The effect of partial substitution of both Nd by Pr and Co by Fe was investigated for illustration of one composition, namely, $Nd_{0.4}Pr_{1.6}Co_{11}Fe_3B$. It may be noted that introduction of Fe in the compound, with x = 1.6, was observed to expand the a and c parameters by $\sim$ 0.2% and $\sim$ 1%, respectively. The substitution of Fe has a significant effect in changing the magnetocrystalline anisotropy of the system, which is discussed in the following paragraphs.

A detailed magnetic phase diagram determined for the present system is shown in Fig. 1. The Curie temperature decreases linearly, although slightly, with the increase in Pr concentration. In the $Nd_{2-x}Pr_xCo_{14}B$ series the Co 3d-3d direct exchange interactions are the strongest and mainly determine the high value of the Curie temperature. The relatively small change in T_c upon substituting Pr for Nd may, in fact, be partially attributed to the R-T exchange interaction which is slightly stronger for the Nd-rich system compared to that of the Pr-rich system.

Measurements of M vs T for Nd-rich compounds at low temperature exhibit a spin-reorientation phenomenon. The spin-reorientation temperatures, T_{SR_1}, determined for x = o and x = 0.1 are shown in Fig. 1. This phenomenon, which qualitatively is similar to that of $Nd_2Fe_{14}B$ (at $\sim$ 135 K) is ascribed to higher order terms in the crystal field potential, resulting in a transition of spin from a c-axis to a cone with temperature decreasing [7]. T_{SR_1} progressively decreases as Pr partially substitutes for Nd. Both the decrease

TABLE I

Lattice Parameters, Curie Temperatures, Saturation Magnetic Moments a.
the Theoretical Energy Product of the $Nd_{2-x}Pr_xCo_{14}B$ System

Composition	a (Å)	c (Å)	T_c (K)	M_s (μ_B/f.u.)		$(BH)_{max}$ [a]
				295 K	77 K	MOe
$Nd_2Co_{14}B$	8.651	11,878	1004	23.2	24.7	32.7
$Nd_{1.9}Pr_{0.1}Co_{14}B$	8.650	11.876	1002	--	--	--
$Nd_{1.6}Pr_{0.4}Co_{14}B$	8.649	11.878	998	22.3	24.1	30.3
$Nd_{1.2}Pr_{0.8}Co_{14}B$	8.648	11.880	996	21.7	23.8	28.4
$Nd_{0.8}Pr_{1.2}Co_{14}B$	8.649	11.884	993	21.4	23.6	27.7
$Nd_{0.6}Pr_{1.4}Co_{14}B$	8.650	11.885	991	21.3	23.4	27.4
$Nd_{0.4}Pr_{1.6}Co_{14}B$	8.650	11.885	989	21.2	23.3	27.2
$Nd_{0.2}Pr_{1.8}Co_{14}B$	8.648	11.888	988	21.1	23.2	26.7
$Pr_2Co_{14}B$	8.649	11.890	988	21.0	23.2	26.5
$Nd_{0.4}Pr_{1.6}Co_{11}Fe_3B$	8.665	11.965	970	24.1	26.2	35.3

(a) Theoretical energy product.

TABLE II

Anisotropy Fields and Spin-Reorientation Temperatures of
the $Nd_{2-x}Pr_xCo_{14}B$ System

Composition	H_A (kOe)		T_{SR_1} (K)	T_{SR_2} (K)
	295 K	77 K		
$Nd_2Co_{14}B$	55	105	34	546
$Nd_{1.9}Pr_{0.1}Co_{14}B$	---	---	30	549
$Nd_{1.8}Pr_{0.2}Co_{14}B$	---	---	--	556
$Nd_{1.6}Pr_{0.4}Co_{14}B$	80	132	--	568
$Nd_{1.2}Pr_{0.8}Co_{14}B$	95	172	--	595
$Nd_{0.8}Pr_{1.2}Co_{14}B$	106	207	--	623
$Nd_{0.6}Pr_{1.4}Co_{14}B$	110	224	--	638
$Nd_{0.4}Pr_{1.6}Co_{14}B$	118	258	--	648
$Nd_{0.2}Pr_{1.8}Co_{14}B$	127	295	--	659
$Pr_2Co_{14}B$	140	350	--	668
$Nd_{0.4}Pr_{1.6}Co_{11}Fe_3B$	66	141	--	681

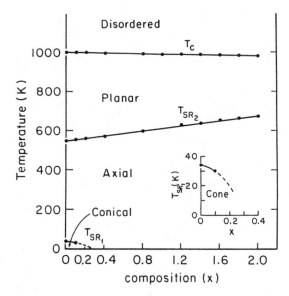

Fig. 1. Magnetic phase diagram of the $Nd_{2-x}Pr_xCo_{14}B$ system. Inset: Composition dependence of T_{SR_1}.

in cone angle and the decrease in temperature can be explained by the smaller R-T exchange field, which results in the increase of the relative importance of the crystal field interaction [7].

The dependence of the saturation magnetization M_s on the Pr concentration at 295 K and 77 K are plotted in Fig. 2 and given in Table I. M_s decreases monotonically with the increase in Pr content. The change in the total moment beyond $x \simeq 1.4$ in $Nd_{2-x}Pr_xCo_{14}B$ is found to be relatively small at both 77 K and 295 K.

Results of magnetic studies observed by powder neutron diffraction for $R_2Fe_{14}B$ (R = Pr or Nd) and $Nd_2Co_{14}B$ [11-14] revealed a ferromagnetic coupling of R-3d magnetic moments. The average of Pr and Nd moments at 77 K was observed to be $\simeq 2.7$ μ_B and $\simeq 2.75$ μ_B. An attempt is made to estimate the change in Co moment, assuming a constant Pr and Nd moment throughout the entire series of $Nd_{2-x}Pr_xCo_{14}B$ and also assuming a collinear ferromagnetic coupling. The estimated value of the Co moment is found to be 1.40 μ_B and 1.27 μ_B in $Nd_2Co_{14}B$ and $Pr_2Co_{14}B$ compounds, respectively. As shown in Fig. 3, a small change in Co moment across the series has been observed. The results are in good agreement with those previously determined in $Y_2Co_{14}B$ [4].

The dependence of the anisotropy field, H_A, on Pr content is given in Table II and shown in Fig. 4. It should be pointed out that H_A values determined at 77 and 298 K for most of the alloys studied are relatively very large. They are to be treated as approximate values, since they were extrapolated from low field measurements. The observed anisotropy field arises from two contributions - the 3d sublattice and 4f sublattice anisotropies. The Co 3d sublattice favors an easy direction of magnetization in the ab plane, as in $Y_2Co_{14}B$, while Nd and Pr both favor an easy direction of magnetization parallel to the c-axis at 77 K and 295 K. However, the strong anisotropy may be attributed to the large 4f sublattice anisotropy which dominates the Co anisotropy. H_A increases significantly from ~ 55 kOe to ~ 140 kOe at 295 K and from 105 kOe to ~ 350 kOe at 77 K, as Pr progressively replaces Nd. A similar extremely

107

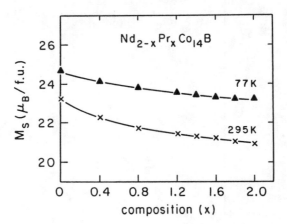

Fig. 2. Concentration dependence of the
saturation magnetization, M_s, of
the $Nd_{2-x}Pr_xCo_{14}B$ system at 295 K
and 77 K.

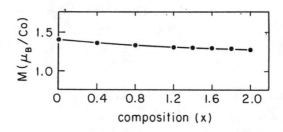

Fig. 3. Concentration dependence of the Co
magnetic moment in the $Nd_{2-x}Pr_xCo_{14}B$
system at 77 K.

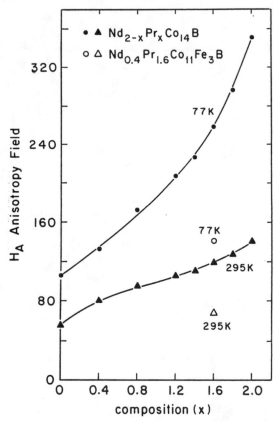

Fig. 4. Concentration dependence of the magneto-crystalline anisotropy field determined at 77 K and 295 K. Results are also shown for the composition $Nd_{0.4}Pr_{1.6}Co_{11}Fe_3B$.

large difference in H_A between $Nd_2Co_{14}B$ and $Pr_2Co_{14}B$ at 4.2 K was also reported from measurements performed at a large external field [6]. This large difference in H_A between Pr-rich and Nd-rich compounds may be attributed to the stronger 4f uniaxial anisotropy of Pr, which persists down to 0°K.

In addition to increasing the magnetization, the substitution of Co by Fe significantly decreases the anisotropy field (for $Nd_{0.4}Pr_{1.6}Co_{11}Fe_3B$ as compared to $Nd_{0.4}Pr_{1.6}Co_{14}B$), as shown in Fig. 4. This behavior is not completely understood at present. It may originate from any or all of the following: 1) different electronic charges on Co and Fe, 2) variation in unit cell dimensions, including changes in the c/a ratio and 3) preferential substitution on the various transition metal sites [12]. A difference in the charges of transition metal ions would significantly alter the crystal field potential experienced by the rare earth and, accordingly, the rare earth anisotropy. Preliminary calculations indicate that a change in 3d charge from +1 to zero would roughly double the leading crystal field term A_2^0. Similarly, a shrinkage of the unit cell and, correspondingly, the interatomic distances, would increase the magnitude of the crystal field parameters. Furthermore, it has recently been shown that the increase in H_A with composition correlates with both the c/a ratio in $Pr_2(Fe,Co)_{14}B$ [15] and the Co occupation of the 4e site in $Nd_2(Co,Fe)_{14}B$ [12].

The effect of competing anisotropies of 3d and 4f sublattices results in another spin-reorientation phenomenon (T_{SR_2}). This effect was observed earlier for $Nd_2Co_{14}B$ and $Pr_2Co_{14}B$ compounds well above room temperature [16–18]. It appears that this effect is also present in the intermediate concentration range of the $Nd_{2-x}Pr_xCo_{14}B$ series. Results of the spin-reorientation temperatures, T_{SR_2}, determined from the thermomagnetic analysis as a function of Pr concentration are shown as part of the phase diagram (Fig. 1) and also given in Table II. T_{SR_2} shows a linear dependence with Pr content. The increase in T_{SR_2} is $\simeq 67$ K per one Pr atom substituted for Nd. The observed behavior indicates a transition between uniaxial and basal plane anisotropy, which originates from the obvious competing effects of the Co 3d and Nd and/or Pr 4f sublattices [19].

The measurements of magnetization at temperatures close to T_{SR_2} performed in different applied fields exhibit a peak or cusp-like shape. An illustrative observation of T_{SR_2} for a selected compound, $Nd_{0.2}Pr_{1.8}Co_{14}B$, is shown in Fig. 5. It can be seen that as the applied field is lowered, the two inflection points converge and the peak becomes a cusp-like feature, coincident with the spin-reorientation temperature.

The increase in T_{SR_2} for $Nd_{0.4}Pr_{1.6}Co_{11}Fe_3B$ upon Fe substitution (see Table II) is consistent with the lowering of the anisotropy fields.

The substitution of Pr for Nd in $Nd_{2-x}Pr_xCo_{14}B$ results in a nearly three-fold increase in anisotropy fields at room temperature and at 77 K. However, this substitution decreases the magnetic moment and theoretical energy product by about 20%. These results thus appear to suggest the importance of these compositions in the fabrication of permanent magnets.

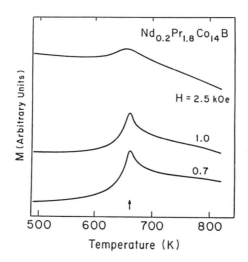

Fig. 5. Magnetization curves near the spin-reorientation temperature region (T_{SR_2}) obtained at different magnetic fields for the selected $Nd_{0.2}Pr_{1.8}Co_{14}B$ compound.

CONCLUSIONS

$Nd_{2-x}Pr_xCo_{14}B$ intermetallics, with x varying from 0 to 2.0, have been investigated. The magnetic phase diagram indicates the occurrence of conical, axial and planar magnetic anisotropies in the composition ranges investigated. These results have been attributed to the competing 3d and 4f sublattice anisotropies. The latter is sensitive to the relative concentrations of the rare earth ions, especially at the praseodymium-rich compositions. The substitution of Pr for Nd results in a relatively huge (nearly 3-fold) increase in the H_A with only a slight decrease ($\sim$ 20%) in the theoretical energy product. These results, therefore, suggest that these compositions may be useful for the fabrication of permanent magnets.

ACKNOWLEDGEMENT

We wish to thank the Lawrence Livermore National Laboratory for their support.

*On leave from Institute of Physics, Jagiellonian University, 30-059 Cracow, Poland.

REFERENCES

1. D. Givord, H. S. Li and J. M. Moreau, Sol. State Commun. 50, 497 (1984).
2. S. Sinnema, R. J. Radwanski, J. J. M. Franse, D. B. de Mooij and K. H. J. Buschow, J. Magn. Magn. Mat. 44, 333 (1984).
3. Y. Matsuura, S. Hirosawa, H. Yamamoto, S. Fujimura and M. Sagawa, Appl. Phys. Lett. 46, 308 (1985).
4. M. Q. Huang, E. B. Boltich, W. E. Wallace and E. Oswald, J. Mag. Mag. Mat. 60, 270 (1986).
5. H. M. van Noort and K. H. J. Buschow, J. Less-Common Met. 113, L9 (1985).
6. K. H. J. Buschow, D. B. de Mooij, S. Sinnema, R. J. Radwanski and J. J. M. Franse, J. Mag. Mag. Mat. 51, 211 (1985).
7. E. B. Boltich and W. E. Wallace, Sol. State Commun. 55, 529 (1985).
8. S. G. Sankar and K. S. V. L. Narasimhan, J. Mag. Mag. Mat. 54-57, 530 (1985).
9. D. Givord, H. S. Li, J. M. Moreau and P. Tenaud, ibid., 445 (1986).
10. Satoshi Hirosawa, Yuaka Matsuura, Hitoshi Yamamoto, Setmo Fujimura and Masato Sagawa, J. Appl. Phys. 59, 873 (1986).
11. J. F. Herbst, J. J. Croat, F. E. Pinkerton and W. B. Yelon, Phys. Rev. B29, 4176 (1984).
12. J. F. Herbst and W. B. Yelon, J. Appl. Phys. 60, 4224 (1986).
13. Ying-chang Yang, Shu-can Fu and W. J. James, MMM Conf., 1987, to appear.
14. J. F. Herbst and W. B. Yelon, J. Appl. Phys. 57, 2343 (1985).
15. F. Bolzoni, J. M. D. Coey, J. Gavigan, D. Givord, O. Moze, L. Pareti and T. Viadieu, J. Mag. Mag. Mat. 65, 123 (1987).
16. A. T. Pedziwiatr and W. E. Wallace, Sol. State Commun. 60, 653 (1986).
17. R. Grossinger, R. Krewenka, S. K. Sun, R. Eibler, H. R. Kirchmayer and K. H. J. Buschow, J. Less-Common Met. 124, 165 (1986).
18. S. Y. Jiang, F. Pourarian, E. B. Boltich and W. E. Wallace, presented to Inter. Mag. Conf., Tokyo, Japan, 1987.
19. E. B. Boltich, A. T. Pedziwiatr and W. E. Wallace, J. Mag. Mag. Mat. 66, 317 (1987).

HIGH FIELD TORQUE MAGNETOMETRY MEASUREMENTS ON $Y_{1.8}Er_{0.2}Fe_{14}B$

C.M. WILLIAMS, N.C. KOON, AND B.N. DAS
U.S. Naval Research Laboratory, Washington, D.C. 20375-5000

ABSTRACT

The magnetocrystalline anisotropy energy has been determined for single crystal $Y_{1.8}Er_{0.2}Fe_{14}B$ in the (001) and (100) planes between 5 K and 300 K using torque magnetometry techniques. The results are compared with a model based on crystal field theory. Excellent agreement was obtained between the model and experiment. Both experiment and model showed a first order spin reorientation between the [001] and an angle 70 degrees from the [001] in the (100) plane.

INTRODUCTION

There has been considerable interest in tetragonal $R_2Fe_{14}B$ compounds as the basis of a new class of permanent magnet materials because of the unusually large energy products they exhibit at room temperature for certain rare earth substitutions. The origin of the high energy products is directly related to the large saturation magnetization and magnetic anisotropy energy. The single crystal magnetic anisotropy energy has been investigated by several groups[1-5]. The origin of the magnetocrystalline anisotropy energy is believed to be directly related to an interaction between the 4f-electrons and the crystal field; however, to date there have been few if any direct comparisons made between the experimental magnetic anisotropy and crystal field theory. The reason being that the large anisotropy energy makes it very difficult to use conventional torque magnetometry techniques to determine the angular dependence of the magnetic free energy , particularly at low temperatures where the higher order terms become more important. Most anisotropy energies measurements have been obtained from magnetization data which do not have the angular fidelity required for a direct comparison with theory. In this investigation we determine the angular dependence of the magnetic free energy as a function of temperature using conventional torque magnetometry techniques and make a direct comparison of the free energy with the energy calculated using a model based on crystal field theory. We circumvent the high anisotropy energy problem to some extent by considering $Y_{1.8}Er_{0.2}Fe_{14}B$. The rationale for using this composition is the large iron sublattice anisotropy favors the c-axis and the erbium anisotropy favors the basal plane; at low temperatures the Fe and Er sublattice anisotropies nearly cancel. The ultimate aim of this investigation is to determine the crystal field constants for this dilute system and use these constants to predict the anisotropy energies of other $R_2Fe_{14}B$ structures which may be potentially useful in high energy product permanent magnet applications.

EXPERIMENTAL

The most direct method of determining the magnetic free energy is by torque magnetometry; however, this method requires well characterized, good quality single crystals. The torque method consist of measuring the torque required to rotate the magnetization away from a principal crystallographic direction in a particular plane. The torque is related to the energy by $L(\theta,\phi) = -\partial E(\theta,\phi)/\partial(\theta,\phi)$, where $E(\theta,\phi)$ is the free energy and θ,ϕ are the angles the

Mat. Res. Soc. Symp. Proc. Vol. 96. ©1987 Materials Research Society magnetization with respect to the z-and x-axes.

The single crystals used in this investigation were prepared by the Czochralski method using a rotating hearth Tri Arc furnance. The samples weighed 2.5 mg and 2.7 mg and were approximately 0.5mm in diameter. The torque was measured using a 6 Tesla high field torque magnetometer. The details of the construction of this magnetometer will be reported elsewhere. The sensitivity of the magnetometer is 5×10^{-3} dyne-cm, far greater than needed for this application.

A typical torque and energy curve in the (001) plane are shown in Figure 1 for a 2.5 mg sample measured at 150 K . The free energy is obtained by a piece wise integration (Simpson's rule) of the torque. At this temperature the [100] and [1$\bar{1}$0] are the easy and hard directions, respectively.   The temperature dependence of the free energy in the (100) plane is shown in Figure 2a and 2b for temperatures between 5.5 K and 300 K. At 300 K the [001] and [110] are the easy and hard directions of magnetization, respectively. As the temperature is decreased spin reorientations occur at about 60 K, where the easy direction changes from the [001] to an angle of about 70 degrees away from the [001] in the (100) plane. The absence of a gradual change in the easy direction of magnetization with temperature suggest that the spin reorientation is first order. The spin orientations are summarized in Figure 3.

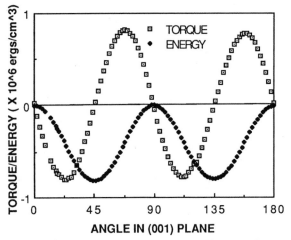

Figure 1 - Torque and energy curves in the (001) plane for
T = 150 K; zero angle corresponds to the [110].

The temperature dependence of the free energy in the (001) plane is shown in Figure 4 for temperatures between 10 K and 200K. As in Figure 1, the [100] and [110] are the easy and hard directions of magnetization, respectively. The increase in free energy with decreasing temperature is the result of an increase in the Er basal plane anisotropy with decreasing temperature. The curves are reasonably symmetric about the [100] and [110] directions between room temperature and 60 K; however, below 60 K the energy curves become increasingly more asymmetric as the temperature approaches 10 K. This asymmetry is characterized by a rather broad energy minima around the [100] and a sharp maxima around the [110]. The origin of this behavior is not clear; however, other data suggest that it may be related to spin fluctuations about the [110].

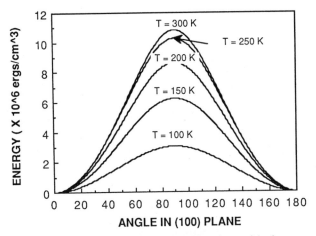

Figure 2a - Experimental temperature dependence of the free
energy in the (100) plane for temperatures
between 100 K and 300 K; zero angle
corresponds to the [001].

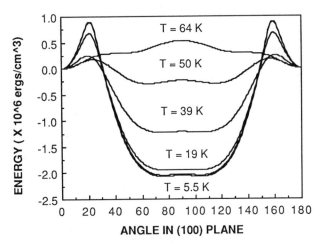

Figure 2b - Experimental temperature dependence of the free
energy in the (100) plane for temperatures
between 5.5 K and 64 K; zero angle
corresponds to the [001].

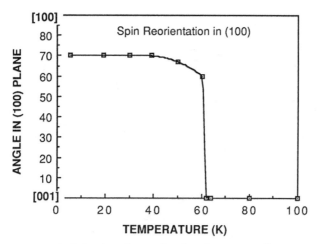

Figure 3 - Spin orientation as a function of temperature for $Y_{1.8}Er_{0.2}Fe_{14}B$.

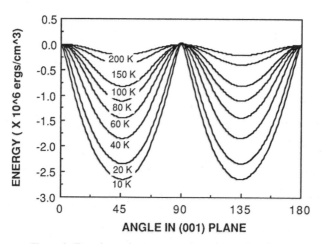

Figure 4 - Experimental temperature dependence of the free energy in the (001) plane for temperatures between 10 K and 200 K; zero angle corresponds to the [110].

MODEL

In principle, the model is rather straightforward. The Hamiltonian is given by

$$H = H_{exch} + H_{cf} \tag{1}$$

where

$$H_{exch} = \sum_i 2 (g_i - 1) \mu_B H_{exch} \cdot J_i \tag{2}$$

and

$$H_{cf} = \sum_{lm} B_l^m O_l^m \tag{3}$$

In these expressions g_i and J_i are the g-factor and the spin operators for the ith ion, while the B_l^m's and the O_l^m's are the usual crystal field parameters and Stevens operator equivalent. The exchange interaction is assumed to scale with the spin part of the rare earth moment which is proportional to $(g_i - 1)$. For different rare earth ions the crystal field parameters are assumed to vary according to the relation

$$B_l^m = \theta_l < r^l > (1 - \sigma_l) A_l^m \tag{4}$$

The A_l^m are parameters assumed to be independent of the rare earth ion and to depend only on the structure. The $< r^l >$'s are the Hartree Fock radial integrals. The σ_l's are the screening parameters taken from Freeman and Watson; the θ_l's are the reduced matrix elements for the rare earth ions.

To calculate the angular dependence of the free energy in a particular plane for a given set of exchange and crystal field parameters the eigenstates are determined for various orientations of the exchange field in the plane. The temperature dependence of free energy is given by

$$F_0 = -kT \sum_i \ln(Z_i) \tag{5}$$

where Z_i is the partition function of the ith ion.

The degree of ease with which the above calculation can be performed depends on the number of crystal field parameters required to describe the free energy. In the case of well characterized crystallographic structures the total number of crystal field constants can be reduced by symmetry arguments. The tetragonal $R_2Fe_{14}B$ structure has been determined by Herbst, et al.[6]. It has a primitive cell containing eight rare earth atoms distributed over two crystallographically inequivalent sites. Each site consist of four atoms located along two mm-axes parallel to the [110] and [110]. This fact permits the number of crystal field parameters to be reduced from 15 to 9 for each rare earth site, leaving a total of 18 parameters. In our model we assume that the number of parameters can be further reduced if we assume that crystal parameters of like order can be averaged for the two inequivalent sites. Under this assumption the allowable crystal field parameters are:

116

$$B_2^0 \; ; \; B_2^{2(s)}$$

$$B_4^0 \; ; \; B_4^{2(s)} \; ; \; B_4^4 \tag{6}$$

$$B_6^0 \; ; \; B_6^{2(s)} \; ; \; B_6^4 \; ; \; B_6^{6(s)}$$

where the (s) indicate complex parameters which arise as a result of rotations of the crystal field. Theoretically, the parameters in (6) can not be further reduced. The relative importance of the remaining crystal field parameters can only be determined by comparison of the calculated and experimental free energies. Figures 5 and 6 shows the temperature dependence of the calculated free energies for the (100) and (001) planes, respectively. The crystal field parameters were determined by choosing A_4^4, A_6^4, and H_{exch} to fit the temperature dependence of the (001) data;

having fixed A_4^4, A_6^4, and H_{exch} ; A_2^0, A_4^0, and A_6^0 were chosen to fit the temperature dependence

of the (100) plane data. Qualitatively the agreement between experiment and the model is excellent. The (001) data scales very well with temperature and some asymmetry about the [100] and [110] directions is observed for temperatures below 60 K; however, the asymmetry is not as pronounced as in the case of the experimental data. The (100) plane data scales reasonably well with temperature. The model predicts a first order spin reorientation between the [001] and an angle about 70 degrees from the [001] which is in agreement with that observed experimentally. The prediction of the onset temperature (70 K) for the spin reorientation is slightly higher than that observed experimentally (60 K). The exchange and crystal field parameters used to fit the experimental data are shown in Table 1. The parameters shown in Table 1 are not necessarily unique; additional data in the (110) are required to uniquely define the parameters.

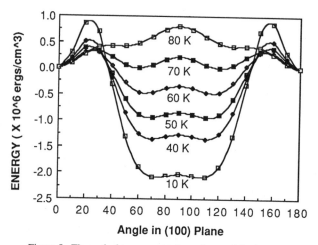

Figure 5 - Theoretical temperature dependence of the free energy in the (100) plane for temperatures between 10 K and 80 K; zero angle corresponds to the [001].

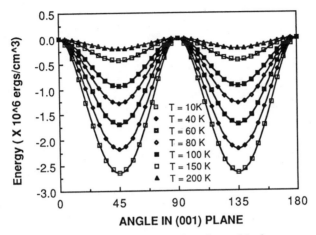

Figure 6 - Theoretical temperature dependence of the free
energy in the (001) plane for temperatures
between 10 K and 200 K; zero angle corresponds
to the [110].

In summary, the experimental and theoretical magnetic free energy has been determined for a dilute $Y_{1.8}Er_{0.2}Fe_{14}B$ compound. Although the agreement between experiment and theory is in good qualitative agreement the crystal field constants determined form this investigation are not necessarily unique. Additional data in the (110) plane will be required to uniquely define the crystal field constants. Once the constants have been uniquely determined for $Y_{1.8}Er_{0.2}Fe_{14}B$, the measurements and calculation will be extended to other dilute systems of the same structure.

TABLE 1

$$H_{exch} = (g - 1)\,\mu_B\,H_o = 35.0 \text{ meV}$$
$$A_2^0 = 76.0 \text{ (meV/}a_0^2)$$
$$A_4^0 = -0.40 \text{ (meV/}a_0^4)$$
$$A_4^4 = -9.50 \text{ (meV/}a_0^4)$$
$$A_6^0 = -0.107 \text{ (meV/}a_0^6)$$

REFERENCES

1. D. Givord, H.S. Li and R. Perrier de la Bathie, Solid State Commun. **51** (1984) 2083.

2. M. Sagawa, S. Fujimura, M. Togawa, H. Yamamoto and Y. Matsuura, J Appl. Phys. **55** (1984) 2083.

3. N.C. Koon, B.N. Das, M. Rubinstein and J. Tyson, J Appl. Phys. **57** (1985) 4091.

4. S. Hirosawa and M. Sagawa, Solid State Commun. **54** (1985) 335.

5. N.C. Koon, B.N. Das and C.M. Williams, J. Magn. Magn. Mat. **54** (1986) 523.

6. J.R. Herbst, J.J. Croat, F.E. Pinkerton W. B. Yelon, Phys. Rev. **B29** (1984) 4176.

SPIN REORIENTATIONS IN $(Er_xY_{1-x})_2Fe_{14}B$ AND $(Er_xPr_{1-x})_2Fe_{14}B$ SYSTEMS

E. B. BOLTICH[a)], A. T. PEDZIWIATR[b),c)] AND W. E. WALLACE
MEMS Department and Magnetics Technology Center, Carnegie-Mellon University,
Pittsburgh, PA   15213

ABSTRACT

The bulk magnetic properties of $(Er_xY_{1-x})_2Fe_{14}B$ and $(Er_xPr_{1-x})_2Fe_{14}B$
systems were studied over the temperature range 4.2 - 1100 K. Lattice parameters, saturation magnetizations, Curie temperatures and spin reorientation
temperatures were determined. Theoretical description of the detailed magnetic
behavior is presented, based on a crystal field model. The $(Er_xY_{1-x})_2Fe_{14}B$
compounds were all found to exhibit plane-to-axis spin reorientations similar
to that observed for $Er_2Fe_{14}B$, with the transition temperature decreasing with
increasing Y content. In contrast, the spin reorientations in the
$(Er_xPr_{1-x})_2Fe_{14}B$ systems appear to be of the cone-to-axis type. Since higher
order crystal field terms appear to be significant only in the cases of Nd^{3+}
and Ho^{3+}, the results are discussed in terms of a crystal field Hamiltonian
involving only 2nd order terms. Using known values of the exchange field, Fe
anisotropy and the ratios of the crystal field coefficients, the multi-ion
crystal field problem was formulated in terms of a single adjustable parameter
$(B_2^0(f))$. It is shown that 2nd order crystal field terms are capable, not only
of explaining the conical anisotropy of the $(Er_xPr_{1-x})_2Fe_{14}B$ systems, but also
the decrease in the Er moment upon passing through the spin reorientation (as
has been observed for $Er_2Fe_{14}B$). The magnetic structure of $Er_{1.5}Pr_{0.5}Fe_{14}B$ is
also predicted.

INTRODUCTION

Two different types of spin reorientations have been observed in the
$R_2Fe_{14}B$ compounds (R = Nd, Ho, Yb, Er and Tm) with increasing temperature[1-12]:
1) transitions attributable to higher order terms in the crystal field
Hamiltonian of the rare earth ions, as in $Nd_2Fe_{14}B$ near 135 K and $Ho_2Fe_{14}B$
near 65 K and 2) transitions resulting from the competition of various sublattices possessing different anisotropies as in $Er_2Fe_{14}B$, $Tm_2Fe_{14}B$ and $Yb_2Fe_{14}B$
near 330, 315 and 115 K, respectively. Both types of spin reorientations are
evidenced experimentally by the appearance of irregularities in the magnetization versus temperature (M vs. T) curves measured on rough chunks in low
applied fields[13]. Transitions from conical to axial structures (and vice
versa) appear as steps in the M vs. T curves, while transitions from planar to
axial structures (and vice versa) appear as spikes.

Spin reorientations in $R_2Fe_{14}B$ systems have been the subject of a great
deal of interest within the magnetics community. Early point charge calculations led to the conclusion that the conical anisotropy observed in $Nd_2Fe_{14}B$
at low temperatures was the result of competing anisotropies of the two rare
earth sites[14]. However, it has since been shown that the conical structure of
this system is due to the presence of higher order terms in the crystal field
potential of the rare earth ions and not to competing anisotropies[2]. In the
present work we report for the first time, an $R_2Fe_{14}B$ system $(Er_xPr_{1-x})_2Fe_{14}B$

a)Supported by the Magnetic Materials Research Group, through the Division of
Materials Research, National Science Foundation, under Grant No. DMR-8613386.

b)Supported by Lawrence Livermore National Laboratory.

c)On leave from Institute of Physics, Jagiellonian University, 30-059 Cracow,
Poland.

which possesses an easy cone that is shown by detailed crystal field theory
to truly be the result of competing rare earth anisotropies, in this case
between Er (positive 2nd order Stevens coefficient) and Pr (negative 2nd
order Stevens coefficient).

In addition to the conical anisotropy referred to above, other implications of crystal field analysis, such as the magnetic structures of the rare
earth sublattices and the anisotropy in the magnetization of $Er_2Fe_{14}B$ will
also be discussed.

EXPERIMENTAL

The samples were produced by melting stoichiometric proportions of the
constituent elements (99.9% purity or better) in a water-cooled copper boat
by induction heating under flowing high purity argon. The as-cast ingots were
annealed at 900°C for two weeks and then rapidly cooled to room temperature.
X-ray diffraction analysis was performed at room temperature on powdered
samples with a Rigaku powder diffractometer using $Cr-K_\alpha$ radiation, both for
randomly oriented samples and those aligned in a 20 kOe magnetic field. By
comparing the x-ray diffraction patterns of the random and aligned samples,
the easy direction of magnetization at room temperature was determined. Lattice parameters were refined by Cohen's least square method. The phase integ-
rity of the alloys was further verified by thermal magnetic analysis (TMA) and
optical metallography. TMA was performed by the Faraday technique, in low
applied fields (about 4 kOe), throughout the temperature range 295 - 1100°C.
The Curie temperatures of the various alloys were also determined by this
method.

Low temperature measurements (4.2 - 295 K) were also done by the Faraday
method. Saturation magnetizations of the various compounds were obtained from
Honda (M versus 1/H) plots. Spin-reorientation temperatures (T_{sr}) were determined from the low field M vs. T curves of rough chunks, according to the
method described in our previous papers [13,15].

CRYSTAL FIELD ANALYSIS

When a rare earth ion is placed in a magnetically ordered crystal, the
free ion electronic wave functions (characterized by the quantum numbers L,
S, J, M_J) are perturbed by a Hamiltonian of the form

$$\mathcal{H} = \mathcal{H}_{CF} + \mathcal{H}_{Exch}$$

where $\mathcal{H}_{CF}$, the crystal field term, represents the electrostatic interactions
with the neighboring ions and $\mathcal{H}_{Exch}$, the exchange term, represents the magnetic interactions with the other ions in the crystal.

For most rare earths (exceptions being Sm and Eu), the excited J multiplets
are well separated from the ground state and the crystal field interaction is
most conveniently treated by the method of operator equivalents[16]. According
to this method, the crystal field part of the Hamiltonian is expressed as

$$\mathcal{H}_{CF} = \sum_{m=0}^{6} \sum_{m=-n}^{n} \sum_{\alpha} B_n^{m\alpha} O_n^{m\alpha} (J)$$

where α takes on the values sine and cosine, the $O_n^{m\alpha}$s are polynomials in the
angular momentum operator, J, and the $B_n^{m\alpha}$s are the crystal field intensity
parameters, characteristic of the rare earth ion and its environment. These
intensity parameters are broken into an ion part and an environmental part
according to the expression

$$B_n^{m\alpha} = <J|\theta_n|J> <r^n> (1-\sigma_n) (-|e|) K_R^m A_n^{m\alpha}$$

where the first three factors are characteristic of the rare earth ion and the last is characteristic of its environment. $<J\|\theta_n\|J>$ is a reduced matrix element, known as the Stevens factor; $<r^n>$ is the expectation value of the nth power of the radius of the $4f$ wave function and σ_n is a shielding factor. The environmental contribution, $A_n^{m\alpha}$, is a sum of the electrostatic potentials due to the surrounding ions. If the surrounding ions are considered as point charges, this factor can be evaluated analytically from the tesseral harmonic expansion

$$A_n^{m\alpha} = (4\pi/(2n+1)) \sum_j Q_j \ (Z_n^{m\alpha}(\theta_j, \ \phi_j)/R_j^{(n+1)})$$

where the index j runs over the surrounding ions. However, in practice, the $B_n^{m\alpha}$s are usually treated as empirical parameters, the present work being no exception.

The exchange term represents a spin-spin interaction and is usually written in a form similar to a molecular field model

$$\mathcal{H}_{Exch} = 2(g-1)\mu_B \ \vec{J}\cdot\vec{H}_{Exch}.$$

In rare earth transition metal systems the R-R exchange interaction is generally much weaker than the R-T interaction, and is usually ignored. In that case, $\mathcal{H}_{Exch}$ is a measure of the R-T interaction.

Since neither Er^{3+} nor Pr^{3+} possess conical anisotropy in the $Nd_2Fe_{14}B$ structure, the spin reorientations were analyzed according to the method employed by Vasquez and Sanchez in their study of $(Er_xGd_{1-x})_2Fe_{14}B$ systems, using a crystal field Hamiltonian containing only 2nd order terms [17]. In this model, the Hamiltonian of each ion is of the form

$$\mathcal{H} = B_2^0 O_2^0 + B_2^{2c} O_2^{2c} + 2(g-1)\mu_B \ \vec{J}\cdot\vec{H}_{Exch},$$

the sine terms being zero due to the choice of the local X-axes (vide infra). This Hamiltonian is justified by the observation that higher order terms are only significant in the cases of Nd^{3+} and Ho^{3+} [18].

The exchange fields, Fe anisotropy constant and the ratios of the various crystal field coefficients were taken from the literature (ref. 17).

The ratios of the various crystal field coefficients [$B_2^0(f)$, $B_2^{2c}(f)$, $B_2^0(g)$ and $B_2^{2c}(g)$] were those estimated from the ^{155}Gd quadrupole splitting [19]. The signs of the B_2^{2c} terms for the particular local axes chosen (Z parallel to [001] and X parallel to [110], for the atoms in the basal plane) were determined by point charge calculations. The crystal field coefficients of Pr^{3+} were determined by scaling the Er values with α and $<r^2>$[20]. These values are given in Table I.

The Fe anisotropy was linearly interpolated from the K_1 values of $Y_2Fe_{14}B$ at 4.2 (7.05×10^6 erg/cm^3) and 300 K (10.2×10^6 erg/cm^3).

The exchange field acting on the rare earth ions was assumed to be due solely to the R-T exchange interaction. The temperature dependence of the exchange field was taken to be the same as that of the Fe hyperfine field, i.e.,

$$|\vec{H}_{Exch}(T)| = |\vec{H}_{Exch}(0)| \ [1 - b(T/T_c)^2]$$

with b = 0.5. $|\vec{H}_{Exch}(0)|$ was taken as the average of the two values reported by Boge et al. [19] and Berthier et al. [21], $2\mu_B \ |H_{Exch}| = 370$ and 385 K, respectively.

After adopting these parameters, the multi-ion crystal field calculation reduces to a straightforward, one-parameter problem (which was chosen as $B_2^0(f)$). The value of $B_2^0(f)$ was determined to be that which produced the best fit to the composition dependence of the spin reorientation in the $(Er_xY_{1-x})_2Fe_{14}B$ systems (vide infra), with the other crystal field coefficients being fixed by the ratios given in Table II.

TABLE I

Relationships Among Crystal Field Intensity Parameters as
Determined by ^{155}Gd Mössbauer Spectroscopy[a]

$$B_2^0(f) = X$$

$$B_2^{2c}(f) = - (.38/.85)\ X^b$$

$$B_2^0(g) = (.79/.85)\ X$$

$$B_2^{2c}(g) = + (1.56/.85)\ X^b$$

$$B_2^m(Pr) = - 13.49\ B_2^m(Er)^c$$

---

[a]as reproduced in ref. 17

[b]The signs of the non-axial terms refer to the basal plane atoms, with the X axis taken along [110].

[c]The scaling factor was obtained using the α's and $\langle r \rangle$'s given in ref. 20.

TABLE II

Magnetic and Structural Data for $(Er_xY_{1-x})_2Fe_{14}B$ and $(Er_xPr_{1-x})_2Fe_{14}B$ Systems

Compound	a [A]	c [A]	T_c (K)	T_{sr} (K)	D.O.M. (4.2 K)	M_s (μ_B/f.u.) (4.2 K)
R = Y						
x=1.0	8.741	11.949	554	330	plane	14.1
0.75	8.745	11.952	557	280	plane	19.6
0.50	8.764	11.990	561	235	plane	24.6
0.25	8.768	12.005	565	169	plane	28.2
0.00	8.771	12.035	565	---	axis	31.6
R = Pr						
x=1.0	8.741	11.949	554	330	plane	14.1
0.875	8.759	11.987	557	248	cone	----
0.75	8.780	12.005	559	155	cone	22.2
0.625	8.791	12.030	561	38	cone	----
0.50	8.799	12.076	563	---	axis	28.5
0.25	8.812	12.165	564	---	axis	34.1
0.00	8.838	12.270	566	---	axis	36.3

The direction of the spontaneous magnetization at a given temperature was determined by varying the direction of the exchange field (in 5° increments) and observing which orientation leads to the lowest total free energy.

The total free energy for a given direction of the exchange field was determined by diagonalizing the Hamiltonian matrices of the various rare earth ions, obtaining the eigenvalues and computing the total free energy.

$$F = -kT \sum_i \ln Q_i + K_1(\text{Fe})\sin^2\theta$$

where Q_i is the partition function of the ith rare earth ion

$$Q_i = \sum e^{-E_{ij}/kT},$$

E_{ij} being the jth eigenvalue of the ith ion.

It should be noted that the 8 rare earth ions in the unit cell fall into four magnetically inequivalent sets: 1) f site ions in the basal plane, 2) f site ions in the midplane, 3) g site ions in the basal plane and 4) g site ions in the midplane. Sets 1) and 2) possess the same crystal field parameters, but for differently oriented local axes (they are related to each other by a 90° rotation). Hence, relative to these symmetry-related local axes, the Hamiltonians for these two sets of atoms differ only in the exchange term ($\vec{J}\cdot\vec{H}_{\text{Exch}}$). Similar considerations hold for the two sets of ions derived from the g site.

Having obtained the spontaneous direction of the exchange field, the magnetic moments of the individual rare earth ions in their respective local reference frames were calculated from the formula

$$\vec{M}_i = \sum_j \vec{\mu}_{ij}\, e^{-E_{ij}/kT} / \sum_j e^{-E_{ij}/kT}$$

with $\vec{\mu}_{ij}$, the magnetic moment of the jth eigenstate of the ith ion being calculated from the eigenvectors according to the expression

$$\vec{\mu}_j = -\langle \Gamma_j | g\vec{J} | \Gamma_j \rangle\, \mu_B$$

The magnetic moments were then transformed to the crystal-physical axes ($\vec{x} \parallel \vec{a}$, $\vec{y} \parallel \vec{b}$, and $\vec{z} \parallel \vec{c}$) by taking the appropriate dot products.

RESULTS AND DISCUSSION

The experimental results for both the $(\text{Er}_x\text{Y}_{1-x})_2\text{Fe}_{14}\text{B}$ and $(\text{Er}_x\text{Pr}_{1-x})_2\text{Fe}_{14}\text{B}$ systems are summarized in Table II and Fig. 1. In both series the lattice constants increase with decreasing Er content, as expected from the lanthanide contraction. These variations in the lattice constants are reflected in the Curie temperatures (T_c), which also increase with decreasing Er content. Because Y is non-magnetic, this increase in T_c must be ascribed to the increase in the interatomic distances, and the corresponding decrease in the negative Fe-Fe interactions. As shown in Table I, the saturation moments at 4.2 K increase with decreasing Er content, due to the antiferromagnetic coupling between the heavy rare earth (Er) and Fe. The more rapid increase in the case of the Pr-substituted systems is a result of the ferromagnetic coupling between the light rare earths and Fe.

Although the inclusion of higher order crystal field terms is essential to account for the low temperature behavior of $\text{Nd}_2\text{Fe}_{14}\text{B}$ and $\text{Ho}_2\text{Fe}_{14}\text{B}$, many of the phenomena observed for the other $R_2\text{Fe}_{14}\text{B}$ compounds, particularly those containing Er and Tm, can be explained by a crystal field Hamiltonian consisting of only the 2nd order terms [2,6,17,18].

As shown in Fig. 1, the spin-reorientation temperature (T_{sr}) of both the $(\text{Er}_x\text{Y}_{1-x})_2\text{Fe}_{14}\text{B}$ and $(\text{Er}_x\text{Pr}_{1-x})_2\text{Fe}_{14}\text{B}$ systems decreases monotonically with decreasing Er content. This behavior is easily understood, since it is the Er which gives rise to the planar anisotropy. The observation that T_{sr} decreases more rapidly in the Pr systems than in the Y systems may be explained by the fact that, in the 2:14:B structure, Pr^{3+} is an axial ion. The value of $B_2^0(f)$ which best fit the T_{sr} versus composition data for the $(\text{Er}_x\text{Y}_{1-x})_2\text{Fe}_{14}\text{B}$ systems was determined to be 0.37 K, in excellent agreement with that obtained by Vasquez and Sanchez [17] for the $(\text{Er}_x\text{Gd}_{1-x})_2\text{Fe}_{14}\text{B}$

124

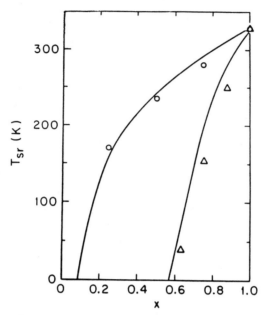

Fig. 1.   Spin-reorientation temperature versus
composition for $(Er_xY_{1-x})_2Fe_{14}B$ (o)
and $(Er_xPr_{1-x})_2Fe_{14}B$ (Δ) alloys.

systems (0.36 K). According to our present model, this value of $B_2^0(f)$ completely determines all of the crystal field parameters with the ratios given
in Table I. These same parameters were then used to explain the various phenomena discussed below. It should also be noted that the present parameters
predict that the spin reorientation occurs over a small temperature range, on
the order of a few degrees, in agreement with the experiments of Hirosawa
et al. [1].

It has previously been observed by thermal magnetic analysis of a freely
rotating single crystal that the net magnetic moment of $Er_2Fe_{14}B$ increases
when heated above the spin-reorientation temperature [1]. The reason for this
increase can be seen by reference to Fig. 2, where the splitting patterns of
the f and g sites immediately below T_{sr} are shown, both for the true exchange
field direction (parallel to [100]) and for a hypothetical state with the exchange field parallel to [001]. It is easily inferred from the figure that
the increase in the next moment on passing through T_{sr} is due to a decrease in
the Er moment, resulting from the significantly lower population of the ground
state when the exchange field is oriented parallel to the c-axis. This decrease in the Er moment is observed in spite of the fact that the ground state
moment is larger when the exchange field lies along the c-axis. It should also
be noted from the figure that, although the B_2^{2c} values are significantly different for the two sites, their splitting patterns, at least for these orientations of the exchange field, are very similar.

On the basis of the crystal field coefficients obtained above, it is also
possible to predict the magnetic structure of $Er_2Fe_{14}B$. The structure predicted
by the present parameters (at 0 K) is described in Table III and illustrated in
Fig. 3. As shown in both the table and the figure, at low temperatures, all of
the magetic moments lie in the basal plane. Each of the rare earth sites is

split into two magnetically inequivalent sublattices, one consisting of the atoms in the z=0 plane, and the other consisting of those in the z=1/2 plane. Furthermore, because these two sublattices are related to each other by reflection in the (010) plane, the resultant moment of each rare earth site lies along the [100] direction and is, therefore, antiparallel to the net Fe moment, as is shown in Table IV. The magnetic moments reported in Table III correspond to angles of 3.5° between the f site moments at z=0 and z=1/2 and 13.9° between the corresponding g site moments. Since this result cannot presently be compared with experiment, due to the lack of single crystal neutron diffraction data for $Er_2Fe_{14}B$, a similar calculation was performed for $Tm_2Fe_{14}B$ (using scaled parameters), for which experimental information is available [6]. In this case, the angles between the f and g site moments were calculated to be 12.0° and 39.9°, respectively. These compare quite well to the experimental

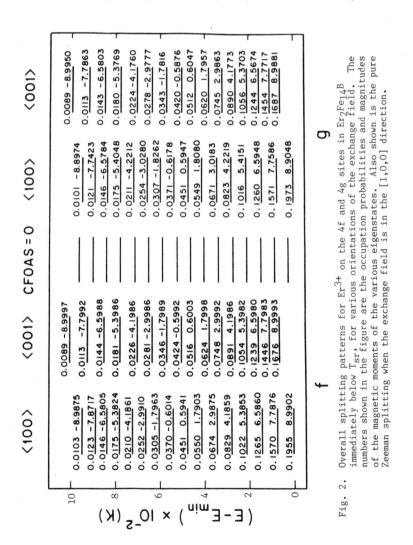

Fig. 2. Overall splitting patterns for Er^{3+} on the 4f and 4g sites in $Er_2Fe_{14}B$ immediately below T_{sr}, for various orientations of the exchange field. The numbers shown in the figure are the occupation probabilities and magnitudes of the magnetic moments of the various eigenstates. Also shown is the pure Zeeman splitting when the exchange field is in the [1,0,0] direction.

TABLE III

Magnetic Moments of Rare Earth Ions in $(Er_xPr_{1-x})_2Fe_{14}B$ Systems at 0 K

Position[a]	Ion Type	M_x [b]	M_y	M_z	$|M|$ [c]
		$Er_2Fe_{14}B$ ($\theta=90$, $\phi=0$)[d]			
f(0)	Er	−8.9931	−0.2717	0.0000	8.9972
f(1/2)	Er	−8.9931	0.2717	0.0000	8.9972
g(0)	Er	−8.9325	1.0867	0.0000	8.9983
g(1/2)	Er	−8.9325	−1.0867	0.0000	8.9984
		$Er_{1.75}Pr_{0.25}Fe_{14}B$ ($\theta=76$, $\phi=0$)			
f(0)	Er	−8.8979	−0.2685	−1.8173	8.9974
	Pr	1.9278	−0.2726	2.5114	3.1777
f(1/2)	Er	−8.8079	0.2685	−1.8173	8.9974
	Pr	1.9278	0.2726	2.5114	3.1777
g(0)	Er	−8.7490	1.0738	−1.8088	8.9983
	Pr	2.3505	1.2081	1.5495	3.0635
g(1/2)	Er	−8.7490	−1.0738	−1.8088	8.9983
	Pr	2.3505	−1.2081	1.5495	3.0635
		$Er_{1.5}Pr_{0.5}Fe_{14}B$ ($\theta=54$, $\phi=0$)			
f(0)	Er	−7.7266	−0.2473	−4.6049	8.9982
	Pr	1.3024	−0.1468	2.9141	3.1953
f(1/2)	Er	−7.7286	0.2473	−4.6049	8.9982
	Pr	1.3024	0.1488	2.9141	3.1953
g(0)	Er	−7.6797	0.9888	−4.5837	8.9981
	Pr	1.5959	0.7196	2.5982	3.1329
g(1/2)	Er	−7.6797	−0.9888	−4.5837	8.9981
	Pr	1.5959	−0.7196	2.5982	3.1329
		$Er_{1.25}Pr_{0.75}Fe_{14}B$ ($\theta=29$, $\phi=0$)			
f(0)	Er	−5.0985	−0.1788	−7.4136	8.9994
	Pr	0.6888	−0.0680	3.1218	3.1976
f(1/2)	Er	−5.0985	0.1788	−7.4136	8.9994
	Pr	0.6888	0.0680	3.1218	3.1976
g(0)	Er	−5.0818	0.7166	−7.3888	8.9962
	Pr	0.8292	0.3377	3.0249	3.1547
g(1/2)	Er	−5.0818	−0.7166	−7.3888	8.9962
	Pr	0.8292	−0.3377	3.0249	3.1547

[a]The ion positions are (ref. 22):
f(0)   (0.3572, 0.3572, 0.0000); (0.6428, 0.6428, 0.0000)
f(1/2) (0.8572, 0.1428, 0.5000); (0.1428, 0.8572, 0.5000)
g(0)   (0.7698, 0.2302, 0.0000); (0.2302, 0.7698, 0.0000)
g(1/2) (0.2698, 0.2698, 0.5000); (0.7302, 0.7302, 0.5000)

[b]These are the components of the magnetic moment in terms of unit vectors parallel to $\vec{a}$, $\vec{b}$ and $\vec{c}$, expressed in Bohr magnetons.

[c]This is the magnitude of the rare earth moment, in Bohr magnetons.

[d]Theta and phi refer to the orientation of the exchange field which minimizes the free energy.

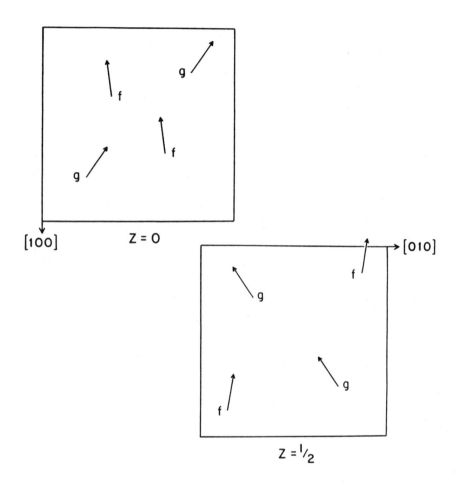

Fig. 3.   Predicted rare earth magnetic structure of $Er_2Fe_{14}B$ at 0 K. The angles between the various moments and the [1,0,0] direction are shown five times actual size.

values of $14.9{+}1.8°$ and $34.1{+}6.4°$ [6]. Our calculations further show that the resultant rare earth moments on each site are antiparallel to the net transition metal moment, both above and below the spin reorientation.

Unlike the $(Er_xY_{1-x})_2Fe_{14}B$ systems, which exhibit planar anisotropy at low temperatures, the M versus T curves of the $(Er_xPr_{1-x})_2Fe_{14}B$ systems (Fig. 4) suggest that, at low temperatures, the easy direction of magnetization in these materials lies at some intermediate angle between the axis and the plane [13]. Although this behavior appears similar to that of $Nd_2Fe_{14}B$, it is of entirely different origin. In $Nd_2Fe_{14}B$, the low temperature anisotropy is dictated by higher order crystal field terms [2,18]. In the present case,

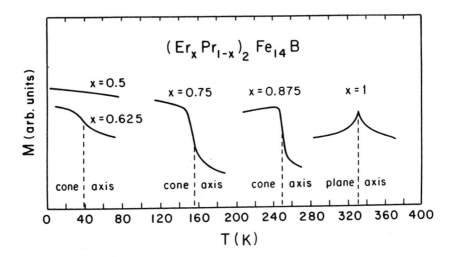

Fig. 4. Magnetization versus temperature curves for the $(Er_xPr_{1-x})_2Fe_{14}B$ compounds in the vicinity of the spin reorientation.

however, the "conical" anisotropy can be attributed to competition between the axis-seeking Pr (and Fe) and the plane-seeking Er (vide infra). This is similar to the intersite competition which was originally thought to be responsible for the low temperature anisotropy of $Nd_2Fe_{14}B$ [14].

In order to characterize the magnetic behavior of the $(Er_xPr_{1-x})_2Fe_{14}B$ systems, the crystal field and exchange parameters reported above were scaled to Pr and the crystal field analysis was performed. In the present calculation, it was assumed that Pr and Er are randomly distributed on both the f and g sites. However, this assumption has little effect on the results, due to the similarity of the two sites (Fig. 2 and related discussion). The fit of the T_{sr} versus composition curve obtained with these parameters (Fig. 1) is quite good, considering the fact that none of them were optimized for this particular system.

In addition to accounting for the composition dependence of T_{sr}, the present crystal field model reveals that, for several of the systems studied, the minimum free energy occurs when the exchange field lies at polar angles other than $0°$ or $90°$. The magnetic moments of the various ions and sublattices are given in Tables III and IV, as are the spontaneous directions of the exchange field for the various systems. The (001) projection of the magnetic moments in the compound $Er_{1.5}Pr_{0.5}Fe_{14}B$ are shown schematically in Fig. 5. Although this figure does illustrate the magnetic structure of this material, the scaled-up angles do tend to obscure one important point, this point being that, for a given ion position, the Er and Pr moments are drawn in the direction of opposite local axes. For example, the moments of the Er ions at (0.3572, 0.3572, 0.0000) are pulled away from [-1,0,0] toward [-1,-1,0] (the local negative X-axis) and the moments of the Pr ions on this site are pulled away from [1,0,0] toward [1,-1,0] (the local negative Y-axis). This behavior is a result of the different signs of the 2nd order Stevens coefficient and the presence of the $B_2^{2c}O_2^{2c}$ terms. These terms give rise to a crystal field potential which is of opposite polarity along the local X and Y axes. Finally, it should be noted from Table III that the predicted net magnetic moments of the systems studied are not necessarily parallel to the exchange field, the angles between them being on the order of 3 to 5°. This observation is important when one attempts to predict anisotropy constants on the basis of crystal field theory [23].

In summary, it has been shown that, although 2nd order crystal field terms are incapable of explaining the low temperature behavior of $Nd_2Fe_{14}B$, they can account quite adequately for many of the phenomena associated with $R_2Fe_{14}B$ systems, including: 1) the composition dependence of T_{sr} in those cases where it is the result of a competition between the rare earth and transition metal sublattices, 2) the increase in the magnetic moment of $Er_2Fe_{14}B$ as it undergoes its spin reorientation, 3) the non-collinear magnetic structure of $Tm_2Fe_{14}B$ (and, presumably, $Er_2Fe_{14}B$) at low temperature and the "conical" anisotropy which occurs in systems containing both axis- and plane-seeking regions. It has further been shown that, with present experimental knowledge, it is possible to account for much of the detailed magnetic behavior of the 2:14:B compounds using a crystal field model with a single adjustable parameter. Needless to say, single crystal neutron diffraction experiments are required to confirm the validity of the proposed magnetic structures of $Er_2Fe_{14}B$ and $Er_{1.5}Pr_{0.5}Fe_{14}B$.

TABLE IV

Magnetic Moments in $(Er_xPr_{1-x})_2Fe_{14}B$ Systems at 0 K, as Computed from Crystal Field Theory

Ion Type	M_x [a]	M_y	M_z	$\lvert M \rvert$ [b]
$Er_2Fe_{14}B$ ($\theta=90$, $\phi=0$) [c]				
Er, 4f	− 35.9726	0.0000	0.0000	35.9726
4g	− 35.7300	0.0000	0.0000	35.7300
RE(net)	− 71.7025	0.0000	0.0000	71.7025
Fe(net)	125.6000	0.0000	0.0000	125.6000
Total	53.8975	0.0000	0.0000	0.0000
$Er_{1.75}Pr_{0.25}Fe_{14}B$ ($\theta=76$, $\phi=0$)				
Er, 4f	− 30.8275	0.0000	− 6.3309	31.4769
4g	− 30.6216	0.0000	− 6.3309	31.2692
Pr, 4f	0.9639	0.0000	1.2557	1.5830
4g	1.1753	0.0000	0.7747	1.4076
RE(net)	− 59.3099	0.0000	− 10.6611	60.2605
Fe(net)	121.8691	0.0000	30.3854	125.6000
Total	62.5592	0.0000	19.7243	65.5950
$Er_{1.5}Pr_{0.5}Fe_{14}B$ ($\theta=54$, $\phi=0$)				
Er, 4f	− 23.1799	0.0000	− 13.8148	26.9844
4g	− 23.0391	0.0000	− 13.7511	26.8308
Pr, 4f	1.3024	0.0000	2.9141	3.1919
4g	1.5959	0.0000	2.5982	3.0492
RE(net)	− 43.3207	0.0000	− 22.0536	48.6112
Fe(net)	101.6125	0.0000	73.8258	125.6000
Total	58.2918	0.0000	51.7722	77.9634
$Er_{1.25}Pr_{0.75}Fe_{14}B$ ($\theta=29$, $\phi=0$)				
Er, 4f	− 12.7463	0.0000	− 18.5340	22.4940
4g	− 12.7045	0.0000	− 18.4719	22.4191
Pr, 4f	1.0332	0.0000	4.5374	4.7953
4g	1.2439	0.0000	4.5374	4.7048
RE(net)	− 23.1738	0.0000	4.5374	4.7048
Fe(net)	60.8921	0.0000	109.8522	125.6000
Total	37.7183	0.0000	82.0664	90.3192

[a]These are the components of the magnetic moment in terms of unit vectors parallel to $\vec{a}$, $\vec{b}$ and $\vec{c}$, expressed in Bohr magnetons.

[b]This is the magnitude of the moment, in Bohr magnetons.

[c]Theta and phi refer to the orientation of the exchange field which minimizes the free energy.

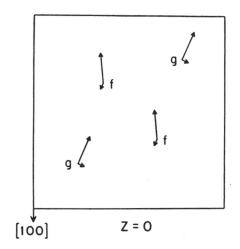

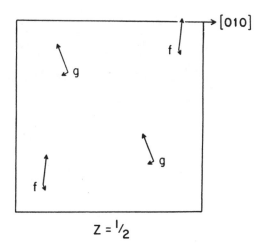

Fig. 5. Predicted rare earth magnetic structure of $Er_{1.5}Pr_{0.5}Fe_{14}B$ at 0 K, projected in the [0,0,1] plane. The angles between the various moments and the [1,0,0] direction are shown three times actual size.

132

REFERENCES

1. S. Hirosawa and M. Sagawa, Solid State Comm. $\underline{54}$, 335 (1985).
2. E. B. Boltich and W. E. Wallace, ibid., $\underline{55}$, 529 (1985).
3. A. Vasquez, J. M. Friedt, J. P. Sanchez, Ph. L'Heritier and R. Fruchart, ibid., 783 (1985).
4. R. L. Davis, R. K. Day and J. B. Dunlop, ibid., $\underline{56}$, 181 (1985).
5. K. Tokuhara, Y. Ohtsu, F. Ono, O. Yamada, M. Sagawa and Y. Matsuura, ibid., 333 (1985).
6. M. Yamada, Y. Yamaguchi, H. Kato, H. Yamamoto, Y. Nakagawa, S. Hirosawa and M. Sagawa, ibid., 663 (1985).
7. N. C. Koon, M. Abe, E. Callen, B. N. Das, H. S. Liou, A. Martinez and R. Segnan, J. Mag. Mag. Mat. $\underline{54-57}$, 593 (1986).
8. N. C. Koon, D. N. Das and C. M. Williams, ibid., 523 (1986).
9. H. Hiroyoshi, N. Saito, G. Kido, Y. Nakagawa, S. Hirosawa and M. Sagawa, ibid., 583 (1986).
10. D. Niarchos and A. Simopoulos, Solid State Comm. $\underline{59}$, 669 (1986).
11. P. Burlet, J. M. D. Coey, J. P. Gavigan, D. Givord and C. Meyer, Solid State Comm. $\underline{60}$, 723 (1986).
12. W. B.Yelon and J. F. Herbst, J. Appl. Phys. $\underline{59}$, 93 (1986).
13. E. B. Boltich, A. T. Pedziwiatr and W. E. Wallace, J. Mag. Mag. Mat., $\underline{66}$,317 (1987).
14. D. Givord, H. S. Li, R. Perrier de la Bathie, Solid State Comm. $\underline{51}$, 857 (1984).
15. A. T. Pedziwiatr and W. E. Wallace, J. Mag. Mat. $\underline{65}$, 139 (1987).
16. K. W. H. Stevens, Proc. Phys. Soc. (London) $\underline{A65}$, 209 (1952).
17. A. Vasquez and J. P. Sanchez, presented at the 17th Rare Earth Research Conference, Hamilton, Canada, June 8-12, 1986, and to appear in J. Less-Common Met.
18. H. Oesterreicher, Phys. Status Solidi $\underline{B131}$, K123 (1985)
19. M. Boge, J. M. D. Coey, G. Czjzek, D. Givord, C. Jeandey, H. S. Li and J. L. Oddou, Solid State Comm. $\underline{55}$, 295 (1985).
20. W. E. Wallace, S. G. Sankar and V. U. S. Rao, Structure and Bonding $\underline{33}$, 1 (1977).
21. Y. Berthier, M. Boge, G. Czjzek, D. Givord, C. Jeandey, H. S. Li and J. L. Oddou, J. Mag. Mag. Mat. $\underline{54-57}$, 589 (1986).
22. D. Givord, H. S. Li and J. M. Moreau, Solid State Comm. $\underline{50}$, 497 (1984).
23. L. Xiaojia, X. Yoy, Y. Guilin, Z. Hongru and L. Yiglie, Proc. 8th Int. Workshop on Rare Earth Magnets, edited by K. J . Strnat (University of Dayton Press, Dayton, OH, 1985), p. 687.

MISCH-METAL AND/OR ALUMINUM SUBSTITUTIONS IN ND-FE-B PERMANENT MAGNETS

B. M. MA AND C. J. WILLMAN
Crucible Research Center, P. O. Box 88, Pittsburgh, PA 15230

ABSTRACT

The effects of misch-metal and/or Al substitutions in the Nd-Fe-B alloy system have been investigated. Selected compositions were processed into magnets using the conventional powder metallurgy technique. As expected, misch-metal substitutions act to degrade the intrinsic magnetic characteristics and thus degrade the properties of the sintered magnet. It was observed that at a level of 25% of the total rare earth being comprised of misch-metal, energy products of the order of 25 MGOe were still able to be achieved. Processing of such material is difficult because higher sintering temperatures are required to fully densify the magnet, and the grain growth that occurs at these temperatures dramatically reduces the coercivity. The effects of misch-metal substitutions are quantified in this discussion.

Al was found to be beneficial for improving the coercivity in both the Nd-Fe-B and misch-metal-Fe-B alloy systems. Unfortunately Al substitutions reduce both the remanence and the Curie temperature. If the Al content is kept to 1 wt.% or less, magnets exhibiting reasonable properties can be made. Al doping as opposed to doping with expensive heavy rare earth elements such as Tb or Dy is a less costly method of improving H_{ci} in the Nd-Fe-B magnets.

INTRODUCTION

Energy products up to 52 MGOe have been reported in the conventional Nd-Fe-B alloy system.[1,2] Such an alloy is characterized by a saturation magnetization, M_s, of 16 kG and a maximum theoretical energy product of 64 MGOe. The base alloy is characterized by a modest anisotropy field of about 65 to 70 kOe.[2] $Dy_2Fe_{14}B$ and $Tb_2Fe_{14}B$ have been observed to have the largest anisotropy fields of the $R_2Fe_{14}B$ compounds. Partial replacement of Nd by these heavy rare earth elements, Dy or Tb, leads to a large improvement in the intrinsic coercivity of sintered RE-Fe-B magnets.[3] Al substitutions into the Nd-Fe-B alloy system were investigated as a less costly means of increasing the anisotropy field in hopes of improving the intrinsic coercivity. While some investigators have observed an increase in anisotropy field at certain Al levels[4], others have observed degradation of anisotropy field upon Al substitution.[5] However, most investigators agree that Al is beneficial to intrinsic coercivity.[4,6,7]

One of the most attractive features of the Nd-Fe-B magnets is their comparatively low $/lb. and $/energy product cost when compared to $SmCo_5$ type magnets.[8] Complete replacement of Nd by misch-metal does not appear very promising for magnets produced by the powder metallurgy technique, however, partial replacement is feasible.[9] The intrinsic coercivity is far too low for most applications. A quantitative investigation on the effects of the least costly form of rare earths, misch-metal, on the properties of RE-Fe-B and RE-Fe-B-Al magnets was conducted.

Mat. Res. Soc. Symp. Proc. Vol. 96. ©1987 Materials Research Society

EXPERIMENTAL PROCEDURE

Alloys were prepared either by vacuum induction melting or arc melting under an argon atmosphere. The intrinsic properties of the alloys were evaluated as described below.

Samples of cast alloy were crushed to -400 mesh then sealed under an argon atmosphere in Vycor capsules. Curie temperatures, T_c, were measured by Thermo-Magnetic Analysis (TMA) at an applied field of 4 kOe. Saturation magnetization values were obtained for cast alloys by measuring the magnetization of a fixed amount of -400 mesh powder using a Vibrating Sample Magnetometer (VSM) at an applied field of 70 kOe; the measurements were conducted at 25 °C. Honda extrapolation was used to determine the true saturation magnetization.

The anisotropy field measurements were conducted on -325 mesh powder which was dispersed in paraffin wax, aligned in a field of 25 kOe, then allowed to harden in the presence of a weak magnetic field. Magnetization curves were measured in both directions parallel and perpendicular to the alignment direction in a field of 25 kOe. The anisotropy fields were determined by extrapolation.

The following alloys, whose compositions are listed in atomic percent, were blended to obtain the desired compositions: $Nd_{15}Fe_{79-x}Al_xB_6$, $Nd_{15}Fe_{77-x}Al_xB_8$, $Nd_{15}Fe_{79}Al_xB_{6-x}$, $Nd_{15}Fe_{79}B_6$, $Nd_{42}Fe_{57}B$, $MM_{15}Fe_{79}B_6$ and $MM_{18}Fe_{69}Al_5B_8$. The alloy blends were processed into magnets utilizing the conventional powder metallurgy technique.[10,11] Second quadrant demagnetization curves were evaluated using a permeameter. The closed circuit technique was utilized to evaluate magnetic properties as a function of temperature.

RESULTS AND DISCUSSION

Nd-Fe-B and Nd-Fe-B-Al

Aluminum can substitute for either Fe or B in the Nd-Fe-B system. In this experiment Al was substituted for Fe in a $Nd_{15}Fe_{79-x}Al_xB_6$ alloy. This substitution results in a decrease in Curie temperature, Table I. Substitution of 2 at.% Al, x = 2, decreased the anisotropy field of the alloy. No further decrease in anisotropy field was observed when x was increased to 10 at.%. In a higher B containing alloy, $Nd_{15}Fe_{77-x}Al_xB_8$, the Al substitutions for Fe were less detrimental to magnetic properties. Little change in anisotropy field was observed even at 10 at.% Al, x=10. The Curie temperature decreased from 312 °C at x=0 to 215 °C at x=10. The magnitudes of the reductions in both anisotropy field and Curie temperature were less for the $Nd_{15}Fe_{77-x}Al_xB_8$ alloy than for the $Nd_{15}Fe_{79-x}Al_xB_6$ alloy.

When Al was substituted for B in a $Nd_{15}Fe_{79}B_{6-x}Al_x$ alloy, the effect of Al on anisotropy field and Curie temperature was much more pronounced. At x = 6, significant reductions in both Curie temperature and anisotropy field were observed, Table I.

Table I.   The Effect of Al Substitutions on Curie Temperature
and Anisotropy Field in the Nd-Fe-B Alloys

Alloy	T_c (°C)	H_a (kOe)
$Nd_{15}Fe_{79-x}Al_xB_6$		
x=0	312	66
x=2	290	46
x=10	215	42
$Nd_{15}Fe_{77-x}Al_xB_8$		
x=0	312	66
x=2	289	65
x=10	230	5(
$Nd_{15}Fe_{79}Al_xB_{6-x}$		
x=0	312	66
x=2	291	58
x=4	108,280	19
x=6	123	9

Magnets were prepared from the Al containing alloys. The Nd content
was maintained at a level of 34.5 ± 0.5 wt.% and B was maintained at 1 wt.%.
The Al contents of the magnets ranged from 0 to 1.7 wt.%. As the Al content
was increased from 0 to 0.6 wt.%, the intrinsic coercivity increased while
the remanence remained fairly constant. Increasing the Al content beyond
0.6 wt.% reduced both remanence and intrinsic coercivity, Table II. The
initial increase in intrinsic coercivity could possibly be attributed to
increased fluidity of the RE-rich phase during sintering as proposed by
Burzo et al.[12] Burzo observed that in $Gd_2Fe_{14-x}Al_xB$ alloys, Al acts to
decrease the temperature of the eutectic reaction between $Gd_2Fe_{14}B$ and Gd
metal by approximately 30%.[12]   The decrease in the eutectic temperature is
thought to improve both the fluidity and wettability of the rare earth-rich
phase during sintering. The net result is a more complete encapsulation of
the hard magnetic 2-14-1 grains by the rare earth-rich grain boundary phase.
Additionally, the grain boundary phase becomes thinner as Al is added to the
alloy. It is believed that the thinner the grain boundary phase the more
effective it is in pinning domain walls.

It is well documented that a post sintering heat treatment is
beneficial for improving the intrinsic coercivity of the conventional Nd-Fe-
B magnets.[13] For the Al containing magnets the optimum aging temperature is
elevated by the order of 50 - 100 °C, Figure 1. The exact nature of the
effect of Al content on aging temperature has not been verified. Possibly
Al acts to increase the dissolution temperature of the platelet phase
commonly observed near the grain boundaries of "as-sintered" magnets.[14-16]
It is suspected that the platelet phase acts as a nucleation site for
reverse domains.[14,15] The post sintering heat treatment is believed to
dissolve these platelets which in turn, results in an increase in
coercivity.[14-16]

136

Table II.   The Effect of Al Substitution for Fe on the
            Magnetic Properties of Sintered Nd-Fe-B-Al Magnets

Aluminum content (wt.%)	B_r kG	H_c kOe	H_{ci} kOe	BH_{max} MGOe	H_k kOe
0	13.2	8.4	9.4	41.0	8.0
0.3	11.9	10.0	11.9	34.8	9.2
0.5	11.4	10.9	14.9	32.5	12.2
0.6	11.3	10.2	16.1	31.8	10.0
0.7	11.6	11.2	15.4	33.4	13.4
0.9	11.7	10.9	14.8	33.4	12.2
1.2	11.2	10.3	14.2	30.8	10.8
1.4	11.2	9.9	13.0	30.4	9.6
1.7	11.3	8.2	9.4	31.2	7.4

Note: Nd content was maintained at 34.5 ± 0.5 wt.%.
        B content was maintained at 1.0 wt.%.

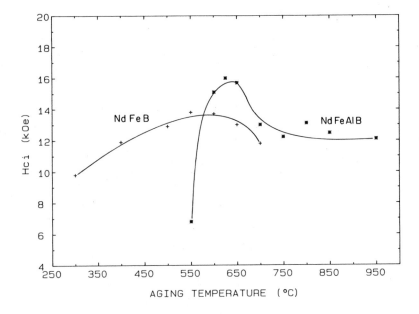

Figure 1.   A comparison of H_{ci} vs Aging Temperature
            for Nd-Fe-B and Nd-Fe-Al-B Magnets.

MM-Fe-B and MM-Fe-B-Al

The misch-metal used for these studies was of the following composition, in wt.% : 43.11 La, 29.14 Ce, 6.84 Pr, 19.17 Nd, 0.44 Gd, 0.23 Y, 0.31 Fe, 0.31 Al, 0.18 Mg, 0.034 Mn and 0.21 Si. Two separate misch-metal containing alloy series were prepared. The alloys had the following compositions, in atomic % : $MM_{15}Fe_{79}B_6$ and $MM_{18}Fe_{69}Al_5B_8$. The final magnet composition was achieved by blending either of these two alloys with a conventional Nd-Fe-B alloy.

Substitution of misch metal for Nd in the $Nd_{15-x}MM_xFe_{79}B_6$ alloy system degrades the intrinsic properties of the magnet alloy. When Nd is completely replaced by misch-metal in such an alloy, the Curie temperature of the main magnetic phase decreases from 312 °C for x=0 to 242 °C for x=15 (B=6 at.%). The anisotropy field is reduced by approximately 75% when Nd is completely replaced by misch metal, Table III. These results are in agreement with those of Grossinger et al.[9]

When Al is added to a MM-Fe-B alloy, $MM_{18}Fe_{69}Al_5B_8$, the alloy exhibits an anisotropy field of 34 kOe, Table III. The nature of the effect of Al on anisotropy field is not well understood at this time. The saturation magnetization and Curie temperature of the main magnetic phase in the two alloys, $MM_{15}Fe_{79}B_6$ and $MM_{18}Fe_{69}Al_5B_8$, were nearly the same, Table III.

Table III. The Efffect of Misch-metal and Al on the Curie Temperature, Saturation Magnetization and Anisotropy Field

Alloy	Ms(kG)	Ha(kOe)	Tc(oC)
$Nd_{15}Fe_{77}B_8$	16.1	68	312
$MM_{15}Fe_{79}B_6$	----	17	242,600,784
$MM_{18}Fe_{69}Al_5B_8$	9.0	34	255,768

For the composition $Nd_{16.3-x}MM_xFe_{77.5-y}Al_yB_{6.2}$, remanence decreased as Nd was replaced by misch-metal, Figure 2. The coercivity also rapidly decreased as Nd was replaced by misch-metal, Figure 3. Coercivity appears to be dependent mainly on rare earth content but small improvements in coercivity over that for the plain MM-Nd-Fe-B can be obtained by partial substitution of Al for Fe, Table IV. The sintering temperature also affected the quality of the MM-Fe-B-Al magnets. The remanence was generally larger for a given composition when the sintering temperature was 1060 °C as opposed to 1050 °C, the remainder of the heat treatment being identical, Figure 2. In most cases, the intrinsic coercivity was less for those magnets sintered at the higher sintering temperature.

For a given sintering temperature, the final grain size of the magnet was much larger for magnets having higher concentrations of misch-metal, Figures 4a - 4d. The decrease in coercivity with misch-metal substitution is most likely due to both a reduction of anisotropy field with increasing misch-metal substitution for Nd and an increase in the grain size of the sintered magnet with increasing misch-metal substitution. Many authors have eluded to the importance of the "grain boundary" phase in RE-Fe-B magnets in regards to coercivity.[13, 16-20] Final grain size, i.e. fraction of grain boundary area, is of utmost importance for obtaining large intrinsic coercivities.

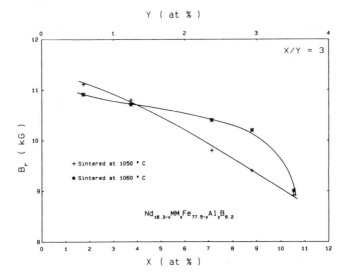

Figure 2. Variation of Remanence with Misch-Metal
Substitution for Nd.

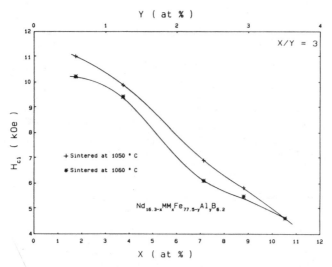

Figure 3. Variation of Intrinsic Coercivity with
Misch-Metal Substitution for Nd.

Table IV. Remanence and Intrinsic Coercivity for Various
Rare Earth-Fe-B Magnets

COMPOSITION	$\underline{B}_r$ (G)	$\underline{H}_c$ (Oe)
$MM_{13.9}Nd_{3.3}Fe_{77.2}B_{5.6}$	4200	900
$MM_{12.8}Nd_{6.8}Fe_{75.2}B_{5.2}$	6200	3900
$MM_{10.4}Nd_{5.8}Al_{3.5}Fe_{74}B_6$	8900	4600
$MM_{8.7}Nd_{7.5}Al_{2.9}Fe_{74.7}B_{6.2}$	9400	5800
$MM_{3.5}Nd_{12.6}Al_{1.2}Fe_{76.5}B_{6.2}$	10800	9900
$Nd_{15}Fe_{79}B_6$	13200	9400

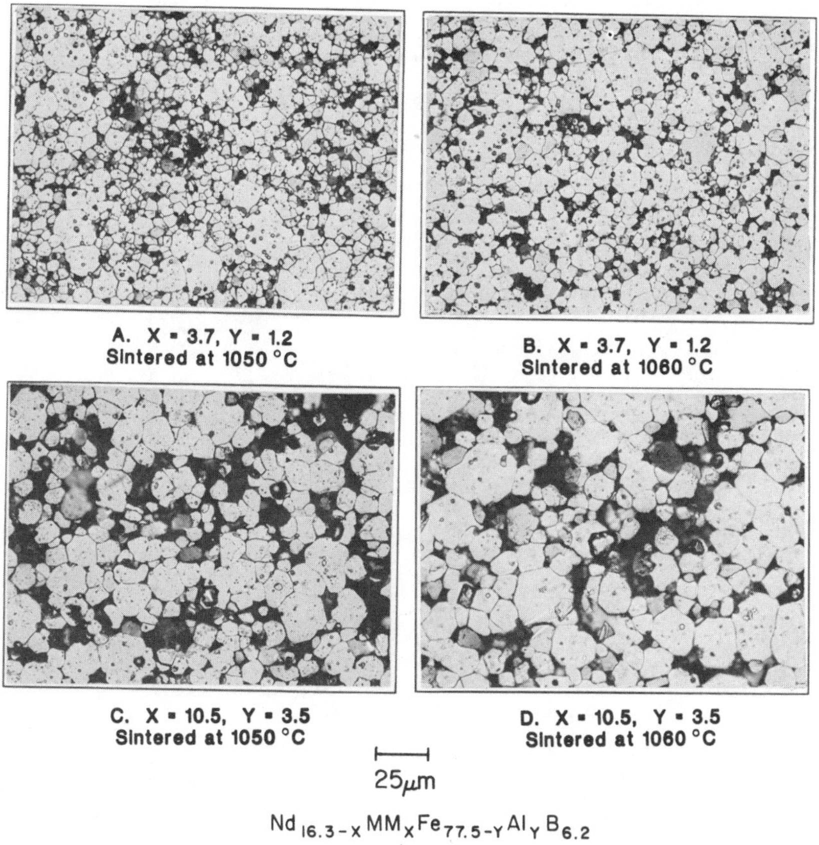

A. X = 3.7, Y = 1.2
Sintered at 1050 °C

B. X = 3.7, Y = 1.2
Sintered at 1060 °C

C. X = 10.5, Y = 3.5
Sintered at 1050 °C

D. X = 10.5, Y = 3.5
Sintered at 1060 °C

$\longmapsto$
$25\mu m$

$Nd_{16.3-X}MM_X Fe_{77.5-Y}Al_Y B_{6.2}$

Figure 4. Microstructures of Misch-Metal Substituted Magnets.

Magnets whose composition was in the range of $Nd_{16.3-x}MM_xFe_{77.5-y}Al_yB_{6.2}$, where x ranged from 1.7 to 10.5 and y from 0.7 to 3.5, exhibited maximum energy products in the range of 15 to 28 MGOe, Table V. For the magnet composition, $Nd_{12.6}MM_{3.7}Fe_{76.5}Al_{1.2}B_{6.3}$, the following properties were obtained: B_r = 10800 gauss, H_c = 8000 Oe, H_{ci} = 10100 Oe, H_k = 5900 Oe and BH_{max} = 25.6 MGOe. This composition, $Nd_{12.6}MM_{3.7}Fe_{76.5}Al_{1.2}B_{6.3}$, had the following temperature coefficients, B_r: α -0.14 %/°C and for H_{ci}: β -0.61 %/°C, Figure 5.

Table V.   Magnetic Properties of $MM_{16.3-x}Nd_xFe_{77.5-y}Al_yB_{6.2}$ Magnets.

x at.%	y at.%	B_r kG	H_c kOe	H_{ci} kOe	H_k kOe	BH_{max} MGOe
5.8	3.5	8.9	4.2	4.6	3.4	16.2
7.5	2.9	9.4	5.1	5.8	4.0	18.5
9.2	2.4	9.8	5.8	6.9	4.1	20.2
12.6	1.2	10.8	8.0	9.9	5.8	25.5
14.6	0.7	11.1	8.1	11.0	6.3	27.8

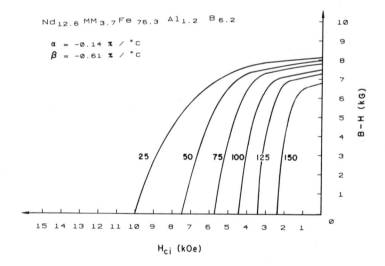

Figure 5.   Second quadrant demagnetization behavior of $Nd_{12.6}MM_{3.7}Fe_{76.5}Al_{1.2}B_{6.3}$ as a function of temperature.

CONCLUSIONS

In the Nd-Fe-B system, Al substitutions for Fe act to increase the
intrinsic coercivity of the sintered magnet. Intrinsic coercivities of the
order of 16.1 kOe are able to be achieved at an Al level of 0.6 wt.% Al.
However, since Al does not carry a magnetic moment, the saturation
magnetization is reduced for larger Al substitutions. Al substitution for
Fe appears to be a cost effective means of improving the intrinsic
coercivity of the sintered Nd-Fe-B magnet without having to dope with heavy
rare earths such as Tb or Dy. Al substitution for B is detrimental to both
Curie temperature and anisotropy field.

Al substitutions tend to alter the phase equilibria in the conventional
RE-Fe-B alloy systems. Generally the optimum aging temperature for
coercivity is elevated by 50 -100°C over that for the conventional alloys.
The exact nature of this phenomenon is not well understood at this time.

The substitution of misch-metal for Nd acts to degrade the intrinsic
magnetic properties of the Nd-Fe-B alloy. Ce or La, which make up the bulk
of the misch-metal, contribute little or no magnetic moment to the alloy and
thus degrade saturation magnetization. Additionally, the replacement of Nd
by misch-metal acts to degrade the anisotropy field because the bulk of the
magnetocrystalline anisotropy in the Nd-Fe-B alloy results from
contributions from the Nd sublattice.

Although substitution of Nd by misch-metal tends to degrade the
intrinsic properties of a Nd-Fe-B magnet alloy, small substitutions of
misch-metal can be tolerated in an Al-containing magnet.

Higher sintering temperatures were required to achieve densification
and remanence in the MM-Nd-Fe-B-Al magnets. Although these higher
temperatures resulted in larger remanence, the coercivity was impaired.

ACKNOWLEDGMENTS

The authors wish to thank Mr. E. J. Dulis for his interest in this
work. Special appreciation is extended to Dr. R. F. Krause for his helpful
discussions and comments. Technical assistance from S. Snyder, T. Sloan and
S. Rendziniak is also greatly appreciated.

REFERENCES

1. K. S. V. L. Narasimhan, J. Appl. Phys., 57, 4081 (1985).

2. M. Sagawa, S. Fujimura, H. Yamamoto, Y. Matsuura, S. Hirosawa and K.
   Hiraga, Proceedings of the 8th International Conference on Rare-Earth
   Magnets and their Applications,(edited by K. J. Strnat, University of
   Dayton, Dayton, Ohio, USA, 1985) 588.

3. M. Sagawa, S. Fujimura, H. Yamamoto and Y. Matsuura IEEE Trans. Magn.,
   MAG-20, 1584 (1984).

4. Y. Yang, W. J. James, X. Li, H. Chen and L. Xu, IEEE Trans. Magn.,
   MAG-22, 758 (1986).

5. E. Burzo, N. Plugaru, V. Pop, L. Stanciu and W. E. Wallace, Solid State
   Commun., 56, 803 (1986).

142

6. T. Mizoguchi, I. Sakai, H. Niu and K. Inomata, IEEE Trans. Magn., MAG-22, 919 (1986).

7. Z. Maocai, M. Deqing, J. Xiuling and L. Shiqian, Proceeding of the 8th International Workshop on Rare-Earth Magnets and Their Application (edited by K. J. Strnat, University of Dayton, Dayton, Ohio, USA 1985), 541.

8. B. C. Semones, Proceedings of the 8th International Conference on Rare-Earth Magnets and their Applications, (edited by K. J. Strnat, University of Dayton, Dayton, Ohio, USA, 1985) 31.

9. R. Grossinger, X. K. Sun, R. Eibler, H. R. Kirchmayr, Proceedings of the 8th International Conference on Rare-Earth Magnets and their Applications (edited by K. J. Strnat, University of Dayton, Dayton, Ohio, USA, 1985), 553.

10. M. Sagawa, S. Fujimura, N. Togawa, H. Yamamoto and Y. Matsuura, J. Appl. Phys., 55, 2083 (1984).

11. K. S. V. L. Narasimhan, MMPA Conference, Atlanta, Georgia, January 1984.

12. E. Burzo, A. T. Pedziwiatr and W. E. Wallace, Solid State Commun. 61, 57 (1987).

13. J. D. Livingston, Proceeding of ASM Materials Week 1986 (edited by K. S. V. L. Narasimhan, ASM 1986), 71.

14. K. Hiraga, M. Hirabayashi, M. Sagawa and Y. Matsuura, Jap. J. Appl. Phys. 24, L30 (1985).

15. K. Hiraga, M. Hirabayashi, M. Sagawa and Y. Matsuura, Jap. J. Appl. Phys. 24, L635 (1985).

16. M. Tokunaga, M. Tobise, N. Meguro and H. Harada, IEEE Trans. Magn., MAG-22, 904 (1986).

17. M. Sagawa, S. Hirosawa, H. Yamamoto, Y. Matsuura, S. Fujimura, H. Tokugara and K. Haraga, IEEE Trans. Magn., MAG-22, 910 (1986).

18. P. Schrey, IEEE Trans. Magn., MAG-22, 913 (1986).

19. R. Ramesh, J. K. Chen and G. Thomas, Presented at the 1986 Intermag Conference, Phoenix, Az, USA, 1986 (unpublished).

20. J. P. Hirth (private communication).

NdFeB MAGNETS WITH IMPROVED TEMPERATURE CHARACTERISTICS

B. M. Ma
Crucible Research Center, P. O. Box 88, Pittsburgh, PA 15230

ABSTRACT

The temperature dependence of magnetic properties for NdFeCoB alloys were investigated. It is well known that dysprosium substitution for neodymium increases the intrinsic coercivity. Aluminum also improves the intrinsic coercivity, H_{ci}, at room temperature but impairs the Curie temperature and the intrinsic coercivity at elevated temperatures. The rate of increase of the H_{ci} with increasing dysprosium in the NdFeCoB system is less than when dysprosium is added to the NdFeB system. A combination of aluminum and dysprosium is more effective than aluminum alone in raising H_{ci}. Molybdenum addition into the NdDyFeCoB alloy was found to reduce the temperature coefficients of B_r and H_{ci}. Temperature coefficients of B_r and H_{ci} obtained on a $Nd_{10}Dy_5Fe_{67}Co_{10}Mo_1Al_1B_6$ alloy are -0.07 %/°C and -0.4 %/°C respectively.

INTRODUCTION

NdFeB permanent magnets exhibit excellent hard magnetic characteristics at room temperature [1]. Unfortunately, because of the low Curie temperature and high temperature coefficients of B_r and H_{ci}, operation is restricted to a maximum temperature of about 100°C. Improvements are required in order for this material to be used at elevated temperatures. Curie temperature, temperature coefficients of B_r and H_{ci}, and the energy product must all be improved in order to increase the maximum operating temperature of these magnets.

There are two approaches one can take to improve the temperature dependence of induction. One can either raise the Curie temperature, for example, by substituting Co for Fe, or one can compensate for the rapid decrease of the Nd moment with increasing temperature by alloying with heavy rare earths [2, 3]. Unfortunately, both of these approaches negatively affect other characteristics of the magnet. Cobalt additions tend to reduce the intrinsic coercivity, while heavy rare earths couple antiferromagnetically with the iron and reduce the saturation magnetization.

Other alloy additions also result in competing positive and negative effects on the magnetic properties. Alloying elements such as aluminum and molybdenum are found to increase the room temperature intrinsic coercivity of NdFeB magnets. Unfortunately, the Curie temperature drops when either aluminum or molybdenum is added to the alloy. Additionally, the intrinsic coercivity at elevated temperature is significantly reduced by aluminum additions.

The combined effects of Al, Mo, Dy and Co additions were investigated. The results are reported in this paper.

EXPERIMENTAL PROCEDURES

Alloys were prepared by vacuum induction melting or arc melting under an argon atmosphere and processed into magnets by the traditional powder metallurgy (P/M) method [4]. Samples used to measure the Curie temperature, the saturation magnetization, and the temperature dependence of magnetization were prepared by crushing the alloy to -400 mesh, placing the powder in argon filled Vycor capsules and sealing. The Curie temperature was measured by Thermo Magnetic Analysis (TMA) with an applied field of either 1.5 or 4.0 kOe. The temperature dependence of magnetization was also measured by TMA over the temperature range of 4 to 1180 K. The room temperature saturation magnetization was determined using a Vibrating Sample Magnetometer (VSM) operating at an applied field of 20 kOe. Honda extrapolation was used to determine the magnetization at an infinite applied field.

The variations of remanence and coercivity with temperature were measured on sintered magnets using the closed circuit technique [5]. The temperature coefficients of B_r and H_{ci} were calculated from the data obtained.

RESULTS AND DISCUSSION

Cobalt substitution for iron in NdFeB compounds

Cobalt substitution for iron in NdFeB magnets increases the Curie temperature, however, the intrinsic coercivity is impaired, Figure 1. The increase of Curie temperature is attributed to the Co-Co exchange interaction, which is stronger than either the Co-Fe or the Fe-Fe interactions. The decrease in H_{ci} involves many factors, intrinsically and metallurgically. Knowing how cobalt substitution occurs in the lattice structure is an important step in understanding the changes in the intrinsic magnetic properties.

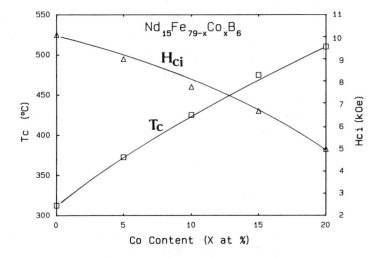

Figure 1. The dependence of Curie temperature, T_c, and intrinsic coercivity, H_{ci}, on Co content for $Nd_{15}Fe_{79-x}Co_xB_6$.

Herbst et al [6, 7] utilized neutron diffraction to analyze the site occupation and magnetic structure of both the $Nd_2Fe_{14}B$ and $Nd_2(Fe_{1-x}Co_x)_{14}B$ systems. They found that in the $Nd_2(Fe_{1-x}Co_x)_{14}B$ system, the j_2 site has a strong preference for Fe while the e site favors cobalt. The other transition-metal sites (k_1, k_2, j_1, and c) are populated randomly according to the amount of Co substitution. The complex variation of the Curie temperature with cobalt concentration depends upon how the Co ions are distributed in the lattice. Cobalt substitution for iron in the NdFeB system also results in a decrease of lattice spacing and a coincident increase in the exchange energy. The saturation magnetization is also increased with minor cobalt substitution [8].

Aluminum substitution in NdFeCoB system

Aluminum and boron are in the same column on the periodic table. Aluminum may either be treated as an alloy element to substitute for Fe or be utilized as metalloid element to replace boron in the bulk composition of NdFeCoB system.

When aluminum was substituted for Fe, $Nd_{15}Fe_{57-x}Co_{22}Al_xB_6$, the intrinsic coercivity of the sintered magnets increased, while the Curie temperature decreased dramatically, Figure 2. The increase in the intrinsic coercivity with increasing aluminum content is due to an increase in the anisotropy field of the alloy. The increase on the intrinsic coercivity with increasing aluminum may also be related to the effect that was proposed by Burzo et al [9]. Burzo et al studied the $Gd_2Fe_{14-x}Al_xB$ system and found that aluminum decreases the eutectic temperature for the reaction of $Gd_2Fe_{14}B$ and Gd metal. The low eutectic temperature increases the fluidity and wettability of the Gd-rich phase during sintering. Similarly, the author claims that the presence of aluminum in the Nd-rich phase of the magnets acts to improve the wettability and thus enables the entire grain of hard magnetic phase to be surrounded by a thin grain boundary layer; thus increasing H_{ci}. Additionally, the higher anisotropy energy that results from the aluminum addition gives rise to a thinner domain wall. Both the thinner wall and the thinner Nd-rich phase contribute to the improved intrinsic coercivity of NdFeCoAlB alloys [9]. The decrease of Curie temperature with increasing aluminum may be due to the excessive conduction electrons of Al, weakening the exchange interactions between transition metals.

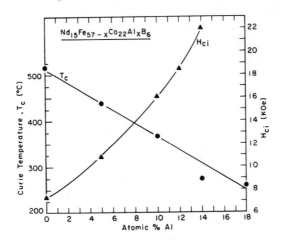

Figure 2. The dependence of Curie temperature, T_c, and intrinsic coercivity, H_{ci}, on Al content for $Nd_{15}Fe_{57-x}Co_{22}Al_xB_6$.

When boron is replaced by aluminum, $Nd_{15}Fe_{59}Co_{20}Al_xB_{6-x}$, the anisotropy field increases and then decreases sharply as the Al content increases to beyond 2 atomic percent [10, 11]. The Curie temperature decreases approximately at the same rate as observed when Al was used to replace Fe in the bulk composition. In the case of high aluminum, 4 atomic percent, the intrinsic coercivity of the sintered magnets is maintained, but the demagnetization curve lacks sufficient squareness. Figure 3 shows the demagnetization curves of the NdFeCoAlB alloys for a boron to aluminum ratio of unity. The skewed demagnetization curves suggest that boron is essential in stabilizing the hard magnetic phase. In spite of the skewness, this magnet still has uniaxial magnetic characteristics.

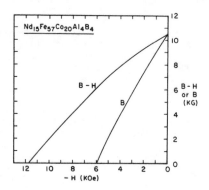

Figure 3. The second quadrant demagnetization curves
of $Nd_{15}Fe_{57}Co_{20}Al_4B_4$.

Figure 4 illustrates the effect of boron and aluminum contents on Curie temperature for a $Nd_{16}Fe_{65-x-y}Co_{20}Al_xB_y$ alloy series, where x varies from 0 to 6 and y varies from 0 to 15. As can be seen, boron does not significantly effect the Curie temperature. However, aluminum lowers the Curie temperature at a rate of approximately 15°C for each atomic percent of Al substitution. The variation of the intrinsic coercivity for NdFeCoAlB alloys, as a function of boron and aluminum content, is illustrated in Figure 5. With no aluminum, x = 0, the H_{ci} of the sintered magnets did not exceed 7500 Oe. At 6 atomic percent of aluminum and y = 12, an H_{ci} of 19 kOe was obtained. In general, H_{ci} increases as both the aluminum and the boron content is increased. When the boron content reaches a value of 13 (y = 15 and x = 4), the addition of aluminum above 4 at% does not significantly increase the H_{ci}.

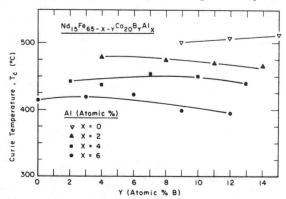

Figure 4. The Curie temperature of $Nd_{15}Fe_{65-x-y}Co_{20}Al_xB_y$ alloys
at various boron and aluminum contents.

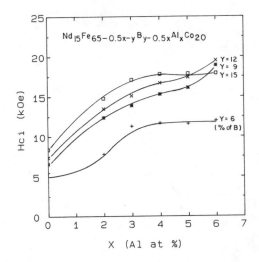

Figure 5. The intrinsic coercivity (H_{ci}) of sintered
$Nd_{15}Fe_{65-0.5x-y}Co_{20}Al_xB_{y-0.5}$ magnets.

Neither aluminum nor boron has any contribution to the magnetic moment.
Therefore, their concentrations must be restricted in order to obtain a
reasonably high saturation magnetization and to moderate the drop of Curie
temperature due to Al substitution. Figure 6 illustrates the variation of
magnetization with temperature for Al-containing NdFeB or NdFeCoB magnets.

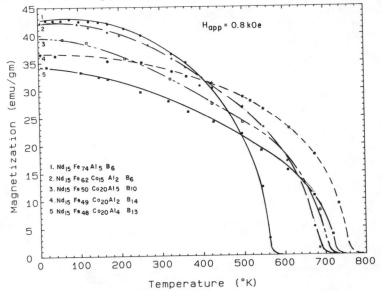

Figure 6. The temperature dependence of magnetization for various
NdFeB or NdFeCoB alloys at an applied field of 0.8 kOe.

The Curie temperature and the slope of induction curves varies with both the Co and Al concentration. The room temperature magnetic properties, B_r, H_{ci}, and BH_{max}, and Curie temperature of these magnets are listed in Table I.

Table I. Magnetic Properties of NdFeCoAlB Alloys Measured at Room Temperature

Alloy	T_c (°C)	B_r (kG)	H_c (kOe)	H_{ci} (kOe)	BH_{max} (MGOe)
$Nd_{15}Fe_{74}Al_5B_6$	290	11.2	9.6	12.2	32.5
$Nd_{15}Fe_{62}Co_{15}Al_2B_6$	410	11.9	10.5	10.9	35.3
$Nd_{15}Fe_{50}Co_{20}Al_5B_{10}$	438	9.4	9.1	18.0	21.6
$Nd_{15}Fe_{49}Co_{20}Al_2B_{14}$	481	9.9	7.4	14.9	21.0
$Nd_{15}Fe_{48}Co_{20}Al_4B_{13}$	448	8.7	7.8	17.9	18.1

When the Curie temperature, saturation magnetization, temperature coefficients and intrinsic coercivities of the sintered magnets are compared, it is seen that optimum magnetic characteristics are obtained when the cobalt concentration is between 5 to 20 atomic percent, the aluminum concentration is less than 6 atomic percent and the ratio of boron to aluminum greater than 1.5. At the optimum aluminum and boron combination, the temperature coefficients of induction for the NdFeCoAlB magnets are not improved by further cobalt additions. This is attributed to the formation of multiple phases. The saturation magnetization, Curie temperature (major and minor phases), and temperature coefficient of induction for the $Nd_{15}Fe_{70-z}Co_zAl_6B_9$ alloys are listed in Table II. The Curie temperatures of 242°C and 860°C for the alloy containing 40 atomic percent Co (z = 40) most likely corresponds to the Nd_2Co_7 and Nd_2Co_{17} phases respectively, while the Curie temperatures of 75°C and 725°C for the alloy containing 60 atomic percent Co (z = 60) correspond to the $NdCo_3$ and $NdCo_5$ phases respectively [12].

Table II. Curie Temperature and Temperature Coefficients of Inductions, α , for $Nd_{15}Fe_{70-z}Co_zAl_6B_9$

Co Content z = at%	Curie Temperatures (°C)	α %/°C (0 - 250 °C)
10	311; 715	-0.1318
20	440; 723	-0.1102
40	242; 513; 860	-0.1436
60	75; 547; 725	-0.1596

As previously indicated, both Curie temperatures, Table I, and the slope of induction versus temperature curves, Figure 6, can be adjusted by altering the aluminum and cobalt concentrations. Listed in Table III are a few NdFeCoAlB alloys that yield high intrinsic coercivities. The alloy with a Curie temperature of 410°C and a maximum energy product of 35 MGOe, $Nd_{15}Fe_{62}Co_{15}Al_2B_6$, was obtained at the cost of low intrinsic coercivity. All other alloys exhibit intrinsic coercivities greater than 18 kOe at room temperature, but show abrupt decreases in H_{ci} with increasing temperature. Figure 7 illustrates the decrease in H_{ci} with increasing temperature for $Nd_{15}Fe_{57}Co_{20}Al_5B_{10}$ alloy, the other three alloys behave in a similar fashion. This severe temperature sensitivity would restrict the use of these alloys to relatively low temperatures.

Although a higher Curie temperature improves the temperature coefficient of B_r, a higher Curie temperature does not improve the temperature dependence of intrinsic coercivity. Heavy rare earth, such as dysprosium, however, can be used to reduce the temperature dependence of H_{ci} [10].

Table III. Magnetic Properties (Room Temperature) of
NdFeCoAlB with Optimized Aluminum, Boron,
and Cobalt Combination

Alloy	T_c (°C)	B_r (kG)	H_c (kOe)	H_{ci} (kOe)	BH_{max} (MGOe)
$Nd_{15}Fe_{62}Co_{15}Al_2B_6$	410	11.9	10.5	10.9	35.3
$Nd_{15}Fe_{50}Co_{20}Al_5B_{10}$	438	9.4	9.1	18.0	21.6
$Nd_{15}Fe_{53}Co_{20}Al_6B_6$	424	10.3	8.5	18.9	25.6
$Nd_{15}Fe_{50}Co_{20}Al_6B_9$	399	8.9	8.1	19.3	18.5

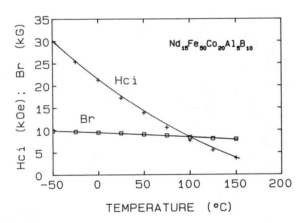

Figure 7. The temperature dependence of intrinsic coercivity,
H_{ci}, for $Nd_{15}Fe_{50}Co_{20}Al_5B_{10}$ from -50°C to 150°C.

Dysprosium substitution for neodymoum in NdFeCoB alloys

The substitution of dysprosium for neodymium increases the anisotropy field of a NdDyFeCoB alloy [13]. As a result, the intrinsic coercivity of NdDyFeCoB magnets increases as the dysprosium content is increased. However, the rate of increase of the intrinsic coercivity is not as large as when the Dy is substituted for Nd in conventional NdFeB alloys, Table IV. Listed in Table IV are the B_r's and H_{ci}'s for $Nd_{16-x}Dy_xFe_{78}B_6$ and $Nd_{16-x}Dy_xFe_{68}Co_{10}B_6$ magnets. The intrinsic coercivity increases at a rate of approximately 5 kOe for each atomic percent of Dy addition in the $Nd_{16-x}Dy_xFe_{78}B_6$ alloy system, and at a rate of 2 kOe for each atomic percent of dysprosium in the $Nd_{16-x}Dy_xFe_{68}Co_{10}B_6$ alloy system. At 4 atomic percent dysprosium, $Nd_{12}Dy_4Fe_{78}B_6$, an H_{ci} of 26 kOe was obtained. This compares to an H_{ci} of 13.6 kOe measured for a $Nd_{12}Dy_4Fe_{68}Co_{10}B_6$ alloy.

Minor aluminum additions appear to enhance the intrinsic coercivity in dysprosium containing alloys, Figure 8. When minor aluminum additions are made to a NdDyFeCoB alloy, the intrinsic coercivity increases more rapidly than what is observed for a series of aluminum free NdDyFeCoB alloys. The cause of this is not clear yet. An investigation of the microstructural changes that occur with the addition of aluminum is required.

Table IV. Dysprosium Effect on the Intrinsic Coercivities of $Nd_{16-x}Dy_xFe_{78}B_6$ and $Nd_{16-x}Dy_xFe_{68}Co_{10}B_6$

Dy Content x at%	$Nd_{16-x}Dy_xFe_{78}B_6$		$Nd_{16-x}Dy_xFe_{68}Co_{10}B_6$	
	B_r (kG)	H_{ci} (kOe)	B_r (kG)	H_{ci} (kOe)
0	13.3	10.5	12.2	6.2
1	12.7	16.9	11.9	8.7
2	11.9	21.0	11.7	10.0
3	11.6	24.1	10.8	11.6
4	11.0	26.0	10.3	13.5

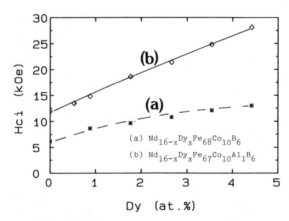

Figure 8. The dysprosium effect on the intrinsic coercivities for
(a) $Nd_{16-x}Dy_xFe_{68}Co_{10}B_6$ and
(b) $Nd_{16-x}Dy_xFe_{68}Co_{10}Al_1B_6$ alloys.

Molybdenum substitution for iron in the NdFeCoB alloys

Molybdenum substitution for iron, $Nd_{15}Fe_{69-x}Co_{10}Mo_xB_6$, also increases the intrinsic coercivity of NdFeCoB alloys, Table V. Listed in Table V is the Curie temperature, saturation magnetization and room temperature magnetic properties of a series of $Nd_{15}Fe_{69-x}Co_{10}Mo_xB_6$ alloys. The Curie temperature drops at a rate of approximately 10°C for each atomic percent of molybdenum substitution. The saturation magnetization also decreases as molybdenum content is increased. Although there is no concrete evidence of the mechanism which causes the increase in H_{ci} with increasing molybdenum contents, microstructural studies do indicate that a not previously observed phase is formed in the grain boundary region. A Mo-rich phase was observed in the grain boundary in addition to the Nd-rich boundary phase(s), Fig. 9.

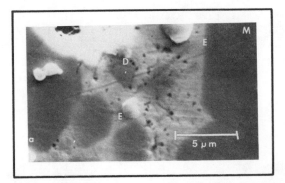

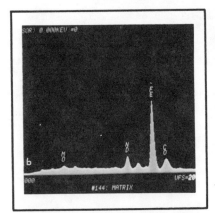

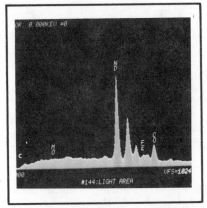

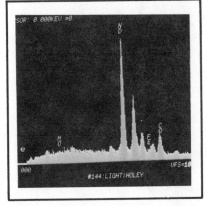

Figure 9.   (a) SEM micrographs of matrix and grain boundary region of
$Nd_{15}Fe_{65}Co_{10}Mo_4B_6$.   The EDAX spectra of (b) matrix phase
(area M), (c) Nd-rich boundary phase (area C), (d) Mo-rich
phase (area D), and (e) neodymium oxide (area E).

This Mo-rich phase is believed to have some contribution to the increase of H_{ci} with increasing molybdenum. The temperature coefficients of both B_r and H_{ci} are also improved with increasing molybdenum [14].

Table V. Curie Temperatures, Saturation Magnetizations, and Magnetic Properties of $Nd_{15}Fe_{69-x}Co_{10}Mo_xB_6$ Alloys

Alloy x =	T_c (°C)	M_s (emu/gm)	B_r (kG)	H_c (kOe)	H_{ci} (kOe)	BH_{max} (MGOe)	H_k (kOe)
0	427	136.8	12.4	6.6	7.0	35.7	6.1
1.4	385	128.0	12.1	7.1	8.6	32.5	6.2
2.8	382	124.0	11.5	7.8	9.9	30.8	7.3
4.3	378	121.0	10.8	10.1	13.1	28.4	10.5

Since Mo improves the temperature coefficient of H_{ci}, and Al results in a high H_{ci} at room temperature and an inferior H_{ci} at elevated temperature, it is of interest to examine the combined effects of Al, Mo, and Dy on the temperature cofficient of H_{ci}. In general, the temperature coefficients of both remanence and intrinsic coercivity are improved by additions of Al, Mo and Dy. Table VI lists the variation of intrinsic coercivity with temperature, normalized to the H_{ci} values at 25°C, for four NdFeCoB magnets. All of these magnets have H_{ci} values in the range of 17 to 19 kOe at 25°C. The normalized value of intrinsic coercivity of a conventional NdDyFeB magnet (Crumax 355) is included for comparison. The alloy compositions and their magnetic properties at 25°C and 150°C are included at the lower portion of Table VI. The intrinsic coercivity at 150°C is significantly improved for the alloy containing 5 atomic percent of cobalt and 7 atomic percent dysprosium, also for the alloy containing 1 atomic percent aluminum, 1 atomic percent Mo and 5 atomic percent of Dy. A temperature coefficient of B_r of -0.07 %/°C and a temperature coefficient of H_{ci} of -0.4 %/°C were obtained on a $Nd_{10}Dy_5Fe_{67}Co_{10}Mo_1Al_1B_6$ sintered magnet. The temperature range for these measurements was -50°C to +150°C.

Table VI. The Normalized Intrinsic Coercivity of NdFeCoB Alloys at Various Test Temperatures

Test Temp. (°C)	A	B	C	D	E
-50	1.75	1.74	1.64	-	1.59
-25	-	1.48	1.40	1.34	1.40
0	1.21	1.23	1.17	1.20	1.17
25	1.00	1.00	1.00	1.00	1.00
50	-	0.80	0.80	0.85	0.86
75	-	0.61	0.61	0.72	0.73
100	0.45	0.45	0.47	0.60	0.59
125	-	0.32	0.36	0.49	0.46
150	0.23	0.21	0.29	0.35	0.34

Alloy Code:

A - Crumax 355
B - $Nd_{15}Fe_{50}Co_{20}Al_5B_{10}$
C - $Nd_{10}Dy_5Fe_{62}Co_{15}Al_2B_6$
D - $Nd_8Dy_7Fe_{74}Co_5B_6$
E - $Nd_{10}Dy_5Fe_{67}Co_{10}Al_1Mo_1B_6$

Alloy Code	Test Temp. (°C)	B_r (kG)	H_c (kOe)	H_{ci} (kOe)	BH_{max} (MGOe)
A	25	12.3	11.3	14.0	35.0
	150	10.2	3.0	3.1	18.7
B	25	9.4	9.1	18.0	21.6
	150	7.9	3.5	3.7	12.3
C	25	10.2	9.4	19.8	24.2
	150	8.8	4.7	4.9	15.1
D	25	9.1	9.0	20.8	20.0
	150	7.8	7.1	7.3	15.1
E	25	9.4	9.4	19.9	22.1
	150	8.3	6.6	6.8	17.0

CONCLUSIONS

The temperature coefficient of induction of NdFeB can be improved by increasing the Curie temperature of the matrix phase. However, the temperature coefficient of intrinsic coercivity must also be considered to qualify this material for elevated temperature application. Higher boron content and the addition of aluminum improve the intrinsic coercivity at room temperature but degrades the energy product and the intrinsic coercivity at elevated temperatures.

Molybdenum improves the thermo-magnetic properties. A temperature coefficient of B_r of -0.07 %/°C and a temperature coefficient of H_{ci} of -0.4 %/°C were obtained for a $Nd_{10}Dy_5Fe_{67}Co_{10}Al_1Mo_1B_6$ sintered magnet.

ACKNOWLEDGMENTS

The author wishes to thank Mr. E. J. Dulis for his interest in this work and Dr. R. F. Krause for helpful discussions and comments. Technical assistance from T. Sloan and S. Rendziniak is appreciated.

REFERENCES

[1] M. Sagawa, S. Fujimura, N. Togawa, H. Yamamoto, and Y. Matsuura, J. Appl. Phys., 55, 2083 (1984).
[2] M. Sagawa, S. Hirosawa, H. Yamamoto, Y. Matsuura, S. Fujimura, H. Tokuhara, and K. Hiraga, IEEE Trans. on Magn., MAG-22, 910 (1986).
[3] B. M. Ma and K. S. V. L. Narasimhan, IEEE Trans. on Magn., MAG-22, 1081 (1986).
[4] K. S. V. L. Narasimhan, J. Appl. Phys. 57, 4081 (1985).
[5] B. M. Ma and K. S. V. L. Narasimhan, J. Magn. Magn. Mat. 54-57, 559 (1986).
[6] J. F. Herbst, J. J. Croat, and W. B. Yelon, J. Appl. Phys. 57, 4086 (1985).
[7] J. F. Herbst, W. B. Yelon, J. Appl. Phys. 60, 4224 (1986).
[8] M. Sagawa, Y. Matsuura, S. Fujimura, H. Yamamoto, and S. Hirosawa, IEEE Translation Journal on Magnetics in Japan, Vol. 1, TJMJ-1, 48 (1985).
[9] E. Burzo, A. T. Pedziwiatr, and W. E. Wallace, Solid State Commun. 61, 57 (1987).
[10] B. M. Ma and K. S. V. L. Narasimhan, IEEE Trans. on Magn., MAG-22, 916 (1986).
[11] T. Mizoguchi, I. Sakai, H. Niu, and K. Inomata, IEEE Trans. on Magn., MAG-22, 919 (1986).
[12] K. H. J. Buschow, Rare Earth Compound in Ferromagnetic Materials, Vol. 1, edited by E. P. Wohlfarth, North-Holland Publishing Co., Amsterdam, p. 385-387, (1980).
[13] H. Fujii, W. E. Wallace and E. B. Boltich, J. Magn. Magn. Mat., 61, 251 (1986).
[14] X. Shen, Y. Wang, Z. Diao, X. Liu, 31st Annual Conference on Magnetism and Magnetic Materials, Baltimore, U. S. A., November 1986.

Elevated Temperature Properties of
Sintered Magnets of (Nd,Dy)-Fe-B
Modified with Cobalt and Aluminum

Y. Xiao, H.F. Mildrum, K.J. Strnat and A.E. Ray
School of Engineering, University of Dayton
Dayton, Ohio 45469

ABSTRACT

The effect of small aluminum additions on the temperature dependence of remanence and intrinsic coercivity in Co- and Dy-containing Nd-Fe-B was studied. Sintered magnets were prepared and demagnetization curves measured at temperatures between -50 to +200°C. Curie temperatures and irreversible flux losses in open circuit were determined. Al increases the coercivity while decreasing remanence, energy product and Curie temperature. Other unfavorable side effects are the increase in temperature coefficients of B_r and, especially, $_MH_c$. Substitution of Al is not beneficial for magnets used at elevated temperature.

INTRODUCTION

The rapid decline on heating of their remanence, B_r, the energy product, $(BH)_{max}$, and especially of the intrinsic coercivity, $_MH_c$, limits the application of ternary Nd-Fe-B magnets to below 100-150°C. To enable such magnets to operate at higher temperatures, different alloy modifications have been tried. Design engineers commonly describe the utility of magnets at elevated temperatures by two parameters: the temperature coefficient, α (B_r), of the remanent induction, and the irreversible flux losses in open circuit during a heating-cooling cycle. The losses must be tied to a specific operating point, e.g. by specifying the unit permeance, $p=-B_d/H_d$.

The large value of $|\alpha|$ for Nd-Fe-B magnets is due to the low Curie temperature, T_c, and the rapid decrease in Nd moment with increasing temperature, T. Hence, two approaches might result in a reduction of $|\alpha|$: raising the Curie temperature of the main phase of the magnet, or partially substituting heavy rare earths (HRE) for Nd. Cobalt is an effective alloying element for raising T_c. [1,2,3] Substitution of 18 to 20 at% Fe by Co enhances T_c to above 500°C and thus lowers α. However, this is at the expense of the intrinsic coercivity. A lower $_MH_c$, in turn, leads to an increase of the irreversible loss.

The irreversible losses are generally inversely related to the coercivity; and since $_MH_c$ usually decreases with increasing operating temperatures, the losses are indirectly related to the temperature coefficient of coercivity, β $(_MH_c)$. The physical parameters affecting the temperature dependence of the coercivity are not well known. One common practical approach to lowering the irreversible losses is simply to increase the intrinsic coercivity at room temperature. Then, even if $_MH_c$ is reduced by the same factor as in the absence of Dy on heating, a higher absolute value is left at T, and the flux loss is lower. A modest dysprosium addition can nearly double $_MH_c$ of a Nd-Fe-Co-B magnet and bring the irreversible loss to below 5% at 200°C [4]. However, the use of the expensive Dy significantly raises the cost of magnets, especially at higher dysprosium contents. Aluminum has also been found effective for enhancing the coercivity in both, Nd-Fe-B [5,6] and Nd-Fe-Co-B magnets.[7] It is much cheaper than Dy. But an unfavorable side effect of aluminum addition is that B_r and T_c decrease with increasing Al content.

Simultaneous substitution of Dy and Al for Nd and Fe, respectively, might be expected to offer a favorable compromise, yielding high coercivity and a good remanence at elevated operating temperatures. Based on the above consideration, a set of experiments was designed, varying the Al content at a constant Dy substitution in the alloy system $(Nd_{0.88}Dy_{0.12})(Fe_{.80-x}Co_{.12}B_{.08}Al_x)_{5.5}$, where x=0.012, 0.024 and 0.036. For control and reference purposes, the properties of a corresponding Dy- and Al-free magnet of composition $Nd(Fe_{.80}Co_{.12}B_{.08})_{5.5}$ were also examined.

EXPERIMENTAL PROCEDURE

Master alloys were prepared either by arc or induction melting, and specific intermediate compositions by blending powders of these. Sintered magnets were made by a conventional powder metallurgy process.[1] All samples were sintered at 1080°C for one hour in vacuum and rapidly cooled in argon gas; they were then reheated to 900°C/one hour in argon, furnace cooled, and they received a final heat treatment at 500°C for one hour. Magnetic measurements were made after the 900° anneal and in the final state.

Demagnetization curves were measured either on ~8mm cubes in closed circuit at temperatures in the range -50 to +200°C with a DC hysteresigraph, or on ~3mm cubes in open circuit in the same temperature range with a magnetometer. The temperature coefficients of B_r and $_MH_c$ were determined from these curve sets. They are defined as follows: [8]
$$\alpha(T_1 \rightarrow T_2) = 100[B_r(T_2)-B_r(T_1)]/B_r(25°C)\cdot(T_2-T_1)$$
$$\beta(T_1 \rightarrow T_2) = 100[_MH_c(T_2)-_MH_c(T_1)]/_MH_c(25°C)\cdot(T_2-T_1) \Big\} \text{ in \% per °C}$$
Before measurement at each temperature, the samples were fully remagnetized in a 100 kOe pulsed field at room temperature.

Irreversible flux losses during cycling to several elevated temperatures, up to 200°C, were measured on 0.25" diameter cylinders of $p=B/H \approx -2$ in open circuit, using a tightly fitting pull coil and a fluxmeter. The Curie temperature of different alloy compositions was determined by a low-field thermomagnetic analysis (TMA) with a 5 kHz field of about 1 Oe amplitude; the average sweep rate was 1°C/min.

RESULTS AND DISCUSSION

Ternary Nd-Fe-B magnets combine excellent room-temperature (r.t.) values of remanence and energy product, with a usefully high coercive force. However, the temperature coefficients, α of B_r, and β of $_MH_c$, in the temperature range 0-150°C are large, -0.126%°C and -0.71%°C, respectively. [9] These values are much higher than those of typical 1-5 and 2-17 Sm-Co magnets. The higher temperature coefficients are usually attributed to the low Curie temperature of $Nd_2Fe_{14}B$, $T_c=312°C$.

In our experiments, introducing 10 at% cobalt increased T_c to 435°C, for $Nd(Fe_{.80}Co_{.12}B_{.08})_{5.5}$ magnets, lowering the α (0→150°C) to 0.09% per °C. By substituting some dysprosium for Nd, in $(Nd_{0.88}Dy_{.12})(Fe_{.80}Co_{.12}B_{.08})_{5.5}$, α could be further lowered to 0.077%, while increasing the coercive force, $_MH_c$, to 13 kOe. No significant change of $\beta(_MH_c)$ occurred compared to Nd-Fe-B.

The variation of remanence, coercive force, energy product and Curie temperature caused by addition of aluminum in $(Nd_{.88}Dy_{.12})(Fe_{.80-x}Co_{.12}B_{0.08}Al_x)_{5.5}$ is shown in Fig. 1. We see that $_MH_c$ rapidly rises with increasing x. For x = 0.036, i.e., 3 at% Al, the $_MH_c$ reaches 19 kOe. Contrary to this, B_r decreases at the rate of 0.5 kG per 1 at% Al addition. Substituting Al initially does not change the Curie point. However, T_c begins to decrease as the Al content exceeds 1 at%. This suggests that the initial small amounts enter only grain boundary phases, while Al in excess of ~1% is partitioned between the main phase, where it depresses T_c, and secondary phases.

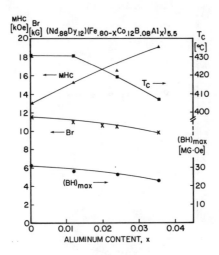

Fig. 1: Variation of magnetic properties with Al content for
$(Nd_{0.88}Dy_{.12})(Fe_{.80-x}Co_{.12}B_{.08}Al_x)_{5.5}$ sintered magnets
(x=0.012, 0.024, 0.036 correspond to 1, 2, 3 at.% Al, respectively).

The influence of Al on the temperature coefficients, α and β, is
shown in Fig. 2. The magnitude of both increases with rising Al content.
Especially β gets rapidly worse. The value of β = -0.89% for 3 at%,
Al even exceeds the β value of -0.71 observed on ternary Nd-Fe-B magnets.
Although the r.t. coercive force is quite high, such a magnet does not
permit a higher operating temperature because $_MH_C$ declines too fast with

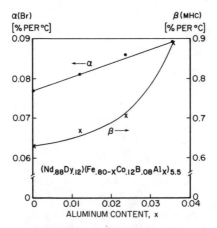

Fig. 2: The influence of Al on temperature coefficients α and β for
$(Nd_{.88}Dy_{.12})(Fe_{.80-x}Co_{.12}B_{.08}Al_x)_{5.5}$ sintered magnets
α and β were calculated for 0→150°C.

158

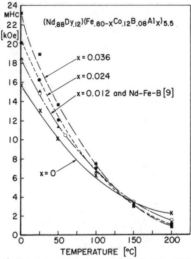

Fig. 3: Variation of the intrinsic coercive force with temperature for
$(Nd_{.88}Dy_{.12})(Fe_{.80-x}Co_{.12}B_{.08}Al_x)_{5.5}$ with x=0, 0.012, 0.024,
0.036 and Nd-Fe-B magnets.

increasing temperature. This can be clearly seen in Fig. 3 where the
temperature dependence of $_MH_C$ of magnets with different Al content is
shown for the temperature range 0-200°C. The curve for Nd-Fe-B is included
for comparison. The following facts can be observed:
    1.   For x=0 (no Al), the coercivity at r.t. is only 13 kOe, but
at 200°C there is still 2.1 kOe left. In contrast, the sample with the
highest Al content, x=0.036, possesses a 40% higher $_MH_C$ at room temperature, but it has only 1/3 as much coercive force left at 200°C than the
x=0 magnet.
    2.   All the curves intersect around 140°C. Up to 140°C, the magnets
with higher r.t. coercivity maintain a higher $_MH_C$.
    We can conclude that, when the operating temperature is below ∼140°C,
the Al-containing magnets may also be expected to show lower irreversible
losses. But above ∼140°C, the Al additions have no benefit for the temperature stability. Direct measurement of irreversible losses for a
$(Nd_{.88}Dy_{.12})(Fe_{.80}Co_{.12}B_{.08})_{5.5}$ magnet and a
$(Nd_{.88}Dy_{.12})(Fe_{.764}Co_{.12}B_{.08}Al_{.036})_{5.5}$ magnet generally confirms this
conclusion. (Fig. 4) The two curves of irreversible loss vs. T intersect
near 150°C. But the gain in stability below this temperature is minimal,
while the irreversible loss near 200°C is severely increased by the Al
addition. Taking into account the concomitant sacrifices in B_r, $(BH)_{max}$
and α , the minor improvement in irreversible loss at lower temperatures
makes Al additions of questionable practical value. According to our
results, substituting Al for some Fe increases the coercive force near
r.t., but it does nothing to extend the useful operating temperature beyond
the 140-150°C level up to which Nd-Dy-Fe-Co-B or even Nd-Dy-Fe-B magnets
are useful. To allow further scrutiny of what is happening at the elevated
temperatures, two complete sets of demagnetization curves measured at
several temperatures for $(Nd_{.88}Dy_{.12})(Fe_{.80}Co_{.12}B_{.08})_{5.5}$ and
$(Nd_{.88}Dy_{.12})(Fe_{.764}Co_{.12}B_{.08}Al_{.036})_{5.5}$ are shown in Fig. 5 and 6.

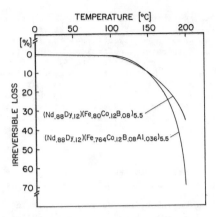

Fig. 4: Irreversible flux losses at p=B/H = -2 during cycling between 25°C and the indicated elevated temperature. Tested were magnets with compositions $(Nd_{.88}Dy_{.12})(Fe_{.80}Co_{.12}B_{.08})_{5.5}$ and $(Nd_{.88}Dy_{.12})(Fe_{.764}Co_{.12}B_{.08}Al_{.036})_{5.5}$.

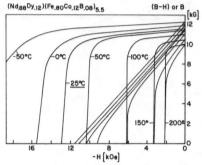

Fig. 5: Demagnetization curves at several temperatures between -50°C and 200°C for a $(Nd_{.88}Dy_{.12})(Fe_{.80}Co_{.12}B_{.08})_{5.5}$ magnet.

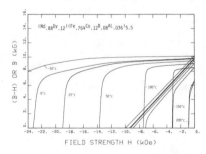

Fig. 6: Demagnetization curves at several temperatures between -50° and 200°C for a $(Nd_{.88}Dy_{.12})(Fe_{.764}Co_{.12}B_{.08}Al_{.036})_{5.5}$ magnet.

Before we summarize, we should briefly consider what these results tell us about the factors which affect the temperature coefficient of coercivity. β does not seem to be directly related to the main-phase Curie temperature. (The temperature coefficient of induction is!) If β were approximately inversely proportional to T_c, our (Co + Al)-containing magnet should have a lower $|\beta|$-value than Nd-Fe-B because its Curie point is higher by more than 100°C. But, on the contrary, it has a much higher $|\beta|$-value! Like $_MH_c$, the β is likely to depend strongly on the temperature variation of the magnetic properties of secondary phases in the grain boundaries. The experiments described cannot yield any direct information about these.

SUMMARY

1. Substituting 10 at.% Co for Fe increases the Curie temperature from $\sim 310°C$ to 435°C and lowers the temperature coefficient of B_r, $\alpha(0 \rightarrow 150°C)$ to 0.09% per °C.
2. Partially substituting Dy for Nd in a Co-containing magnet of composition $(Nd_{.88}Dy_{.12})(Fe_{.80}Co_{.12}B_{.08})_{5.5}$ not only increases the coercivity, but it also decreases (i.e., improves) the temperature coefficient of B_r.
3. Additionally substituting Al for some of the Fe increases the coercivity, but it lowers the remanence and energy product. It raises the temperature coefficients, $\alpha(B_r)$ and, especially, $\beta(_MH_c)$.
4. Such dual substitution of Co and Al is therefore not a good way of raising the useful operating temperature of magnets above 150°C. However, the magnets have attractively high coercive force near room temperature.

ACKNOWLEDGMENT

This work was supported by the U.S. Army Electronics Technology and Devices Laboratory, Ft. Monmouth, New Jersey, under contract DAAK20-84-K-0458.

REFERENCES

1. M. Sagawa et al., IEEE Trans. Magnetics MAG-20, (1984), 1584-1589.
2. T. Shibata et al., IEEE Trans. Magnetics MAG-21, (1985), 1952-1954.
3. Matsuura et al., Appl. Phys. Letters 46, 308 (1985).
4. M. Tokunaga et al., IEEE Trans. Magnetics MAG-22, (1986), 904.
5. M.(Maocai) Zhang et al., Proc. 8th Int'l. Workshop on REPM and their Applications, Dayton, OH, USA, May 1985, 541-552.
6. B.-M.(Bao-Min) Ma and K.S.V.L. Narasimhan, IEEE Trans. Magnetics MAG-22, (1986), 916-918.
7. T. Mizoguchi et al., IEEE Trans. Magnetics MAG-22, (1986), 919-921.
8. H.F. Mildrum and K.M.D. Wong, Proc. 2nd Int'l. Workshop on REPM and their Applications, Dayton, OH, USA, (1976), 35-54.
9. D. Li, K.J. Strnat and H.F. Mildrum, J. Appl. Physics, 57, (1985), 4140.
10. Y. Xiao et al., INTERMAG 1987, to be published in IEEE Trans. Magnetics, MAG-23 (Sep. 1987).

MAGNETIC HARDENING MECHANISM IN SINTERED
R-Fe-B PERMANENT MAGNETS

M. SAGAWA AND S. HIROSAWA
Sumitomo Special Metals Co. Ltd., Mishimagun, Osaka 618, Japan

ABSTRACT

The state-of-the-art description of the magnetic hardening mechanism in the sintered Nd-Fe-B permanent magnet is given. Recent experimental results concerning the coercivity-anisotropy (H_c-H_A) correlation in B-rich Pr-Fe-B and Nd-Fe-B sintered magnets and the influence of the surface conditions of the sintered Nd-Fe-B magnets on the coercivity are reported. These results are interpreted in terms of the $\mu_0 H_c$ versus $c\mu_0 H_A - N I_S$ plot, where I_S is the spontaneous magnetization of $R_2 Fe_{14} B$ (R=Pr or Nd) and N the effective demagnetization field coefficient.

INTRODUCTION

It has been widely accepted that, in the new class of sintered permanent magnets based on the $R_2 Fe_{14} B$ inter-metallic compounds (R stands for rare earths), the coercivity is governed by nucleation of reversed domains [1] and that the intrinsic coercivity, H_{cI}, is related to the anisotropy field, H_A, and the saturation magnetization, I_S, of the matrix as [2-5]

$$\mu_0 H_{cI} = c \mu_0 H_A - N I_S \qquad (1)$$

The first term in the right side of the equation is the nucleation field of the reversed domains, and the second term arises to take into account the local stray fields at the nucleation sites. It is presumed that the constants c and N are related to the microstructure of the sintered magnets. The purpose of this paper is to study how the constants c and N vary with the change in the composition and microstructure of the sintered Nd-Fe-B and Pr-Fe-B magnets. This is the first step necessary for understanding the magnetic hardening mechanism of the new magnets, leading to improvement of their magnetic properties.

COERCIVITY OF $R_{15} Fe_{77} B_8$ MAGNETS

The H_A-H_{cI} relation in $Nd_{15} Fe_{77} B_8$ sintered magnet has been studied from temperature dependences of H_A and H_{cI} and from concentration dependences of H_A and H_{cI} in the $(Nd_{1-x} R_x)$-Fe-B system (R=Tb, Dy and Er)[2]. In the study, the H_A-H_{cI} relationship in the Nd-Fe-B system was compared with the Pr-Fe-B system, because in the latter a complication due to a spin reorientation transition which is observed at 135K [6] in $Nd_2 Fe_{14} B$ does not exist and therefore a more straight-forward conclusion can be drawn from the Pr-Fe-B system than from the Nd-Fe-B system. This method is applied in the present study also.

Figure 1 shows the temperature dependence of H_{cI} of $Pr_{15} Fe_{77} B_8$ and $Nd_{15} Fe_{77} B_8$ sintered magnets [2]. In spite of the distinct difference in the anisotropy behavior of the two compounds mentioned above, the H_{cI}-T curves are very similar to each other. No anomaly is observed at the spin reorien-

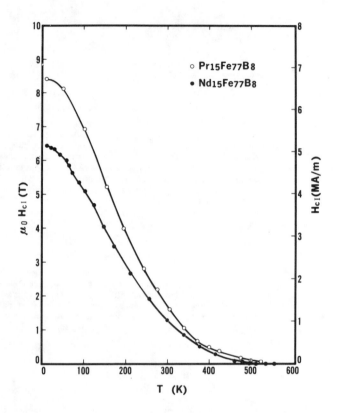

Fig. 1   Temperature dependence of the intrinsic coercivity of $Pr_{15}Fe_{77}B_8$ and $Nd_{15}Fe_{77}B_8$ sintered magnets.

tation temperature (T_{SR}) of $Nd_2Fe_{14}B$ on the H_{cI}-T curve for $Nd_{15}Fe_{77}B_8$. This is an indication that high order terms in the expression of the magnetic anisotropy energy with regard to the angle (θ) between the crystallograpic c-axis and the magnetization direction of $Nd_2Fe_{14}B$ become essentially important in magnetic hardness of the magnet.

Dysprosium has a much larger uniaxial anisotropy in $R_2Fe_{14}B$ than Nd and is known to enhance both H_A and H_{cI} when it replaced Nd in the Nd-Fe-B system [7, 8]. Figure 2 shows the temperature dependence of H_{cI} of $(Nd_{1-x}Dy_x)_{15}Fe_{77}B_8$ sintered magnets [9]. The room temperature values of $\mu_0 H_{cI}$ increases linearly with the Dy concentration and reaches approximately 5T (50kOe in c.g.s.) for x=0.47, while the maximum energy product ((BH)max) is maintained larger than 160 kJ/m^3 (20MGOe). To the authors' knowledge, this is the largest room temperature coercivity ever obtained in sintered magnets. Introduction of Dy exceeding this concentration results, in a decrease in the room temperature H_{cI}. The temperature dependence of H_{cI} of the magnets with x=0.47 and 0.60 is larger than that of the magnets with smaller Dy concentrations. These behaviors of H_{cI} of the highly substituted Dy-rich magnets are presumably due to precipitations of ferromagnetic phases such as (Nd, Dy)Fe$_4$B and DyFe$_2$ as indicated by our electron probe microanalyses (EPMA).

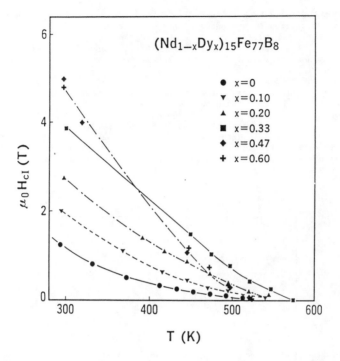

Fig. 2   Temperature dependence of the intrinsic coercivity of
$(Nd_{1-x}Dy_x)_{15}Fe_{77}B_8$ sintered magnets.

The relationship between H_{cI} of the sintered Nd-Fe-B-type magnets and H_A of $R_2Fe_{14}B$ has been expressed as [2]

$$\mu_0 H_{cI} + I_S = c \mu_0 H_A. \qquad (2)$$

In this equation, the effective demagnetization coefficient (N) in Eq. (1) is equated with unity. This is an empirical equation which has been derived from both the temperature and composition dependences of H_A and H_{cI} for $Pr_{15}Fe_{77}B_8$ and $Nd_{15}Fe_{77}B_8$. In our previous work, which is shown in Fig. 3 [2], the value for the constant c was determined as 0.37. In Fig. 3, temperature is the implicit parameter, which varies from 4.2K to the Curie temperature (Tc) for the Pr system whereas it varies from 300K to Tc for the Nd system. The reason for excluding low temperature data from the Nd system is to avoid the complication which arises from the spin reorientation transition in $Nd_2Fe_{14}B$. In fact, contributions of the high order anisotropy constants, expecially K_2, are not negligible even at room temperature as will be shown later. It must be emphasized that the simple linearity between $\mu_0 H_{cI} + I_S$ and $\mu_0 H_A$ remains down to low temperatures in the Pr-Fe-B system. This owes to the fact that the first order anisotropy constant (K_1) predominates in the entire temperature range in $Pr_2Fe_{14}B$. The downward deviation of the data points at the lowest temperatures is presumably due to the K_2 contribution, which is in fact **non-negligible at very low temperatures** as has been indicated in the onset of the first order magnetization process (FOMP) in $Pr_2Fe_{14}B$ [10].

The $\mu_0H_{cI}+I_S$ versus μ_0H_A plot for the highly substituted $(Nd_{1-x}Dy_x)$-Fe-B system is shown in Fig. 4 [9]. Eq.(2) does not seem to describe the H_A - H_{cI} relation in this system except for low Dy concentrations and high temperatures. Our EPMA analysis on annealed ingots of the same compositions as the magnets has shown considerable enrichment of Dy in the matrix $R_2Fe_{14}B$ phase. Therefore, Fig. 4 would have become somewhat different if we take into account of a possible inhomogeneous distribution of Dy in the sintered magnets. However, we do not go further into this subject at this moment, because the coercivity is determined by structural and chemical features of a region in about a few nanometers depth on the surface or interface of the $R_2Fe_{14}B$ granis (See later paragraphs for details) and because the quarternary systems Nd-Dy-Fe-B is already too much complicated to obtain such micro-

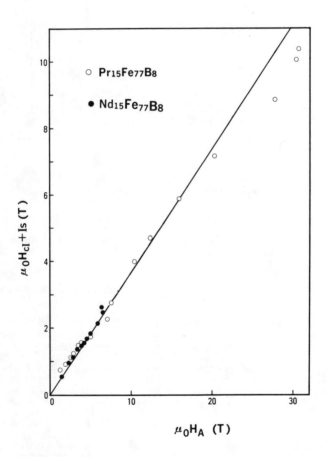

Fig. 3   The $\mu_0H_{cI}+I_S$ versus μ_0H_A plot for $Pr_{15}Fe_{77}B_8$ and $Nd_{15}Fe_{77}B_8$ sintered magnets.

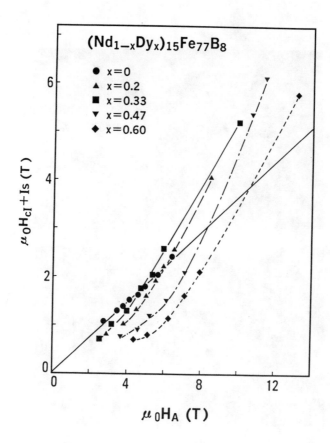

Fig. 4   The $\mu_0 H_{cI} + I_S$ versus $\mu_0 H_A$ plot for $(Nd_{1-x}Dy_x)_{15}Fe_{77}B_8$ sintered magnets.

scopic information as mentioned above. For the moment, the correlation between the intrinsic properties and the extrinsic properties in the Nd-Fe-B type magnet should be examined in the simplest systems, Pr-Fe-B and Nd-Fe-B.

COERCIVITY OF BORON-RICH $R_{17}Fe_{83-x}B_x$ MAGNETS

The ideal system on which the H_{cI}-H_A correlation may be studied is a single particle system. However, complication due to surface contamination and surface defects is another problem which always becomes the largest factor in real determination of coercivity of single particles [11]. This makes the coercivity of single particles a different matter from that of real sintered magnets.

In the ternary Nd-Fe-B system, the Nd-rich phase and the B-rich phase are both nonmagnetic and therefore it can be hoped that by increasing the amount of these phases one may be able to examine the magnetization process

166

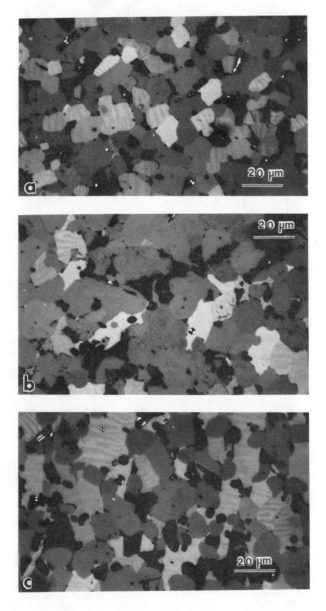

Fig. 5   Kerr micrographs of (a) $Nd_{18}Fe_{53}B_{30}$, (b) $Nd_{17}Fe_{46}B_{37}$ and (c) $Pr_{17}Fe_{53}B_{30}$ sintered magnets.

of $R_2Fe_{14}B$ grains which are separated from each other by the nonmagnetic
phases. We propose that B-rich sintered Pr-Fe-B and Nd-Fe-B magnets are the
system on which we can investigate the magnetization reversal process of
more-or-less isolated grains. Pr- and Nd-rich alloys are too aggressive
chemically and are inappropriate for our purpose. The same procedures as
were reported in our previous paper [1] are applicable for the B-rich
sintered magnets. It was found that the optimum temperatures for post-
sintering heat-treatments are common regardless of the B content, 870K for
the Nd-Fe-B magnets and 850K for the Pr-Fe-B magnets.

Figures 5(a) through 5(c) show Kerr micrographs of the B-rich sintered
magnets. The alignment direction is parallel to the plane of the micrographs. The $R_2Fe_{14}B$ grains can be identified by the domain patterns observed on the grains. The dominant nonmagnetic phase (grey) is $R_{1+\epsilon}Fe_4B_4$.
The Nd-magnets with x=30 and x=37 are of several $Nd_2Fe_{14}B$ grains gathered
together and of mainly isolated $Nd_2Fe_{14}B$ grains, respectively. On the other
hand, in the Pr-magnets, the anticipated strucutre has not been obtained.
The $Pr_2Fe_{14}B$ grains in $Pr_{17}Fe_{53}B_{30}$ are of very irregular shapes and are
interconnected by narrow channels of $Pr_2Fe_{14}B$, as is shown in Fig. 5(c).

The virgin magnetization process of the $Nd_{17}Fe_{83-x}B_x$ magnets at room
temperature are compared in Figs. 6(a) through 6(d). The numbers written in
the figures indicate the maximum magnetizing field (μ_0H_M) expressed in units
of T (teslas). Each curve was obtained after thermal demagnetization of the
samples.

The room temperature values of μ_0H_{cI} of the B-rich Nd-magnets increases
substantially with increasing the B content and reaches 2.1 T for x=37. The
magnets become gradually difficult to be magnetized when the B content
becomes large, which is recognized by comparing the demagnetization curves
recorded after magnetizing the magnet with the same field (e.g., 1T).
It has been recognized that, although the coercivity is controlled
undoubtedly by nucleation of inversed domains, a large magnetizing field is
required to obtain the full squareness of the hysteresis loops. The present
result that separating the $Nd_2Fe_{14}B$ grains does not alter this tendency
suggests that the difficulty in magnetizing the magnet to the saturation is
intrinsic to each grain in the magnet. This is due either to heterogeneous
pinning of domain walls near the grain boundaries or to large local
demagnetization fields.

To clarify the origin of the difficulty in saturating the Nd-Fe-B sintered magnet, we have examined the dependence of the remanence (B_r) of the
$Nd_{17}Fe_{83-x}B_x$ magnets on the magnetizing field (H_M). The demagnetization
curves of these magnets (Figs. 6(a) through 6(d)) can be considered to be
composed of two parts, the first of which is of a low coercivity and the
second is of a large coercivity. The fraction of the larger coercivity part
increases with increasing H_M. Therefore, it may be assumed that a $Nd_2Fe_{14}B$
grain has a negligible coercivity when H_M is not enough to eliminate inverse
domains from the grain and that this grain gains the full coercivity when
H_M is large enough to eliminate inverse domains from it. Then, the dependence of B_r on H_M should reflect a distribution of a magnetic field which is
required to eliminate inverse domains from a $Nd_2Fe_{14}B$ grain in the magnets.
An examination of temperature dependence of the B_r-H_M correlation, therefore, will throw light on this issue.

Figure 7 shows the dependence of B_r on H_M at various temperatures. In
this figure, B_r is normalized by the maximum value of B_r (B_r(max)) obtained
for μ_0H_M=1.5T and H_M is normalized by the spontaneous magnetization (I_s) of
$Nd_2Fe_{14}B$. By doing this, it is found that the B_r/B_r(max) versus H_M/I_s plot
falls on the same curve except for the data for T=77K at which the magnetization is no longer on the c-axis because of the spin reorientation transition. First, we concentrate on the universal relation between B_r/B_r(max)
and H_M/I_s obtained at temperatures for which $Nd_2Fe_{14}B$ has uniaxial anisotropy. The dependence of B_r on H_M itself may suggest that there are several
pinning sites of different pinning strength. The fact that H_M can be scaled

168

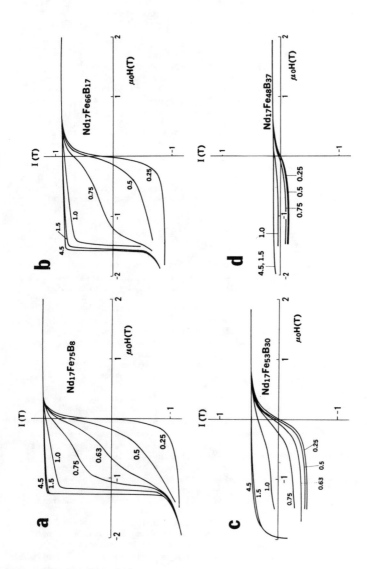

Fig. 6 Demagnetization curves of $Nd_{17}Fe_{83-x}B_x$ sintered magnets after magnetized by magnetic field of various strength (shown in the figures in the unit of teslas).

not by H_A but by I_S clearly indicates that distribution of the local demagnetization field is the main origin of the observed B_r-H_M relation because the local demagnetization fields vary as I_S. Previously, the dependence of H_{cI} or B_r on the magnetizing field has been taken as an indication for local pinning caused by anisotropy energy. However, this not the case because the change of magnetic anisotropy of $Nd_2Fe_{14}B$ is much greater than that of I_S and would give rise to much larger change in the B_r-H_M relation than actually observed. The reason for the large magnetizing field required to obtain full squareness on the sintered R-Fe-B magnets is therefore concluded to be the large local demagnetization field in the $R_2Fe_{14}B$ grains which is presumably created by the irregularity of the shapes of the grains and nonmagnetic inclusions such as oxides.

Secondly, we speculate that the deviation of the $B_r/B_r(max)$ versus H_M/I_S plot at 77K (solid circles in Fig. 7) is related to the spin reorientation and the misalignment of the grains in the magnet. At 77K, the magnetization of $Nd_2Fe_{14}B$ is distributed on four equivalent directions which make a certain angle (approximately 28°[8]) with the c-axis. It is well expected that a weak applied field can move the magnetization direction to one of the four directions which is closest to the direction of the applied field. This implies a slight misorientation of the $Nd_2Fe_{14}B$ crystal with respect to the applied field, which is always present in the sintered magnet. Such a process will make B_r for a small H_M a little larger than a process which is operative when all the spins are on the c-axis, shifting the entire $B_r/B_r(max)$ versus H_M/I_S curve toward low values of H_M.

The temperature dependence of H_{cI} of $Nd_{17}Fe_{83-x}B_x$ is shown in Fig. 8. The values of H_{cI} of the magnets with x=17 and 30 show monotonous increase with decreasing temperature, which somehow resembles to that shown in Fig.

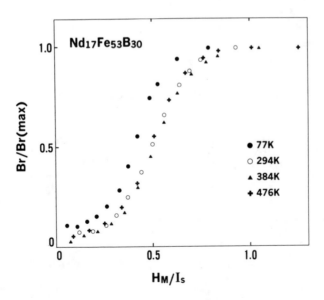

Fig. 7   Dependence of the normalized remanence of sintered $Nd_{17}Fe_{53}B_{30}$ on the magnetizing field normalized by the spontaneous magnetization of $Nd_2Fe_{14}B$.

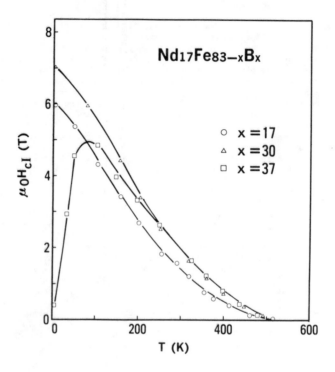

Fig. 8   Temperature dependence of the intrinsic coercivity of·sintered
$Nd_{17}Fe_{83-x}B_x$ magnets.

1. The H_{cI} value of the magnet with x=37 exhibits anomalous downfall at low
temperatures. This is due to a combination of two effects, namely, magnetic
ordering of the B-rich phase ($Nd_{1.1}Fe_4B_4$) at 15K [12] and an increasing
paramagnetic susceptibility of this phase when temperature approaches to
this temperature from above. The latter of these contributions twists the
hysteresis loops anticlockwise even at temperatures higher than Tc of the B-
rich phase and makes the apparent H_{cI} values (defined at zero magnetic
induction) smaller. The hysteresis loop of $Nd_{17}Fe_{46}B_{37}$ at 4.2K is shown in
Fig. 9 which shows a typical two-stage demagnetization process for a mixture
of a soft and a hard magnetic phase. An attempt to estimate the fraction of
the two phases from the obtained hysteresis loops at low temperatures was
unsuccessful because of another contribution to the stepwise decrease of the
magnetization which is characteristic to a system with a tilted spin direction. This contribution occurs in the similar process which has been presented for the magnetization process of $Nd_{17}Fe_{53}B_{30}$ at 77K. In the present
case of demagnetization, a reverse field moves the magnetization toward one
of four equivalent easy directions which is closest to the demagnetizing
field, resulting in a sharp downfall of the magnetization of the magnet[13].
    The correlation between H_{cI} and H_A in the $Nd_{17}Fe_{83-x}B_x$ sintered magnets
is examined according to Eq.(1) which is rewritten as

$$\mu_0H_{cI}/I_S = c\mu_0H_A/I_S - N \qquad (3)$$

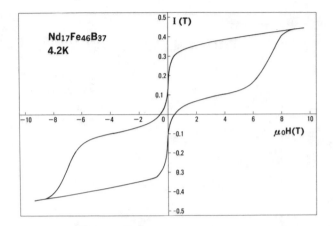

Fig. 9   Hysteresis loop of $Nd_{17}Fe_{46}B_{37}$ at 4.2K.

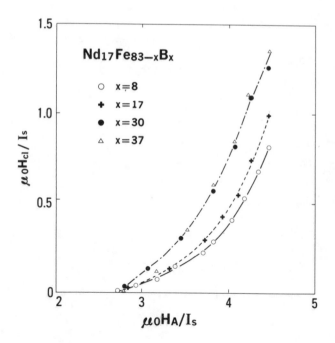

Fig. 10   The $\mu_0 H_{cI}/I_S$ versus $\mu_0 H_A/I_S$ plot for the sintered $Nd_{17}Fe_{83-x}B_x$ magents.

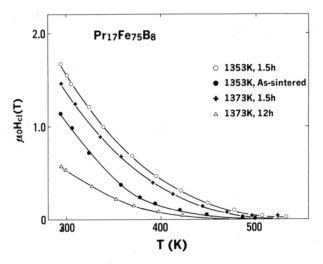

Fig. 11   Temperature dependence of the intrinsic coercivity of sintered
$Pr_{17}Fe_{75}B_8$ magnet prepared under various conditions.

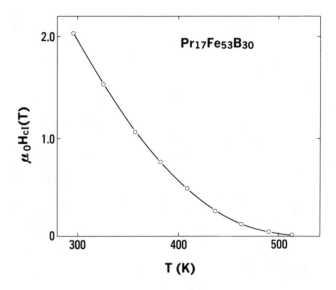

Fig. 12   Temperature dependence of the intrinsic coercivity of $Pr_{17}Fe_{53}B_{30}$.

Accordingly, values for c and N can be determined by plotting $\mu_0 H_{cI}/I_S$ against $\mu_0 H_A/I_S$ if the plot lies on a straight line. Figure 10 shows the $\mu_0 H_{cI}/I_S$ versus $\mu_0 H_A/I_S$ plot for the $Nd_{17}Fe_{83-x}B_x$ magnets. It is clearly recognized that there does not exist a linearity between the two quantities. This is an indication that the contribution of the K_2 term is already non-negligible in the temperature range investigated. Such a contribution of the K_2 term to the magnetization reversal process in real magnets can not be derived from theoretical prediction for a critical field for a magnetization reversal via a coherent magnetization rotation even if the K_2 term is taken into the calculation [3,4]. One may suspect the plausibility of using Eq. (1). However, the validity of using Eq. (1) can be verified clearly when we compare the Nd-Fe-B system with the Pr-Fe-B system.

In the Pr-Fe-B magnets, one can put aside the difficulty in treating K_2 and higher order terms of the anisotropy, because in $Pr_2Fe_{14}B$, K_1 is dominant against K_2 over wide range of temperature [14]. The temperature dependences of H_{cI} for the $Pr_{17}Fe_{75}B_8$ magnet and the $Pr_{17}Fe_{53}B_{30}$ magnet in the temperature range between room temperature and the Curie temperature of the magnets are shown in Figs. 11 and 12. To clarify the effects of crystal grain size and heat treatment on coercivity, measurements were made for different sintering temperatures and time as well as for the magnets with and without a post-sintering heat treatment.

The average grain size in the direction parallel to the c-axis weighed by volume is 14 μm for the sample sintered at 1353K for 1.5h, 16 μm for 1373K, 1.5h and 20 μm for 1373K, 12h, as estimated from Kerr micrographs of the sintered magnets. With coarsening the $R_2Fe_{14}B$ grains, the coercivity decreases significantly. The coercivity of the magnet without a post sintering heat treatment is low and drops very rapidly with increasing temperature. Isolation of the magnetic particles of $Pr_2Fe_{14}B$ with the nonmagnetic $PrFe_4B_4$ enhances the coercivity of the magnet.

Figures 13 and 14 show the $\mu_0 H_{cI}/I_S$ vs. $\mu_0 H_A/I_S$ plots for the $Pr_{17}Fe_{75}B_8$ and $Pr_{17}Fe_{53}B_{30}$ magnets. It is to be noted that a very good linearity holds for the optimumly processed Pr-Fe-B magnets in spite of the irregular shapes of the grains. Comparing the line for the $Pr_{17}Fe_{75}B_8$ magnet sintered at 1353K followed by the post-sintering heat treatment at

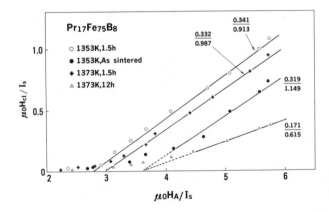

Fig. 13   The $_0H_{cI}/I_S$ versus $_0H_A/I_S$ plot for $Pr_{17}Fe_{75}B_8$ prepared under different conditions. The numbers indicate values for c (the upper row) and N (the lower row) in Eq. (3).

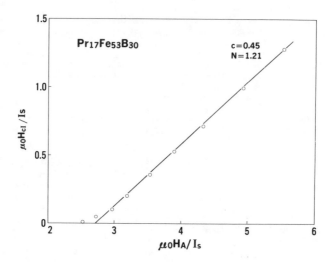

Fig. 14 The $\mu_0 H_{cI}/I_S$ versus $\mu_0 H_A/I_S$ plot for $Pr_{17}Fe_{53}B_{30}$.

850K for one hour (open circle in Fig. 13) and that for the $Pr_{17}Fe_{53}B_8$ magnet, we can see that the isolation of the magnetic particles with nonmagnetic substance increases both the values of c, from 0.34 to 0.45, and N, from 0.913 to 1.21. Coarsening of the magnetic particles results in decreasing both the vlaues of c, from 0.332 to 0.171, and N, from 0.987 to 0.615. The post sintering heat treatment increases the value of c, from 0.319 to 0.341, and decreases the value of N from 1.149 to 0.913.

RELATION BETWEEN COERCIVITY AND MICROSTRUCTURE OF THE R-Fe-B SINTERED MAGNETS

The nucleation of reversed domains takes place in a magnetic field much smaller than H_R, the field for a homogeneous rotation of magnetization when an inverse field is applied to a fully-magnetized perfect crystal. This must be caused by the existence of regions with smaller magnetic anisotropy compared with the perfect $Nd_2Fe_{14}B$ phase. For instance, preferential oxidation of Nd on the surface grains of a sintered Nd-Fe-B magnet leaves highly oxidized region on the surface in which the magnetic anisotropy may change gradually because of a diffuse concentration profile [15]. Or, magnetically soft phases such as R_2Fe_{17} and α-Fe may be formed by local oxidation. The bcc boundary phase, although suspected by some researchers [18, 19], is a soft magnetic phase [16, 17]. Because of such numerous nucleation sources dispersed in the alloy especially along grain boundaries, the nucleation fields H_N is distributed from almost zero to the field of H_R.
Each nucleation site is exposed to a local stray field such as an inverse field created by the adjacent grain whose magnetization is already reversed, or the demagnetization field caused by the adjacent nonmagnetic substances like oxides, B-rich phase (RFe_4B_4) and R-rich phase. The coercivity of a grain is determined by a nucleation site whose H_N is small and the inverse stray field is large.

It was shown that the c value is large in the magnets in which the hard magnetic grains are isolated by the non-magnetic B-rich phase, and in which the grain size of the magnetic grains is small. This is explained as follows: The smaller the volume connected magnetically in a magnet is, the smaller the volume which a site with a small H_N governs is. Inversely, in a magnet in which each magnetic grain is imperfeclty insulated each other or

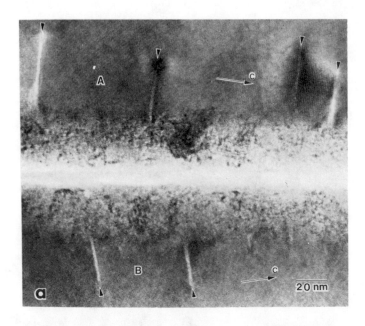

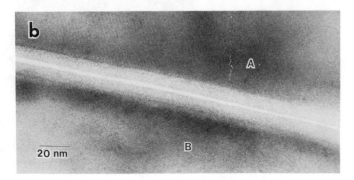

Fig. 15 Electron micrographs of sintered $Nd_{15}Fe_{77}B_8$ [16]: (a) before the heat treatment (low coercivity state), (b) after the heat treatment (high coercivity state).

in a magnet having a large grain size, the sites having a small H_N control a big volume. As a result, the c value is small as a whole body in such a magnet, and vice versa.

The result that the value of N increases with the isolation of the magnetic grains with the nonmagnetic RFe_4B_4 phase is contrary to the expectation that N would decrease with decreasing the magnetic interactions among neighboring grains. In the ordinary magnets in which $R_2Fe_{14}B$ predominates as the majority, the value of N, being approximately unity, is due principally to the magnetic field generated by the adjacent grain whose magnetization has already been reversed. It is plausible that the large value of N in the magnets containing much boron is related to the shape of the $R_2Fe_{14}B$ grains. As shown in Fig. 5(c), grains in the $Pr_{17}Fe_{53}B_{30}$ magnet have very rugged shapes with many sharp projections and depressions. In such an alloy, strong local demagnetization fields act at sharp corners. If one can prepare a magnet consisting of the $R_2Fe_{14}B$ grains with a small grain size, maintaining the grain shape more round and more smooth than that actually obtained in the B-rich magnets, a very high coercivity may be realized.

The post-sintering heat treatment affects the values of both c and N, increasing c and decreasing N. It enhances the value of c presumably by making a thin Nd-rich layer more clear-cut at grain boundaries between the $R_2Fe_{14}B$ grains, insulating each grain magnetically.(Fig. 15) Another effect of the heat treatment may be to diminish the sharp plunges of the impurity phases as shown in Fig. 5(a), leading to reduction of the value of N. The decrease in N in the magnet with increasing the grain size may be attributed to the change in the shape of the grain; sintering at high temperatures for a long time makes grains round and smooth.

We have observed with a transmission electron microscope (TEM) a non-equilibrium bcc phase forming a thin layer along grain boundaries in the

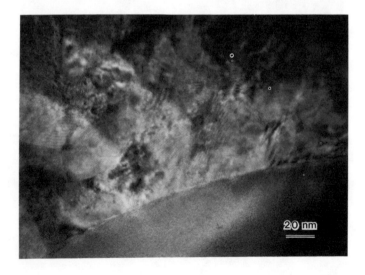

Fig. 16 Electron micrograph of the boundary region between the Nd-rich phase (the upper part) and the $Nd_2Fe_{14}B$ phase (the lower part) in sintered and heat treated $Nd_{15}Fe_{77}B_8$ showing no bcc boundary phase.

true structure of the grain boundary. We have confirmed that, as shown in Fig. 16, the matrix and the Nd-rich pocket contact directly with a sharp-cut boundary; no bcc layer is observed here. For the time being, we have not yet reached a conclusion as to whether the bcc phase exist or not at thin grain boundaries formed between the $R_2Fe_{14}B$ matrix grains. These grain boundaries are easy to be etched out preferentially during the ion thinning process. The study is now in progress and the result will be reported in the near future.

Surface of the sintered magnet is another place where the coercivity has a strong dependence on its condition and where one may be able to control its feature artificially. It has been recognized that the surface envelop of a sintered Nd-Fe-B magnet extending to approximately 10 μm depth does not have coercivity as large as that of the inner grains [13,20]. We have studied the dependence of the coercivity of the surface grains of sintered $Nd_{15}Fe_{77}B_8$ magnets on the surface condition which has been changed by Ar-sputtering and sputter-deposition of various metallic substances [21].

The dashed line in Fig. 17 is the hysteresis loop for a thin slab magnet, 100 μm thick and 4x5 mm^2 large with its magnetization direction parallel to the slab surface. Because the grains exposed to the surfaces have much smaller coercivities than the grains interior, the demagnetization curve is deteriorated compared to that with a perfect squareness of the bulk magnet from which the thin slab magnet was cut out. The squareness is dramatically recovered by the 3 μm thick Nd deposition onto both slab surfaces by sputtering, followed by annealing at 873K for one hour in vacuum.

It is interesting to note that the heat treatment is indispensable in generating coercivity on the surface because this implies that nucleation of inverse domains on the surface takes place very easily even if the surface has been sputtered by Ar ions in order to remove oxidized layers (this procedure was employed before depositing Nd metals on the surface). An

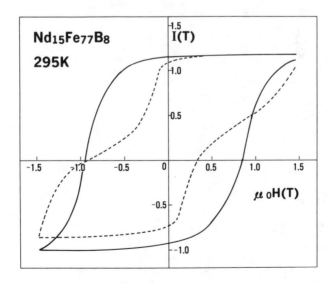

Fig. 17   Hysteresis loops of a 100 μm slab magnet of $Nd_{15}Fe_{77}B_8$ with the Nd-coating (the full curve) and without the Nd-coating (the dotted curve).

Auger electron spectroscopy has revealed that the concentration modulation
of Nd and Fe on the surface of $Nd_2Fe_{14}B$ which has been sputtered and then
exposed to 10^{-2} Pa oxygen gas does not extend deep into the matrix, the depth
being less than 6 nm [21]. Threfore, it may be surmised that even a slight
diffuseness of the surface of $Nd_2Fe_{14}B$ leads to a significant decrease of
the coercivity. The result of this experiment may support a hypothesis that
existence of structurally and chemically difuse layers on outer envelope of
each $Nd_2Fe_{14}B$ grain in sintered magnets is one of the reasons why the value
of c in $H_N=cH_A$ is smaller than unity.

CONCLUDING REMARKS

In this paper, we have examined the dependence of the parameters c and
N in Eq. (1) upon chemical compositions and heat treatments using boron-rich
Pr-Fe-B and Nd-Fe-B sintered magnets. Within the temperature range where
the first order anisotropy constant K_1 of the $R_2Fe_{14}B$ compound is dominant
over the second order constant K_2, the coercivity is related to the intrinsic magnetic properties of the respective compound as is described by Eq.
(1). Because K_1 is dominant against K_2 over a wide temperature range in
$Pr_2Fe_{14}B$, the Pr-Fe-B magnet is more convenient than Nd-Fe-B for a study of
the magnetic hardening mechanism in the R-Fe-B magnets. By comparing the
Pr-system with the Nd-system, it has been inferred that the K_2 contribution
is essential even at room temperature in the magnetization reversal process
in the Nd-Fe-B magnet. In Ref. [2], we proposed to replace K_1 in the
difinition of H_A with ΔE_A which is the difference of the anisotropy energy
between the easy direction and the hard direction of magnetization
($\Delta E_A=K_1+K_2$ for $T \geq T_{SR}$ and $\Delta E_A=K_1+K_2+K_1^2/4K_2$ for $T < T_{SR}$). This does
straight up the μ_0H_{cI}/I_S versus μ_0H_A/I_S plot of Nd-Fe-B magnets shown in
Fig. 10 as in the case of $Nd_{15}Fe_{77}B_8$. Such a result is contradictory to the
theoretically estimated dependence of nucleation field upon the anisotropy
constants based upon a coherent rotation of the magnetization [3]. For
further clarification of this issue, a calculation of domain wall energy and
domain wall width for the case of a large K_2 is necesary.
The values of the parameters c and N relate to the difficulty of
nucleation of inverse domains and the outer shape of each $R_2Fe_{14}B$ grain in
real sintered magnets, respectively. The change of the values of c and N
for the R-Fe-B magnets upon varying chemical composition and heat treatment
conditions are attributable to structural changes in the sintered magnets.
The value of c tends to increase with increasing the degree of isolation of
the grains of the hard magnetic phase, and decreases with increasing its
grain size. The c value is presumably related to the nature of the magnetically soft region formed mainly by preferential oxidation of rare earth
elements on surfaces and grain boundaries. The value of N, on the other
hand, is related to the surface roughness and shape of the grains. NI_S
should be regarded not as the average demagnetization field of the grain but
as the local demagnetization field at the nucleation site around grain
boundaries. The value of N is large, sometimes exceeding unity in the
magnets containing very rugged grains with sharp projections and depressions.
It has been emphasized from the temperature dependence of the B_r-H_M
correlation that the local demagnetization field is the principal origin of
the observed dependence of the coercivity on the magnetizating field of the
sintered Nd-Fe-B-type magnet. Previously, the fact that a large magnetizing
field is needed to obtain the full squareness of a sintered Nd-Fe-B magnet
has been taken as an indication of local heterogeneous pinning of domain
walls.
The study of the surface coercivity of $Nd_{15}Fe_{77}B_8$ has revealed that
even a subtle diffuseness of the surface can cause a significant decrease of
the coercivity on the surface. It has been inferred that the sharpness of

the surface feature of a $Nd_2Fe_{14}B$ grain, where the term "surface" includes the interface with the grain boundary phases, does determine the coercivity of the grain.

Further experimental and theoretical works are necessary for a deeper and quantitative understanding of the coercivity of the Nd-Fe-B-type magnets.

REFERENCES

1. M. Sagawa, S. Fujimura, N. Togawa, H. Yamamoto, and Y. Matsuura, J. Appl. Phys. 55, 2083 (1984).
2. S. Hirosawa, K. Tokuhara, Y. Matsuura, H. Yamamoto, S. Fujimura and M. Sagawa, J. Magn. Magn. Mater. 61, 363 (1986).
3. G. Herzer, W. Fernengel and E. Adler, J. Magn. Magn. Mater 58, 48 (1986).
4. K. -D. Durst and H. Kronmüller, inProceedings of the 8th International Workshop on Rare Earth Magnets and Their Apllications, Eddted by K. J. Strnat (University of Dayton, 1985) pp. 725.
5. E. Adler and P Hamann, ibid, pp. 747.
6. D. Givord, H.S. Li, R. Perrier de la Bathie, Solid State Commun. 51, 857 (1984).
7. M. Sagawa, S. Fujimura, H. Yamamoto, Y. Matsuura, and K. Hiraga, IEEE Trans. Magn. MA6-20, 1584 (1984).
8. S. Hirosawa, Y. Matsuura, H. Yamamoto, S. Fujimura, M. Sagawa and H. Yamauchi, J. Appl. Phys. 59, 873 (1986).
9. M. Sagawa, S. Hirosawa, K. Tokuhara, H. Yamamoto, S. Fujimura, Y. Tsubokawa and R. Shimizu, J. Appl. Phys. (to appear).
10. H. Hiroyoshi, H. Kato, M. Yamada, N. Saito and Y. Nakagawa, Solid State Commun. (to appear).
11. H. Zilstra, J. Appl. Phys. 41, 4881 (1970).
12. D. Givord, J. M. Moreau and P. Tenaud, Solid State Commun. 55, 303 (1985).
13. D. Givord, P. Tenaud and T. Viadies, J. Appl. Phys. 60, 3263 (1986).
14. S. Hirosawa, Y. Matsuura, H. Yamamoto, S. Fujimura, M. Sagawa and H. Yamauchi, Japanese J. Appl. Phys. 24, L803 (1985).
15. H. K. Smith and W. E. Wallace, Lanthanide Actinide Res. 1, 217 (1985-6).
16. K. Hiraga, M. Hirabayashi, M. Sagawa and Y. Matsuura, Jpn. J. Appl. Phys. 24, 699 (1985).
17. M. Sagawa, S. Hirosawa, H. Yamamoto, Y. Matsuura, S. Fujimura, H. Tokuhara and K. Hiraga, IEEE Trans. Magn. MAG-22, 910 (1986).
18. R. Ramesh, J. K. Chen and G. Thomas. Digest of INTERMAG 85 FB-5 (1985).
19. P. Schrey, IEEE Trans. Magn. MAG-22 913 (1985).
20. M. Sagawa, S. Fujimura, H. Yamamoto, Y. Matsuura and S. Hirosawa, IEEE Translation J. Magn. Japan, TJMJ-1, 979 (1985).
21. S. Hirosawa, K. Tokuhara, H. Yamamoto, S. Fujimura and M. Sagawa, to be submitted.

ANALYTICAL ELECTRON MICROSCOPE STUDY OF HIGH- AND
LOW-COERCIVITY
SmCo 2:17 MAGNETS

JOSEF FIDLER , J. BERNARDI AND P. SKALICKY
Institute of Applied and Technical Physics, University of Technology,
Karlsplatz 13, A-1040 Vienna, Austria.

ABSTRACT

Sintered, precipitation hardened SmCo 2:17 magnets contain a multiphase
microstructure. Our electron microscopic investigations reveal that the size
of the rhombic, cellular precipitation structure and the formation of cell
interior and cell boundary phases is determined by the nominal composition
of the alloy and the postsintering heat treatment conditions and primarily
control the intrinsic coercivity of the magnet. Selected area electron
diffraction together with high resolution electron microscopy showed a high
density of basal stacking faults (microtwinning) of the cell interior phase
of low coercivity ($_iH_C$ < 700 kA/m) magnets with a (c/a)*- ratio of the basic
structural unit of > 0.843. High coercivity magnets ($_iH_C$> 1000 kA/m),
containing a high density of the platelet phase perpendicular to the c-axis,
exhibit cell diameters up to 200 nm with a (c/a)*-ratio of the basic
structural unit of the cell interior phase of < 0.843.

INTRODUCTION

Commercially available permanent magnet materials may be divided into
hardferrites, AlNICo-magnets and rare earth (RE)-permanent magnets (PM).
There exist two groups of REPM, the RE-cobalt magnets and the recently
developed RE-iron magnets. The RE-cobalt magnets can be divided into five
types depending on whether the magnet has a single-phase or a two-phase
microstructure. The ideal microstructure of the single phase magnets
consists of aligned single-domain grains with a $SmCo_5$- or Sm_2Co_{17}-crystal
structure. Two types of precipitation hardened magnets can be distinguished:
the one type contains 2:17-precipitates in a 1:5-matrix, the other type
forms 1:5-precipitates in a 2:17-matrix. Besides these magnets there are the
bonded magnets, in which the single domain particles are embedded in a non
magnetic phase. Rare earth-cobalt magnets are produced by a powder
metallurgical process with complicated post-sintering heat treatment
procedures in the case of precipitation hardened SmCo 2:17-magnets. In fact,
the nominal composition, the sintering and annealing parameters determine
the intrinsic coercivity of SmCo 2:17 magnets. To understand the
microstructure and the reasons for high coercivity of sintered SmCo 2:17
magnets transmission electron microscopy has widely been used [1-5]. The
purpose of this work is to compare the microstructure of low coercivity
($_iH_C$<700 kA/m) and high coercivity ($_iH_C$>1000 kA/m) precipitation hardened
SmCo 2:17-magnets, containing a cellular microstructure.

EXPERIMENTAL

Besides optical metallography and microprobe analysis we have used
transmission electron microscopy together with STEM X-ray microanalysis to

identify the precipitation and to study the magnetic domain structure and their interaction with the precipitation structure. The investigations have been carried out on a JEOL 200CX microscope equipped with a high take-off angle energy dispersive detector for X-ray microanalysis. The specimens for transmission electron microscopic investigations were prepared by a combination of electropolishing and ionbeam thinning procedures.

The investigated samples were sintered magnets with a nominal composition close to $Sm(Co_{.68}Fe_{.22}Cu_{.08}Zr_{.02})_{7.5}$.

They were prepared and given a systematically varying heat treatment after sintering to achieve maximum magnetic hardness by different producers, such as KRUPP WIDIA GmbH., FRG, TDK Corp. and SHIN-ETSU Chem.Corp., Japan. For some unknown reason different coercivities were obtained under the same preparation conditions. Magnets with low coercivities (< 700 kA/m) and magnets with high coercivities (> 1000 kA/m) were choosen for transmission electron microscopy. Commercial sintered SmCo 2:17 magnets can also be divided into high $_iH_C$ and low $_iH_C$ magnets (fig.1).

In order to study the influence of the effect of the nominal Samarium content on the microstructure and coercivity a set of samples was prepared and given a systematically varying heat treatment by THYSSEN Edelstahlwerke, FRG with a Sm-content of 23.9 wt% ($_iH_C$=250 kA/m), 24.8 wt% ($_iH_C$ > 1700 kA/m) and 25.7 wt% ($_iH_C$=800 kA/m). Thin slices of these magnets were cut parallel to the alignment direction and different discs were prepared near the edge and the centre of the magnet for electron microscopy as shown in fig.2.

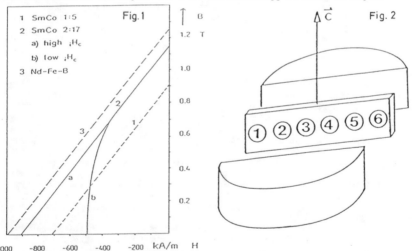

Fig.1: Room temperature demagnetization curves of rare earth permanent magnets.

Fig.2: Schematic diagram showing the preparation of discs for transmission electron microscopy of a cylindrical magnet.

CRYSTALLOGRAPHY OF Sm_2Co_{17}

The compounds $SmCo_5$ and Sm_2Co_{17} may be regarded as layered structures with different stacking sequences of a basic structural layer, consisting of a Co-layer and a mixed Sm- and Co-layer. The stacking sequence of the hexagonal Sm_2Co_{17} structure is

$$.....ABABABABA.....$$ (1)

and the one of the rhombohedral Sm_2Co_{17} structure is

$$.....ABCABCABCA..... \qquad (2)$$

Besides these perfect layer structures structures containing stacking faults in the basal plane also occur in real magnets. In rhombohedral SmCo 2:17 crystals the close packed planes are the (0001)-basal planes, which are also the glide planes and coherent twin planes. The twin orientation corresponds to a mirror plane reflection about {0001}. The stacking sequence for a coherent twin with the dashed line and asterisk as the centre of the fault is

$$\begin{array}{c} * \\ \vdots \\ .....ABCABCBACBA..... \qquad (3) \\ \vdots \\ * \end{array}$$

A coherent twin is formed by a <u>multiple</u> shearing operation on the (0001)-plane, if the displacement is $1/3 \cdot [10\bar{1}0]$. On the other hand an intrinsic stacking fault is formed when <u>one</u> layer of the basic structural unit is removed from the normal sequence, leaving

$$\begin{array}{c} *\vdots* \\ ....ABCABC\vdots BCABCA.... \qquad (4) \\ *\vdots* \end{array}$$

It should be mentioned that this fault can also be regarded as being formed by two twinning operations (marked by asteriks) separated by <u>one</u> basic layer (microtwinning).

An extrinsic stacking fault is formed by the addition of <u>one</u> layer of the structural unit to the normal sequence, giving

$$\begin{array}{c} *\vdots* \\ ....ABCABCBABCABCA.... \qquad (5) \\ *\vdots* \end{array}$$

Again this can also be regarded as being formed by two twinning operations seperated by <u>two</u> basic layers (microtwinning).

A combination of an intrinsic stacking fault with the formation of a coherent twin gives the stacking sequence

$$\begin{array}{c} * \\ \vdots \\ ...ABCABCABACBACBA.... \qquad (6) \\ \vdots \\ * \end{array}$$

All the above faults preserve close packing and exhibit low surface energies compared with other higher order planar faults which are summarized for the case of perfect and imperfect modifications of Sm_2Co_{17} compounds by C.W. Allen et al. [6].

From the stacking sequence of the planar faults (4) and (6) within the rhombohedral Sm_2Co_{17}-crystal it can be seen that due to the crystallographic shear a platelet with one unit cell thickness of the hexagonal Sm_2Co_{17}-phase with the stacking sequence BC and AB are formed.

$RECo_5$ compounds crystallize in the ordered $CaCu_5$ structure type. Substitution of rare earth atoms by Co-atoms lead to the formation of the disordered $TbCu_7$-structure type in which rare earth atoms are replaced by pairs of Co-atoms [6]. When nearly 22% of the RE atoms have been substituted by pairs of Co atoms, the disordered $TbCu_7$-structure is no longer stable and a two-phase microstructure is formed. Substituting 33% of the RE atoms by Co-atom pairs leads to RE Co [10]. In the case of Sm-(Co,Fe) alloys the rhombohedral Sm_2Co_{17} modification is more stable than the hexagonal modification [7]. The $(c/a)^*$ - ratio of the reduced basic structural unit of the rhomb. Sm_2Co_{17}-structure is found to be close to 0.843 (± 0.006) [8]. Only in hyperstoichiometric Sm_2Co_{17}-alloys, such as $Sm(Co,Fe)_{9-12}$, the hexagonal modification with a ratio $(c/a)^* > 0.85$ is found as stable room temperature phase [9]. Table I summarizes the crystal structure - relationships, the range of stability and $(c/a)^*$-ratios of phases occurring in $Sm(Co,Fe)_{5+x}$ alloys {0 ≤ x ≤ 7}.

TABLE I: CRYSTAL STRUCTURE RELATIONSHIPS OF $Sm(Co,Fe)_{5+x}$ - ALLOYS

x = 0	$CaCu_5$	$P6/mmm$	$(c/a)^* = 0.795$	ordered
0<x<2	$TbCu_7$	$P6/mmm$	$0.795 < (c/a)^* < 0.837$	disordered
x=3.5	Th_2Zn_{17}	$R\bar{3}m$	$0.837 \le (c/a)^* \le 0.349$	ordered
3.5≤x≤7	Th_2Ni_{17}	$P6_3/mmc$	$(c/a)^* > 0.85$	ordered
	"Th_2Ni_{17}"hyperst.	$P6_3/mmc$ $P6/mmm$		partly disordered

$(c/a)^*$ is the ratio of the basic structural unit; data taken from [9,10]

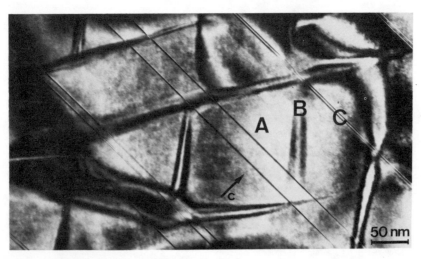

Fig.3: Rhombic cellular microstructure of a sintered precipiation hardened SmCo 2:17 magnet. A is the cell interior phase, B the cell boundary phase and C the platelet phase perpendicular to the alignement direction.

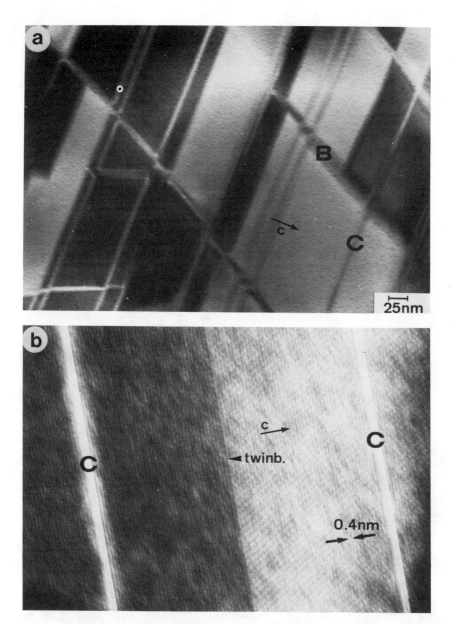

Fig.4: Electron micrographs showing coherent twinning in high coercivity ($_iH_c$>1000 kA/m) magnets. C is the platelet phase and B the cell boundary phase.

186

RESULTS

Microstructure of high coercivity SmCo 2:17-magnets

Copper containing RE-cobalt magnets with a nominal composition of
$Sm(Co,Fe,Cu,TM)_{6-8}$ with TM=Zr,Ti or Hf show a fine scale, cellular
microstructure [5]. Rhombic cells of the type $Sm_2(Co,Fe)_{17}$ rhomb. - phase A
- are separated by a $Sm(Co,Cu,TM)_{5-7}$ cell boundary phase - phase B. In
magnets with high coercivities thin platelets - phase C - are found
perpendicular to the hexagonal c-axis (fig.3). Our high resolution electron
microscope investigations [5] show that the crystal structure of the
platelet phase C is close to the hexagonal Sm_2Co_{17} structure with a
c-crystal parameter of 0.8 nm. This is in agreement with metallurgical
considerations on the formation of various phases in sintered SmCo 2:17
magnets by A.E. Ray [9,12,13]. Remarkable for high coercivity SmCo 2:17
magnets is also the observed twinning within the rhombohedral 2:17-cell
interior phase. The electron micrographs of fig.4a and b, taken under
different reflection conditions, show the cellular microstructure and the
twinning with the basal plane as twin boundary. Twin boundaries may coincide
with the platelet phase but also occur within the rhomb. 2:17-phase
(fig.4b), corresponding to planar faults of the type (6) and (3),
respectively.

Microstructure of low coercivity SmCo 2:17 magnets

From our investigations it is obvious that there are different reasons for
low coercivities. Since in precipitation hardened SmCo 2:17 magnets the the
cellular precipitation structure acts as attractive pinning centre for
magnetic domain walls during the magnetization reversal (fig.5) [14,15], the

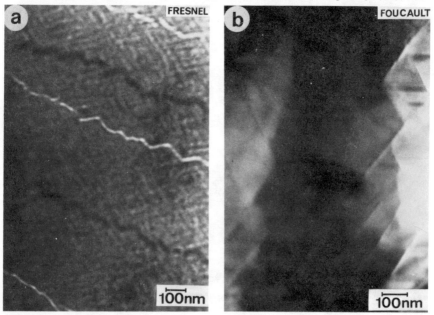

Fig.5: Lorentz electron micrographs showing the attractive domain wall
pinning at the continuous cell boundary phase in the demagnetized state.

size, the composition and the completeness of the continuus cellular precipitation structure controll the coercivity. The platelet phase (C) predominately acts as diffusion path for the chemical redistribution process during the post-sintering heat treatments and is favourable for the formation of the cell boundary phase of the 1:5-ordered type or 1:7-disordered type. Figure 6 shows two electron micrographs of the magnet with 23.9 wt% nominal Sm-content. A high density of the platelet phase is shown, but due to the lack of samarium the formation of the cell boundary phase is

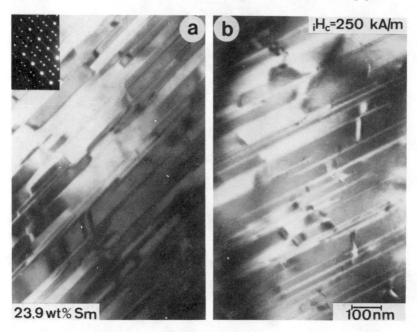

Fig.6: Bright field electron micrographs of a low coercivity magnet. Due to the lack of samarium the cell boundary phase is only incompletely formed. A high density of the platelet phase is observed.

poor leading to an intrinsic coercivity of 250 kA/m. Impurities, primarily such as oxygen and carbon, lead to the formation of macroscopic precipitates of Sm_2O_3, ZrC, TiC etc. A high amount of these phases in the magnet impedes the formation of the platelet phase and the cellular precipitation structure. It has been shown [16] that small amounts of Zr, Ti or Hf are necessary for the formation of the platelet phase (C).

The high resolution electron micrograph of fig.7 shows the cell interior of a magnet with an intrinsic coercivity of $_iH_c$ = 600 kA/m. No platelet phase (C) is found and X-ray microanalysis shows that this region of the magnet is depleted from zirconium [11]. The contrast of this image correspond to the atomic positions along the [11$\bar{2}$0]-zone axis. A high density of stacking faults, and therefore also microtwinning is observed. It should be noted that the distance between atomic layers in the basal plane varies, especially near stacking faults. The development and growth of the cellular precipitation structure occurs during the isothermal aging procedure and involves diffusion of samarium, whereas the chemical redistribution of the transition metals during the step aging procedure following the isothermal aging increases the coercivity [16]. Possible

188

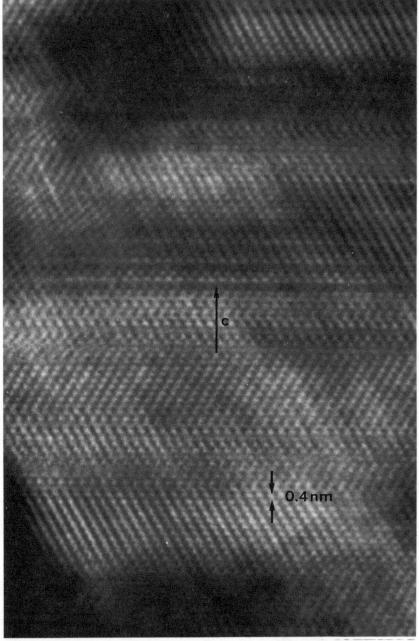

Fig.7: High resolution electron micrograph of the cell interior phase of a low coercivity ($_jH_c$=600 kA/m) magnet showing a high density of stacking faults in the basal plane (microtwinning).

diffusion paths for this diffusion process are the platelets of phase (C) and the disordered regions near planar stacking faults and twin boundaries in the basal plane as observed in fig.7.

Our electron microscopic investigations of magnets with low coercivity revealed also a strong dependence of the size of the cellular microstructure on the position of the magnet where the samples were taken. The Kerr-optical micrograph of fig.8 shows the magnetic domain structure perpendicular to the

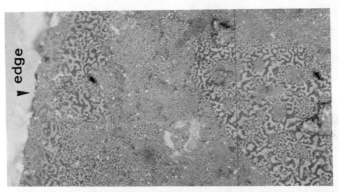

Fig.8: Kerr-optical micrograph of a polished surface perpendicular to the alignment direction showing the magnetic domain structure and chemical inhomogeneities from the edge to the centre of the sintered SmCo 2:17 magnet.

alignment direction of a cylindrical magnet near the edge of the magnet with about 20 mm in diameter. From differences in the domain structure chemical inhomogeneities within the material are concluded. Especially regions with large domain widths in the demagnetized state as in fig.8 are (Co,Fe)-enriched regions with low magnetocrystalline anisotropy.

DISCUSSION AND SUMMARY

Selected area electron diffraction, high resolution and analytical transmission electron microscopy of low ($<$ 700 kA/m) and high ($>$1000 kA/m) coercivity magnets revealed different microstructures. The electron diffraction patterns are different for magnets containing the platelet phase (C) -high coercivity- (fig.9.a) and magnets containing a high density of stacking faults and microtwins, respectively -low coercivity- (fig.9.b and c). Due to the narrowness of the rhombohedral Sm_2Co_{17} regions the rhombohedral $\langle n0\bar{n}n \rangle$ with (n=1,2)- diffraction spots are streaked. As shown earlier, due to microtwinning under certain contitions thin layers of a hexagonal Sm_2Co_{17} phases are formed. Fig.9c was taken from such a region and shows also the hexagonal $\langle 10\bar{1}1 \rangle$ reflections. The c/a-ratio which is sensitive to the chemical composition (see Table I) of the cell interior phase can directly be determined from fig.9. Table II summarizes the reduced (c/a)* - ratios of the basic structural unit determined from magnets with different intrinsic coercivities. It is obvious that a ratio larger than the ideal value of the rhombohedral 2:17-type (0.843) is due to a hyperstoichiometric concentration of the cell interior phase and leads to the formation of a high density of stacking faults which is linked with the transformation to the hexagonal Sm_2Co_{17} structure type. On the other hand a ratio less than the ideal value is an indication for the formation of the platelet phase (C) and leading to high intrinsic coercive forces.

190

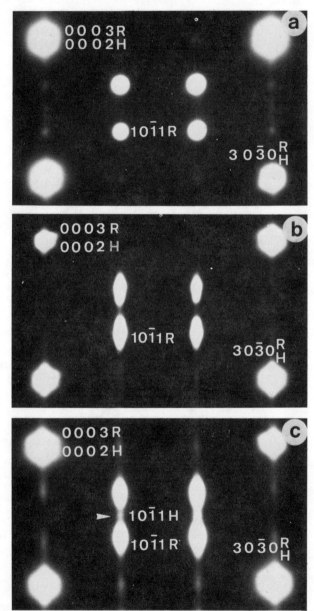

Fig.9: Selected area electron diffraction patterns taken parallel to [1̄210]
of a high coercivity magnet (a) and low coercivity magnets (b) and (c) .
The streaking of the rhombohedral 2:17 reflections in (b) and (c) is due
to microtwinning of the cell interior phase.

TABLE II: (c/a)* - RATIOS OF THE BASIC STRUCTURAL UNIT OF THE CELL INTERIOR
PHASE OF MAGNETS WITH DIFFERENT INTRINSIC COERCIVITIES

magnet	(c/a)*	$_iH_c$ [kA/m]	remarks
# 1	.842	400	microtwinning
# 2	.853	600	microtwinning, hex.$\langle 10\bar{1}1\rangle$
# 3	.814	> 2000	platelets
# 4	.824	1200	platelets
# 5	.820	1000	platelets, overaged
# 6	.846	42	as sintered
# 7	.835	250	nominal 23.9 wt% Sm, platelets
# 8	.829	800	nominal 25.7 wt% Sm, platelets
# 9	.838	> 1700	nominal 24.8 wt% Sm, platelets

In sintered precipitation hardened SmCo 2:17 magnets the coercivity primarily depends on the multiphase microstructure. The degree of completeness of the rhombic cellular precipitation structure determines the effectiveness of the domain wall pinning and therefore the intrinsic coercivity. The shape of the rhombic, cellular structure is determined by the direction of zero deformation strains due to the crystal lattice misfit between the different cell boundary and cell interior phases and the acute angle of the of the rhombic cells is an indication for the compositional difference between these phases [17]. We also found earlier that the postsintering heat treatment is very sensitive to the observed cell size and coercivity of the magnet [5,17].

In contrast to sintered SmCo 1:5 and Nd-Fe-B based permanent magnets in the case of SmCo 2:17 magnets the grain size is not changed during the post sintering heat treatment and does not influence the coercivity [18]. We found that maximum coercivities (> 2000 kA/m) occur in magnets with cell diameters of about 200 nm and containing a high density of the platelet phase (C). As result of our investigations we revealed several factors limiting the intrinsic coercivity. Chemical inhomogeneities especially near the edges of a magnet influence the Sm-rich precipitation structure with whether an ordered $CaCu_5$ or disordered $TbCu_7$ structure type. A high content of oxygen and carbon impurities locally deplete the magnet from samarium. An optimum postsintering heat treatment is necessary for the nucleation and growth of the continuus cellular precipitation structure.

Acknowledgements

One of the authors (J. Fidler) would like to thank Profs. D.E. Laughlin and W.E. Wallace for stimulating discussions during his stay at the Carnegie Mellon University and the partial support from the Magnetic Materials Research Group, through the Division of Materials Research, National Science Foundation, under Grant No. DMR-86133386. The authors are grateful to Drs. K. Kuntze and H. Nagel from Thyssen Edelstahlwerke AG,FRG for providing samples with different nominal composition.

References

1.   J.D. Livingston and D.L. Martin, J. Appl. Phys. <u>48</u>, 1350 (1977).
2.   R.K. Mishra, G. Thomas, T. Yoneyama, A. Fukuno and T. Ojima,
      J.Appl.Phys. <u>52</u>, 2517 (1981).
3.   G.C. Hadjipanayis, E.J. Yadlowsky and S.H. Wollins, J. Appl. Pys.
      <u>53</u>, 2386 (1982).
4.   J. Fidler and P. Skalicky, J. Magn. Magn. Mat. <u>27</u>, 127 (1982).
5.   J. Fidler, P. Skalicky and F. Rothwarf, IEEE Trans. Magn. <u>MAG-19</u>,
      2041 (1983).
6.   C.W. Allen, D.L. Kuruzar and A.E. Miller, IEEE Trans. Mag. <u>MAG-10</u>,
      716 (1974).
7.   Y. Khan, Acta Cryst. <u>B29</u>, 2502 (1973).
8.   Q. Johnson, G.S. Smith and D.H. Wood, Acta Cryst. <u>B25</u>, 464 (1969).
9.   A.E. Ray, Proc. of Soft and Hard Magnetic Materials with
      Applications, Lake Buena Vista, Florida, Oct.1986, to be
      published by ASM.
10.   Y. Khan, Pys. Stat. Sol.(a) <u>21</u>, 69 (1974).
11.   J. Fidler, P. Skalicky and F. Rothwarf, Mikrochimica Acta [Wien]
      <u>Suppl. 11</u>, 371, (1985).
12.   A.E. Ray, J. Appl. Phys. <u>55</u>, 2094 (1984).
13.   A.E. Ray, IEEE Trans. Magn. <u>MAG-20</u>, 1614 (1984).
14.   J. Fidler, J. Mag. Magn. Mat. <u>30</u>, 58 (1982).
15.   H. Kronmüller, K.D. Durst, W. Ervens amd W. Fernengel, IEEE Trans.
      Magn. <u>MAG-20</u>, 1569 (1984).
16.   L. Rabenberg, R.K. Mishra and G. Thomas, J. Appl.Phys. <u>53</u>, 2389
      (1982).
17.   J.Fidler, R. Grössinger, H. Kirchmayr and P. Skalicky,   ERO -
      Report of U.S. Army, Pr.No. DAJA (1983)   37-82-C-0050.
18.   J. Fidler and P. Skalicky, Radex-Rundschau <u>2/3</u>, 63 (1986).

MICROSTRUCTURAL STUDIES OF HIGH COERCIVITY $Nd_{15}Fe_{77}B_8$ SINTERED MAGNETS

M. H. GHANDEHARI* AND J. FIDLER**
* Unocal Corporation, Science and Technology Division, P.O. Box 76,
  Brea, CA 92621 USA
**Institute of Applied and Technical Physics, Technical University
  of Vienna, Vienna, Austria

ABSTRACT

Microstructures of $Nd_{15-x}Dy_xFe_{77}B_8$ prepared by alloying with Dy, and by using Dy_2O_3 as a sintering additive, have been determined using electron microprobe and transmission electron microscopy. The results have shown a higher Dy concentration near the grain boundaries of the 2-14-1 phase for magnets doped with Dy_2O_3, as compared to the Dy-alloyed magnets. A two-step post sintering heat treatment was also studied for the two systems. The resultant concentration gradient of Dy in the 2-14-1 phase of the oxide-doped magnets is explained by the reaction of Dy_2O_3 with the Nd-rich grain boundary phase and its slow diffusion into the 2-14-1 phase. Increased Dy concentration near the grain boundary is more effective in improving the coercivity, as domain reversal nucleation originates at or near this region.

INTRODUCTION

$Dy_2Fe_{14}B$ displays a much higher single crystal anisotropy than the corresponding Nd compound [1,2]. Therefore, a Dy-substituted sintered $Nd_{15}Fe_{77}B_8$, which embodies the 2-14-1 phase displays a higher practical coercivity, as first reported by Sagawa and coworkers [2]. As a low-cost approach in introducing Dy, the reaction of Dy_2O_3 as a sintering additive with $Nd_{15}Fe_{77}B_8$ powder was investigated by one of the authors [3]. This resulted in Dy_2O_3-doped magnets with coercivities similar to those prepared by substitution of Dy for Nd in the alloying step [3]. In this study, we compare the resultant microstructures by scanning transmission electron microscopy (STEM) and by electron microprobe. A two-step post-sintering heat treatment, employed to optimize the coercivity, is also compared for the two systems.

EXPERIMENTAL

Powder samples were prepared from a nominal $Nd_{15}Fe_{77}B_8$, and an alloy in which 3.5 wt% (approximately 10 atomic %) of Nd was replaced by Dy in the alloying step. Both of these powder samples were prepared under identical conditions and contained typically 0.5 to 0.6 wt% oxygen prior to pressing and sintering. 4.0 wt% Dy_2O_3 (3.5 wt% Dy equivalent) was hand-mixed with the former, and both types of magnets were sintered at 1070°C for an hour and annealed in two steps [4]. The first step of the annealing process consisted of 900°C for 2 hours and the second step varied between 550 and 680°C for an hour. Samples were rapidly quenched in a cooler part of the furnace in between the two steps. After the magnetic measurements, the sintered samples were analyzed for O_2 by a vacuum fusion technique. Boron, Nd, Dy and Fe were analyzed chemically by inductively coupled plasma (ICP) spectroscopy. A wavelength dispersive spectrometer was employed with the electron microprobe to scan for Dy.

Samples for TEM analyses were mechanically thinned to 200 micron and electropolished using a solution of perchloric acid in methanol. The surface of the thin foils were cleaned by ion milling and were examined in a JEOL 200 CX microscope at 200 keV equipped with a high-take off angle X-ray detector. The quantitative X-ray analysis was carried out using the k-factors and standards from the software system.

RESULTS

Magnetic Properties & Chemical Composition

Tables I and II show the intrinsic coercivity of the magnets prepared by Dy-alloying and by Dy_2O_3-doping, respectively, as a function of the second step post-sintering heat treatment described above. Table III compares a chemical analysis for two samples selected from Tables I and II. The optimum second step post-sintering heat treatment for the Dy_2O_3-doped magnets is lower by about 50 to 90°C while affording a higher intrinsic coercivity (see Tables I and II). As expected, the Dy_2O_3-doped sample (Table III) has a higher oxygen concenctration than the Dy-alloyed magnet. About 0.5 wt% oxygen is introduced in the sample as a result of adding 4 wt% Dy_2O_3 to the magnet powder in these experiments. It is also noted that the Dy_2O_3-doped sample of Table III has a slightly higher Dy content. This, however, does not fully account for the observed difference in the ^iHc values between the two systems. The control samples in the absence of Dy_2O_3 showed an ^iHc increase of about 2,000 Oe for 1 wt% Dy. The chemical analysis of Table III are accurate within a relative error of ±3%.

Remanence, Br, and the energy product, BHmax, were not affected significantly by the added oxide concentration. A typical demagnetization curve for the heat-treated Dy_2O_3-doped $Nd_{15}Fe_{77}B_8$ magnets is shown in Figure 1. This indicates a good curve squareness and a characteristic energy product, for the magnets of this type.

Microstructural Determinations

Dy line scans of the samples shown in Figures 2a and b indicate that Dy is distributed unevenly in the Dy_2O_3 doped magnet. Consistent with this result, Figure 2c shows a TEM of a Dy_2O_3-doped magnet indicating a higher Dy concentration near the grain boundaries. An X-ray spectrum (not shown) near the region A_4 showed a remarkably low ratio of Fe:Nd, indicating lack of Dy, whereas region A_5 near the grain boundary showed a high Fe:Nd ratio. These results are consistent with uneven distribution of Dy in the electron microprobe line scans. X-ray analyses of Dy-alloyed magnets, under transmission electron microscope showed a much more homogeneous distribution of Dy in the 2-14-1 grains.

DISCUSSION

The sintered Nd-Fe-B magnets display a multiphase microstructure which consists of hard magnetic 2:14:1 grains, a B-rich and a Nd-rich phase [5]. Because of its lower melting point, the latter phase is preferentially formed along the grain boundaries. Compared to the alloyed NdDyFeB magnets, a higher Dy concentration is found near the grain boundary of the 2-14-1 phase in the Dy_2O_3-doped magnets (see Figure 2c). This may be explained by considering the reaction of Dy_2O_3 with the liquid, Nd-rich layer, and its subsequent displacement of Nd in the $Nd_2Fe_{14}B$ phase, as depicted in Figure 3. Such a diffusion process is apparently slow enough to create a Dy gradient, with a higher Dy near the reaction zone at the grain boundary.

TABLE I

INTRINSIC COERCIVITY OF $Nd_{13.5}Dy_{1.5}Fe_{77}B_8$ AS A FUNCTION OF POST-SINTERING HEAT TREATMENT

(ALL SAMPLES WERE SINTERED AT 1070°C FOR 1 HOUR)

SAMPLE	POST-SINTERING HEAT TREATMENT		$_iH_C$(Oe)
	T (°C)	t (HOURS)	
1	900	2	13,500
1	900	2	
	630	1	13,800
1	900	2	
	630	1	15,100
	680	2	
2	900	2	13,600
2	900	2	
	680	1	15,000

TABLE II

INTRINSIC COERCIVITY OF Dy_2O_3-DOPED $Nd_{15}Fe_{77}B_8$ AS A FUNCTION OF POST-SINTERING HEAT TREATMENT

ALL SAMPLES WERE DOPED WITH 4 WT% Dy_2O_3 AND SINTERED AT 1070°C FOR 1 HOUR)

SAMPLE	POST-SINTERING HEAT TREATMENT		$_iH_c$(Oe)
	T (°C)	t (HOURS)	
1	900	2	17,000
	610	1	
2	900	2	17,500
	590	1	
3	900	2	15,400
	570	1	
4	900	2	16,300
	630	1	
5	900	2	17,000
	570	1	
6	900	2	14,800
	550	1	
7	900	2	16,200
	600	1	
8	900	2	16,900
	610	1	
9	900	2	15,200
	590	1	
10	900	2	15,600
	620	1	

TABLE III

CHEMICAL ANALYSES OF A Dy-ALLOYED AND A DY$_2$O$_3$-DOPED Nd$_{15}$Fe$_{77}$B$_8$ SINTERED MAGNETS

Sample	i_{Hc} (Oe)	ELEMENTS (wt%)				
		Nd	Dy	Fe	B	O$_2$
Dy-ALLOYED	15,000	27.7	3.13	70.0	1.36	0.5
Dy$_2$O$_3$-DOPED	17,500	29.8	3.51	65.2	1.22	1.2

FIG. 1 - DEMAGNETIZATION CURVES OF A Dy_2O_3- DOPED $Nd_{15}Fe_{77}B_8$
(SAMPLE 2 FROM TABLE I)

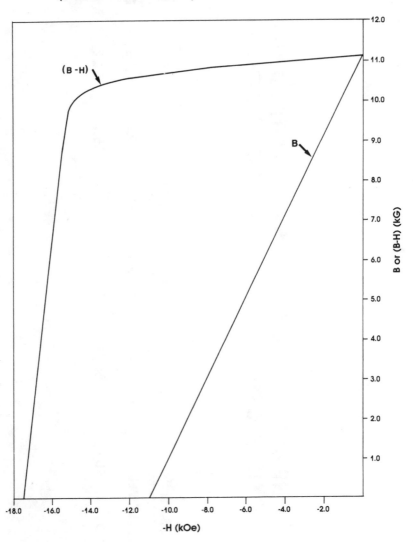

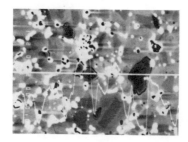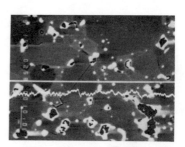

FIG. 2 a, b – Dy LINE SCAN OF; a) $Dy_2O_3^-$ DOPED $Nd_{15}Fe_{77}B_8$ MAGNET & b) $Nd_{13.5}Dy_{1.5}Fe_{77}B_8$ MAGNET

A) SCAN LINE
B) Dy INTENSITY PROFILE FROM X-RAY SPECTROMETER

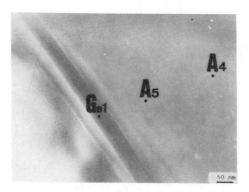

FIG. 2c – TRANSMISSION ELECTRON MICROGRAPH OF A $Dy_2O_3^-$ DOPED $Nd_{15}Fe_{77}B_8$. REGION A_5 NEAR THE GRAIN BOUNDARY SHOWS A HIGHER CONCENTRATION OF Dy THAN REGION A_4 IN THE MIDDLE OF THE GRAIN

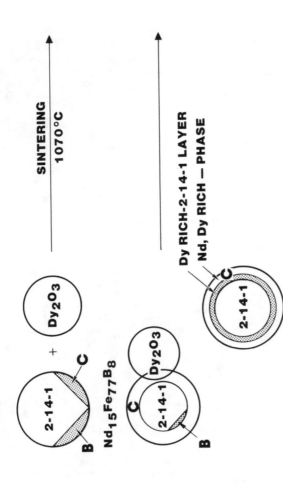

FIG. 3 — SCHEMATIC DIAGRAM SHOWING THE REACTION OF Dy_2O_3 WITH $Nd_{15}Fe_{77}B_8$ DURING THE SINTERING. REGIONS B AND C DESIGNATE THE B-RICH AND Nd-RICH PHASES.

The nucleation of reversed domains takes place at regions with high local demagnetization fields and low magnetocrystalline anisotropies, such as soft magnetic precipitates and disturbed regions near the grain boundaries. As the grain boundary is concentrated with the Dy-substituted 2-14-1 phase, the region's anisotropy increases, translating to a higher intrinsic coercivity as indicated in Tables I & II for the Dy_2O_3-doped magnet. The expected high coercivity may be partially offset by undissolved, and large size Nd_2O_3 particles by-product, or unreacted Dy_2O_3. At the same time, dissolved oxide at the grain boundary could lower the melting point of the Nd-rich phase, thus accounting for the lower second-step post sintering heat treatment for optimum coercivity (see Tables I and II). A uniform NdDy-rich phase along the boundaries of the 2:14:1 grains leads to a better separation of these grains. This can delay the expansion field of the reversed nuclei across the non-magnetic grain boundaries which in turn can account for a higher intrinsic coercivity.

ACKNOWLEDGMENT

The authors would like to thank K. L. McNutt for assistance in magnet preparation and Unocal's management for permission to publish this work. Useful discussion with Dr. R. Ramesh of the Department of Materials Science, University of California, Berkeley, is also acknowledged.

REFERENCES

[1] S. Hirosawa, Y. Matsuura, H. Yamamoto, S. Fujimura and M. Sagawa, J. Appl. Phys. 59 (3), 873 (1986).

[2] M. Sagawa, S. Fujimura, H. Yamamoto, H. Matsuura, and K. Hiraga, IEEE Trans. Mag. MAG-20, 1584 (1984).

[3] M. H. Ghandehari, Appl. Phys. Lett. 48 (8), 548 (1986).

[4] M. Tokunaga, M. Meguro, M. Endoh, S. Tanigawa, and H. Harada, IEEE Trans. Mag. MAG-21, 1964 (1985).

[5] J. Fidler, IEEE Trans. Magn., MAG-21, 1955 (1985).

EFFECT OF Dy ADDITIONS ON MICROSTRUCTURE AND MAGNETIC PROPERTIES OF Fe-Nd-B MAGNETS

R.Ramesh[#], G.Thomas[#] and B.M.Ma[*]

Department of Materials Science and Mineral Engineering and Materials and Molecular Research Division, Lawrence Berkeley Laboratory, University of California, Berkeley, Ca 94720
* Crucible Research Center, Pittsburgh Pa 15230.

ABSTRACT :

This paper addresses the effect of Dy addition upon the magnetic properties, microstructure and microcomposition of Fe-Nd-B magnets. It is shown that increasing additions of Dy causes the remanence, B_r to decrease linearly. The intrinsic coercivity, iHc, increases sharply for small additions of Dy, but the increase is not proportional for higher Dy contents. The iHc increases almost linearly with the effective anisotropy field of the $RE_2Fe_{14}B$ phase until the Dy content is about 10% of the total rare earth content. Above this concentration, there is a strong deviation from linearity. Various types of possible concentration profiles of the substituted rare earth are suggested. It is also argued that preferential segregation of Dy to the interfaces could be beneficial in increasing the nucleation field. Morphologically there is no apparent effect of Dy on the microstructure. However, in the 5 atomic % Dy sample, Dy rich oxides were observed. It is shown through Energy Dispersive Xray Spectroscopy (EDXS) line profiling that Dy partitions preferentially into the $RE_2Fe_{14}B$ phase in all the cases. No segregation of Dy to the interphase interfaces has been detected.

INTRODUCTION :

Current industrial and academic interest in Fe-Rare Earth-B permanent magnets is well evidenced by the number of conference symposia and workshops being held on this class of magnets. Since the discovery of these magnets[1-3], considerable advances have been made in the understanding of their structure, microstructure and properties [4,5], although some questions still remain regarding the actual site for initiation of magnetization reversal. One of the efforts now is in improving the temperature dependent properties through alloying additions. These efforts are being made to offset the high temperature coefficient of $_iH_c$ and Br, due to the low Curie temperature, T_c, of the $RE_2Fe_{14}B$ phase. Thus, Dy,Tb,Al additions have led to significant improvements in the intrinsic coercivity, while Co addition has helped in raising the Curie temperature of the $RE_2Fe_{14}B$ phase[6-12]. It has also been shown that addition of small amounts of Zr can still improve the anisotropy field[13]. However, heavy rare earths such as Dy or Tb decrease the magnetization due to the ferrimagnetic coupling between the rare earth atom and the Fe atoms. Thus, the effect of these additions in the case of single crystal samples, on the magnetic properties are well understood. Recently, Hirosawa et.al[14] have examined the relationship between $_iH_c$ and H_A of heavy rare earth (Dy and Tb) substituted Fe-RE-B magnets and also the influence of temperature and concentration of heavy rare earth on $_iH_c$ and H_A.

In the case of sintered magnets, however, the effects may be more complex due to the polycrystalline nature as well as due to microstructural influences. Thus, the addition of a quaternary element, such as Dy, can bring about two possible changes:(1) a change in the intrinsic properties (such as H_A,M_S) of the 2-14-1 phase with no change in the microstructure and composition;(2) a change in the microstructure (such as the appearance of a new phase) and/or microcomposition of the magnet. Both these potential changes will influence the $_iH_c$, B_r and $(BH)_{max}$ of the magnet. Thus, a systematic study of the influence of these additions on the magnetic properties, microstructure and microcomposition is essential. The microstructure and magnetic properties are sensitive to changes in alloy composition and processing conditions.Thus it is also important to study the effect of each element independent of the influence of other such additions. Hence, controlled samples have to be made and their properties measured and microstructure characterized.

The distribution of the substituted rare earth in the microstructure will depend upon the processing conditions adopted. Figs.1a(i-iv) show schematically a few of the possible concentration profiles for the quaternary element. In this case, a microstructure similar to that observed in the case of sintered Fe-RE-B magnets is examined. The microstructure, shown in Fig.1b, consists of grains of the $RE_2Fe_{14}B$ phase, at the junctions of which exists a RE rich fcc phase which often extends through the two-grain boundaries[19]. Fig 1a(i) is a case where there is no partitioning of the additive to the grain boundary phase or the 2-14-1 phase. In (ii) the additive element partitions preferentially into the 2-14-1 phase while the converse is the case for (iii). In (iv) is shown a concentration profile wherein the additive segregates to the interface

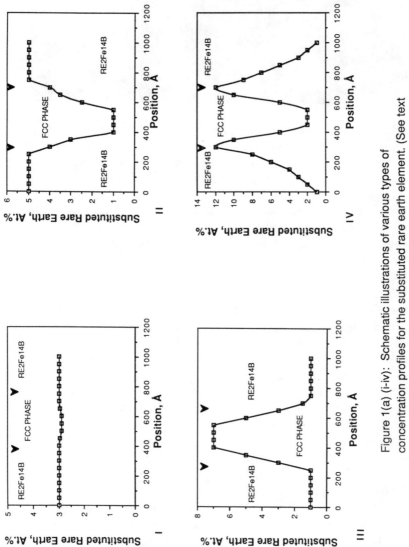

Figure 1(a) (i-iv): Schematic illustrations of various types of concentration profiles for the substituted rare earth element. (See text for description of each plot.)

Figure 1(b) : Transmission electron micrograph of the typical microstructure of the sintered magnet consisting of the matrix $RE_2Fe_{14}B$ phase at the junctions of which a rare earth rich oxygen stabilized phase exists and frequently extends into the two-grain boundaries.

between the two phases. This type of segregation could also occur at grain boundaries. In all the cases, the arrows indicate the interface between the matrix and the grain boundary fcc phase. Among all these hypothetical concentration profiles, (iv) is most interesting, if it is practically achieved in the case of a substitutional element such as Dy. $Dy_2Fe_{14}B$ has a very high anisotropy field (158 kOe) as compared to 71kOe for $Nd_2Fe_{14}B$[15-17]. In such a case, it would be possible to obtain a very localized increase in the magnetocrystalline anisotropy (and hence a high local anisotropy constant), thus increasing the nucleation field required to nucleate a reverse domain wall at the grain boundary or two-phase interface. Such interfaces are generally accepted to be the sites of domain wall nucleation in this class of magnets. Note that this does not appreciably alter the bulk magnetization of the matrix $RE_2Fe_{14}B$ phase, as is the case when the heavy rare earth element is homogeneously distributed in the matrix. Thus, it is clear that by altering the microstructure and/or the composition close to the grain boundaries (or two phase interfaces), considerable changes in the magnetic properties could be potentially brought about.

With this in mind, a systematic investigation of the microstructure and magnetic properties of substituted Fe-Nd-B sintered magnets was initiated. The quaternary addition was systematically varied, keeping the total rare earth content constant. The influence of such changes upon the magnetic properties was investigated with respect to the microstructure and microanalysis of the magnets. Here the results of the Dy addition are reported.

EXPERIMENTAL PROCEDURES:

(1) Alloys : The alloys were prepared by vacuum induction melting and processed into aligned magnets by traditional Powder Metallurgy(P/M) technique[18]. Magnets were sintered in the temperature range of $1000^{\circ}C$ and $1100^{\circ}C$ for three hours and then cooled to room temperature. An isothermal treatment in the temperature range of $500^{\circ}C$ to $700^{\circ}C$ for three hours was then applied to obtain the optimum intrinsic coercivity (iHc).

(2) Magnetic Measurements: All magnets were magnetized in a pulsed magnetic field of 35kOe prior to each measurement. Magnetic properties were measured using a Hysteresigraph at a maximum applied field of 25kOe.

(3) Transmission Electron Microscopy and Analytical Electron Microscopy: Electron transparent thin foils were made by Argon ion -milling discs which had been mechanically thinned down to a thickness of about 50 micrometers. In some cases, the sample was dimpled to reduce the milling time. Argon ion-milling was carried out at 6kV with a gun current of 0.3 mA, corresponding to a specimen current of 20-30microamperes. The milling was completed in about 12-15 hours.

Transmission Electron Microscopy(TEM) imaging was carried out using a Philips 400 TEM/STEM at 100kV. Analytical electron microscopy was carried out using a JEOL 200CX microscope at 200kV. This microscope is fitted with an ultra-thin window Energy Dispersive Xray(EDX) detector as well as a regular Be window EDX detector. The use of the ultra-thin window detector enables the detection and quantification of light elements such as oxygen. Spectral deconvolution and quantification was carried out on the KEVEX 8000 system software using theoretically generated thin foil k-factors. Care was taken to ensure statistical significance of the data by acquiring a minimum of 10^5 counts. Microanalysis line profiles were carried out at different regions in order to check the correctness of the profiles as well as the reliability of the data.

RESULTS AND DISCUSSION:

(1) Magnetic Properties : The addition of increasing amounts of Dy causes the remanence, B_r, to decrease progressively, as is shown in Fig.2. The trend exhibited can be explained as due to the ferrimagnetic coupling

208

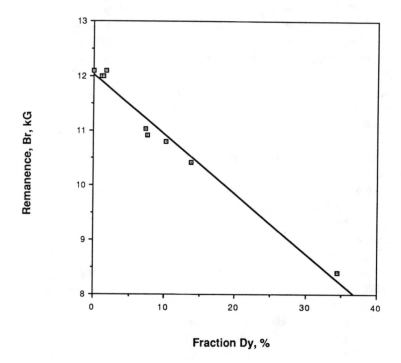

Figure 2 : Effect of Dy content on the Remanence, B$_r$ of Fe-Nd-B sintered magnet. The total rare earth content in all the cases is 35.1 wt.%.

between the Fe spins and the Dy spins. As the Dy content of the alloy is
increased, the Dy content of the $RE_2Fe_{14}B$ phase also increases (see
section on Microstructure and Microanalysis) and hence the remanence
drops. The intrinsic coercivity, iHc, on the other hand, shows a sharp rise
for small additions of Dy. However, above about 10%Dy (i.e., corresponding
to $Nd_{.9}Dy_{.1}Fe_{14}B$)the rate of increase in iHc drops, Fig.3. Since both the
iHc as well as the anisotropy field increase with increasing additions of
Dy, a correlation between iHc and the anisotropy field can be expected. In
an earlier paper, Hirosawa et. al have shown that the iHc increases
linearly with anisotropy field for values of the anisotropy field in the
range 70-80 kOe. In this investigation, a larger range of Dy content has
been examined. In Fig.4 the iHc is plotted against the effective anisotropy
field. The effective anisotropy field is calculated as follows :
Effective Anisotropy Field $(H_A$ eff.$) = X_{Nd}.H_A{}^{Nd} + X_{Dy}.H_A{}^{Dy}$,
where X_{Nd} is the relative atomic fraction of Nd and X_{Dy} is the relative
atomic fraction of Dy ; $H_A{}^{Nd}$ and $H_A{}^{Dy}$ are the anisotropy fields of
$Nd_2Fe_{14}B$ and $Dy_2Fe_{14}B$ respectively. The assumption here is that the
$RE_2Fe_{14}B$ phase is made up of a homogeneous mixture of $Nd_2Fe_{14}B$ and
$Dy_2Fe_{14}B$ with the fractions X_{Nd} and X_{Dy} respectively.
In the initial portion of the curve (i.e., for low Dy contents) the
dependence is linear, as was observed by Hirosawa et. al. However, at
higher concentrations, the slope of the curve drops. This indicates that
for higher Dy concentrations, some changes in the microstructure may be
occurring, which may be limiting the achievement of a higher value of iHc.

MICROSTRUCTURE AND MICROANALYSIS :

   In an earlier paper, the microstructure of Nd-Fe-B magnets has been
discussed in detail[19]. It was shown that the grain boundary phase is fcc
and is stabilized by significant quantities of oxygen. The overall
microstructure (morphology) was the same for all the magnets examined
in this investigation and is described in [19].   Hence, microanalytical
characterization was carried out with two main objectives : (1) to
examine the partitioning of Dy between the grain boundary phase and the
matrix, or segregation of Dy to the grain boundaries or two phase
interfaces ; (2) to examine whether Dy rich second phase particles appear

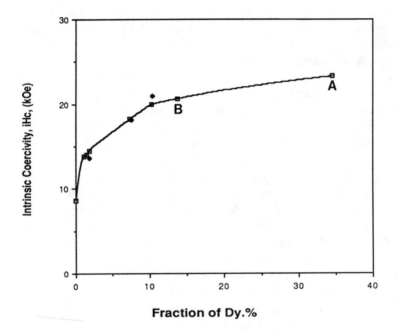

Figure 3 : Effect of Dy content on the Intrinsic Coercivity, iHc, of Fe-Nd-B
sintered magnet. The total rare earth content in all the cases is 35.1 wt.%.

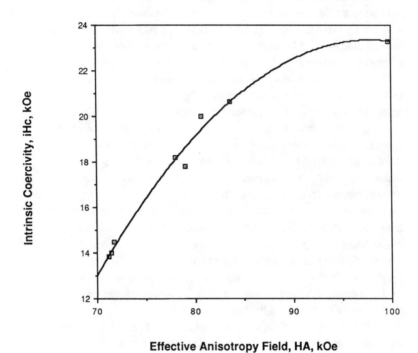

Figure 4 : Dependence of Intrinsic Coercivity upon the effective
Anisotropy Constant, $_{eff}H_A$. The total rare earth content is constant at
35.1 wt.%.

for the highest Dy concentration sample, which shows a marked deviation from linearity. Note that since these measurements are made at room temperature only, a linear relationship between iHc and the anisotropy field can be expected. Hence,xray microanalysis was carried out on the two samples marked A and B in Fig.3, to see whether there was any difference in the microanalytical details at high Dy concentrations.

The distribution of Dy in the microstructure was examined by line profiling, which is shown schematically in Fig.5. In this case, line profiling has been carried out across the triple grain junction. Fig.6(a) shows a typical profile in sample A, which is the high Dy sample. The plot indicates that Dy partitions preferentially to the matrix phase,as shown more clearly in Fig.6(b). As can be seen from Fig.6(b), within the experimental limits of EDXS analysis there is no significant partitioning of Dy to the grain boundary phase, although one might expect Dy, as with the other rare earth elements, to be associated with the oxygen stabilized grain boundary phase[19]. On the contrary, the trends indicate that Dy is preferentially segregated away from the grain boundary phase and the interfaces. This means that the increase in iHc observed is mainly due to the increase in the Dy content of the matrix phase. However, in sample A the Dy concentration of the matrix (3.7-4.0 at.%) was consistently and repeatably found to be less than the nominal Dy content of the alloy, i.e., 5 at.%. This indicated that some of the Dy could exist as inclusions, oxides, etc. Examination of the microstructure of sample A did reveal the presence of such inclusions. One such example is shown in Fig.7(a), while Fig.7(b) shows the corresponding EDX spectrum from this particle. Quantification yielded a composition of 59 at.% Nd and 41at.% Dy. Since this spectrum was obtained with a Be window detector, the possible presence of oxygen is not revealed. However, it is quite possible that it is an oxide phase. The presence of such Dy-rich inclusions justifies the drop in iHc observed for sample A(see Figs. 3 and 4), since it not only reduces the Dy content of the matrix phase (thereby reducing the effective anisotropy field), but also provides potential sites for the nucleation ofreverse domains.   Fig. 8(a) shows the composition profile for alloy B, which is the lower Dy content sample. In this case Dy partitions completely into the matrix and the grain boundary phase does not contain any detectable amount of Dy, as seen from Fig.8(b). Thus if Dy is added at the induction melting stage, as in this case, it partitions almost completely into the matrix (corresponding to case (ii) in Fig.1(a)). Thus a

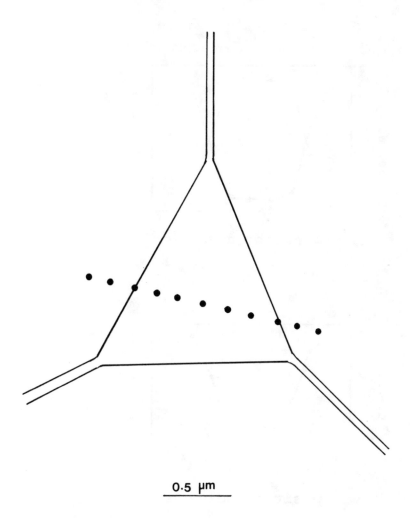

0·5 μm

Figure 5 : Schematic illustration of the line profiling method across the interface of the $RE_2Fe_{14}B$ phase- grain boundary phase . The diameter of the spot corresponds approximately to the probe size.

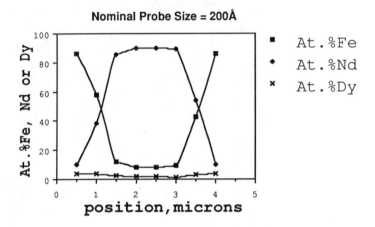

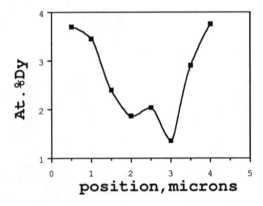

Figure 6(a) : A typical elemental concentration profile in the case of Sample A (high Dy content); (b) An enlarged view of the Dy concentration profile. (The precision in quantification (2 sigma values) is within the width of the symbols).

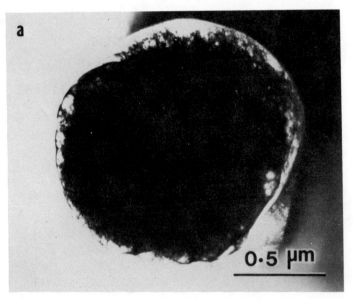

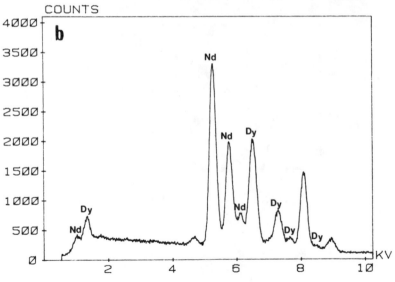

Figure 7(a) : Bright field transmission electron micrograph of an inclusion in sample A ; (b) EDX spectrum from the inclusion in (a) , indicating that the inclusion is Dy-rich.

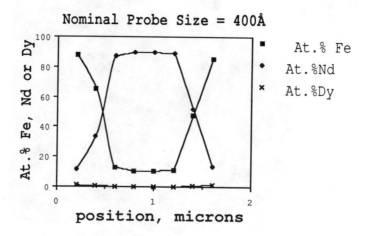

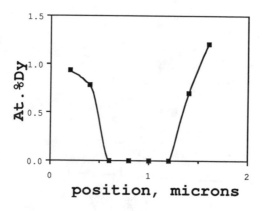

Figure 8(a) : A typical elemental concentration profile for the case of Sample B (low Dy content) ; (b) An enlarged view of the Dy concentration profile. (The precision in quantification (2 sigma values) is within the width of the symbols).

Dy concentration profile, such as that shown in Fig.1(a)(iv), may not be possible to achieve by the processing technique adopted in the preparation of the magnets examined in this study. However, it is possible that such a concentration profile can be achieved by the addition of: (i) pure Dy metal before the sintering stage such that Dy diffuses along grain boundaries and interfaces and into the matrix during sintering; (ii) a compound of Dy such as Dy_2O_3, which could react with the $Nd_2Fe_{14}B$ phase , displacing Nd with the resulting enrichment of Dy in the matrix phase.

The first method has the drawback that Dy has a higher melting point ($1412^{\circ}C$) than the sintering temperatures generally employed. The second method has been adopted by Ghandehari[20] to produce magnets with high iHc (comparable to that of Dy-substituted magnets). However this method also has the drawback that the presence of the oxide causes a decrease in the remanence and possibly provides easy nucleation sites for the nucleation of domain walls.

CONCLUSIONS :

The effect of Dy addition upon the magnetic properties, microstructure and microcomposition of Fe-Nd-B magnets has been investigated. It has been shown that there is a linear decrease in the remanence of the magnet with increasing Dy addition. The intrinsic coercivity, on the other hand, increases sharply for small additions of Dy. Additions of Dy above 10% of the total rare earth content does not produce a concomitant increase in the iHc. The intrinsic coercivity increases linearly with the effective anisotropy field for Dy contents upto about 10% of the total rare earth content. Above this concentration, there is a sharp deviation from linearity. This is attributed to the incomplete ingestion of Dy into the $RE_2Fe_{14}B$ phase leading to the formation of other Dy-rich phases. Various types of possible concentration profiles of the substituted rare earth element have been suggested. In all the cases studied experimentally, it appears from the EDXS analyses (within the limits of experimental error) that Dy partitions preferentially into the matrix phase. No segregation of Dy to the interphase interfaces has been detected by EDXS. It is suggested that in order to preferentially segregate Dy to the interphase interfaces, non-conventional processing techniques such as addition of Dy_2O_3 or low

218

melting Dy-rich alloys be examined. In this way, it may be possible to increase the anisotropy constant locally, near the grain boundaries and two phase interfaces.

ACKNOWLEDGEMENTS :

The authors wish to acknowledge the technical assistance of Mr. C. J.Echer, National Center for Electron Microscopy, Lawrence Berkeley Laboratory, University of California, Berkeley in carrying out the microanalyses. This work was supported by the Director, Office of Basic Energy Research, Office of Basic Energy Sciences, Materials Sciences Division of the US Department of Energy under contract No. DE-AC03-76SF00098.

REFERENCES

1. N.C.Koon and B.N.Das, J. Appl. Phys. 55, 2063(1984).
2. J.F.Herbst, J.J.Croat, F.E.Pinkerton and W.B.Yelon, Phys. Rev.B 29, 4176 (1984).
3. M.Sagawa, S.Fujimura, H.Yamamoto, Y.Matsuura and K.Hiraga, IEEE Trans. on Magnetics, 20, 1584 (1984).
4. J.D.Livingston, in Proc. of the Eighth International Workshop on Rare Earth Permanent Magnets and their Applications, edited by K.J.Strnat (University of Dayton Press, 1985),pp423-443.
5. J.D.Livingston in Proc. of ASM Symposium on Soft and Hard Magnetic Materials, edited by J.A.Salsgiver et.al(Published by ASM, 1986), pp71-79.
6. H.Fujii, W.E.Wallace and E.B.Boltich, J.Mag. and Magn. Matls., 61, 251(1986).
7. Y.C.Yang, W.J.James, X.D.Li, H.Y.Chen and L.G.Xu, IEEE Trans. on Mag.,22, 7579(1986).
8. B.M.Ma and K.S.V.L.Narasimhan, IEEE Trans. on Mag., 22, 916(1986).
9. B.M.Ma and K.S.V.L.Narasimhan, J.MMM, 54-57, 5599(1986).
10. T.Mizoguchi, I.Sakai, H.Niu and K.Inomata, IEEE Trans. on Mag., 22,

919(1986).

11. W.Rodewald, in Proc. of the Eighth International Workshop on Rare Earth Permanent Magnets and their Applications, edited by K.J.Strnat (University of Dayton Press, 1985), pp737-746.

12. L.Wei, J.Long, W.D.Wen, S.T.Diao and Z.Jinghan, J. Less Common Metals, 126, 95(1986).

13. M.Jurczyk and W.E.Wallace, J.MMM,59,L182-L184(1986).

14. S.Hirosawa, K.Tokuhara, Y.Matsuura, H.Yamamoto, S.Fujimura and M.Sagawa, J.MMM., 61, 363(1986).

15. E.B.Boltich, E.Oswald, M.Q.Huang, S.Hirosawa and W.E.Wallace, J. Appl. Phys., 57, 4106(1985).

16. M.Sagawa, S.Fujimura, H.Yamamoto, Y.Matsuura, S.Hirosawa and K.Hiraga in Proc. of Eighth International Workshop on Rare Earth Permanent Magnets and their Applications edited by K.J.Strnat, (University of Dayton Press, 1985), pp587-612.

17. C.Abache and H.Oesterreicher, J. Appl. Phys., 57, 4112(1985).

18. M.Sagawa, S.Fujimura, N.Togawa, H.Yamamoto, Y.Matsuura, J. Appl. Phys., 55, 2083(1984).

19. R. Ramesh, J.K.Chen and G. Thomas, J. Appl. Phys. (in Press).

20. M.H.Ghandehari, Appl. Phys. Lett. 48, 548(1986).

ALUMINUM SUBSTITUTIONS IN ND-FE-B SINTERED MAGNETS

J. K. CHEN* AND G. THOMAS**
*Department of Materials Science and Mineral Engineering, University of
California, Berkeley, CA 94720
**Department of Materials Science and Mineral Engineering, and
the National Center for Electron Microscopy, Materials
and Chemical Sciences Division, Lawrence Berkeley Laboratory, University ot
California, Berkeley, CA 94720.

ABSTRACT

A microstructural and microanalytical study of aluminum substituted
Nd-Fe-B sintered permanent magnets were carried out to determine the
effect, if any, of aluminum on structure, composition, and magnetic
properties, particularly on the observed increase in coercivity. It was found
that Al enrichment occurred in the Nd-rich phase at the grain boundaries. The
possible role(s) of this enrichment on the observed coercivity increase are
discussed.

INTRODUCTION

Ever since its introduction about three years ago, the Nd-Fe-B ternary
system has made a big impact on the permanent magnet industry. [1] Since
then, many alloying substitutions have been carried out in hopes of increasing
this system's magnetic properties. One of the most prominent substitutional
elements for increasing the coercivity is aluminum. [2,3] In an effort to
explain this effect, two Nd-Fe-B permanent magnet alloys were prepared in
identical fashion, except that 2 at.% Al was substituted for B in the second
alloy (see Table 1). Both alloys were also given the same, optimizing
post-sintering heat treatments, after which the magnetic properties were
measured. As clearly shown, the Al-substituted alloy has a 25% increase in
coercivity, with only a 3% decrease in saturation magnetization, relative to
the standard Nd-Fe-B alloy.

This dramatic coercivity increase can be partially attributed to the
increase in anisotropy field (H_A) in Al-substituted Nd-Fe-B magnets, where
Al substitutes for Fe in the $Nd_2Fe_{14}B$ hard magnetic phase. [2] If the
expected coercivity increase is scaled by the increase in anisotropy field

Table I

Processing Conditions and Magnetic Properties of Alloys Studied

Alloy Composition	Post-Sintering Heat Treatment	B Sat [Tesla]	Coercivity [kA/cm]
$Nd_{14}Fe_{78}B_8$	950°C/1 hr, slow cooling to 630°C/1hr, rapid cooling to RT	1.28	6.9
$Nd_{14}Fe_{78}B_6Al_2$	"	1.24	8.6

alone, using values of H_A from Young et. al., it is possible to determine to a first approximation the contribution of the anisotropy field to the coercivity increase. As shown in Fig. 1, these expected coercivities are far below the measured coercivity, leading to the conclusion that Al-induced <u>structural</u> or

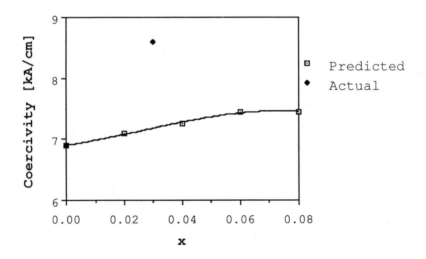

Fig. 1.   Comparison of measured coercivity increase with predicted increase from variations in anisotropy field with x in $Nd_2(Fe_{1-x}Al_x)_{14}B$. H_A values are based on measurements by Yang et. al. [2]

<u>compositional</u> changes are the primary reason for the coercivity increase. In order to investigate such possible changes, the structure of the two alloys in Table 1 were carefully compared. The microstructure of the standard Nd-Fe-B alloy has been determined previously. [4,5] This microstructure consists primarily of two phases, viz, large nearly perfect grains about 10 μm in diameter of the $Nd_2Fe_{14}B$ hard magnetic phase, surrounded by a highly defective fcc Nd-rich phase, which extends from the grain boundaries to the grain junctions. This Nd-rich phase also contains about 20 at.% oxygen, 10 at.% iron, and no boron. Trace amounts of the tetragonal boride phase $NdFe_4B_4$ was also found intragranularly. The Al-substituted alloy was then compared to this "standard" alloy, from both microstructural and microanalytical viewpoints.

EXPERIMENTAL

The magnets were prepared from powdered alloys which were magnetically aligned , pressed, then sintered at 1070° C for one hour. After sintering, they were cooled rapidly and annealed as shown in Table 1. The magnetic properties were measured at room temperature with a Foner sample magnetometer. X-ray diffraction with Cu K_α radiation was carried out on bulk magnets, which were mechanically polished with sub-micron diamond paper.

Transmission electron microscope (TEM) specimens were prepared by mechanically thinning a 250 μm disk to thicknesses of around 100 μm, followed by dimpling to around 20μm. They were then argon ion-milled to electron transparency. Precautions were taken to reduce and identify specimen preparation artefacts as discussed in an earlier paper. [5] The specimens were examined in a Philips 400 TEM/STEM fitted with a Be window energy dispersive (EDS) detector operating at 100 kV, and a JEOL 200CX TEM/STEM fitted with an ultra-thin window EDS detector operating at 200 kV.

In order to detect the low aluminum concentrations, each EDS spectra contained a minimum of 100,000 counts, and over 50 spectra were taken to confirm the observed trends. Quantification was carried out using the system k-factors, which were checked against the standard $Nd_2Fe_{14}B$ matrix phase.

RESULTS/DISCUSSION

As shown in Fig. 2, there is essentially no difference between the x-ray diffractograms of the standard Nd-FE-B and the Al-substituted Nd-Fe-B

224

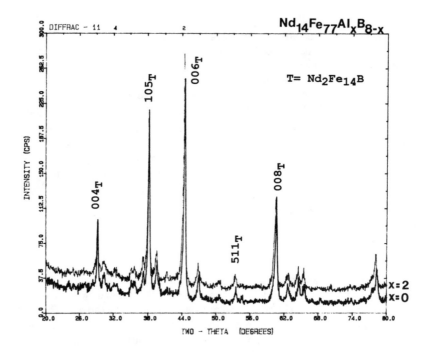

Fig. 2.   X-ray diffractograms from both the standard Nd-Fe-B and
Al-substituted Nd-Fe-B sintered magnets.

alloys. The similarity of these diffractograms has three important
implications:

a) No precipitation of a new phase has been induced by the
   Al-substitution.
b) The hard magnetic phase 2:14:1 is essentially the same in both
   alloys, with no detectable changes in lattice parameter.
c) The similarity of the minor peaks which may be indexed as arising
   from the Nd-rich and $NdFe_4B_4$ phases indicate that both phases are
   still present, and are not altered by the Al-substitution.

As shown in Fig. 3, the general microstructure of the Al-substituted
magnet is also essentially the same: namely, large nearly perfect grains
about 10 μm in diameter of the tetragonal 2:14:1 hard magnetic phase,
surrounded by an fcc grain boundary phase that extends into the grain

Fig. 3.   General microstructure of the Al-substituted Nd-Fe-B sintered magnet.

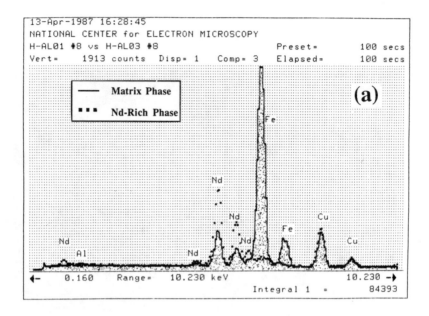

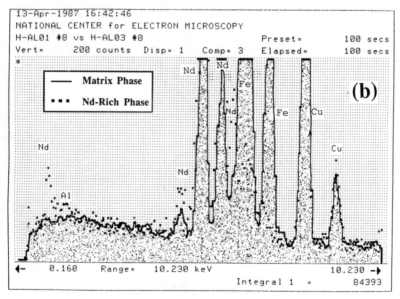

Fig. 4. (a) Superimposed EDS spectra from both the Nd-rich and matrix phases in the Al-substituted Nd-Fe-B sintered magnet. (b) Same as (a), except for a vertical magnification.

junctions. To compare the composition of the grain boundary and matrix phase in the Al-substituted alloys, typical EDS spectra from the two regions are superimposed as shown in Fig. 4. The usual Nd-enrichment of the grain boundary phase is also observed in these alloys, along with an Al enrichment, as shown in Fig. 4(b). If a line analysis is done across a grain junction region by taking point analyses along a line with a 200 Å probe, as schematically illustrated in Fig. 5, a composition profile across the Nd-rich phase is obtained as shown in Fig. 6. This profile clearly illustrates the Al-enrichment at the grain boundary phase.

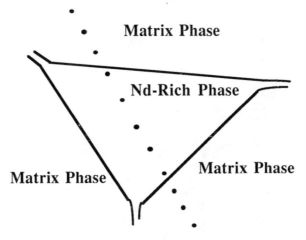

Fig. 5.   Schematic illustration of an EDS concentration profile across a Nd-rich phase.

Some comments on the EDS data need to be made here. First, when trying to detect an element that is less that 1 wt.% of the nominal composition, great care must be exercised in carrying out the experiment. A large number of counts must be collected so that the signal peak can be clearly distinguished from the background. This requires high count rates, long counting times, and an extremely low system background. Secondly, for highly absorbed x-rays such as Al, quantification must be carefully done. Factors such as the relative absorption of the different phases due to differences in composition and thickness must be accounted for. Finally, the detected Al concentration was fairly inhomogenous throughout each specimen in both the Nd-rich and matrix phases. Consequently, many specimens and many regions within each specimen were analyzed before the observed Al enrichment in the Nd-rich phase was acknowledged.

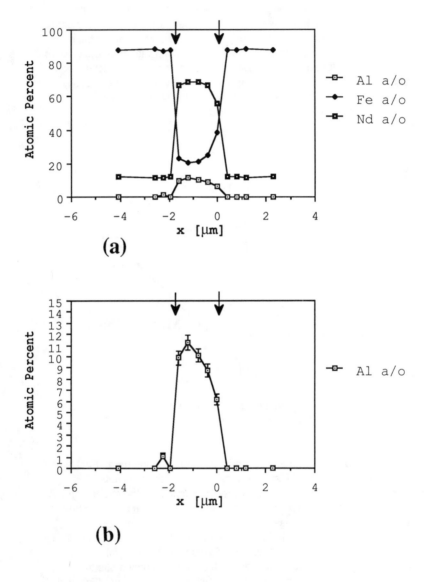

Fig. 6. (a) Concentration profile across the Nd-rich phase of the
Al-substituted Nd-Fe-B sintered magnet. Arrows mark the
the interface.
(b) Same as (a), except for a vertical magnification.

Given that the primary microstructural change induced by the Al-substitution is the enrichment of Al in the Nd-rich phase, the next logical consideration is how the Al enrichment increases the coercivity. Burzo et. al. have considered the possibility that Al increases the fluidity of the Nd-rich phase during sintering, leading to greater magnetic isolation, and consequently higher coercivities through more effective coating of the matrix grains. The resulting thinner Nd-rich phase would also be more effective at pinning the domain walls. [3] Alternatively, Ramesh et. al. have proposed that oxygen de-adsorption at the matrix/grain boundary interface during post-sintering heat treatments is important for high coercivity. This de-adsorption process may be greatly facilitated by Al enrichment in the Nd-rich phase. [6] Although both suggestions are plausible, much more work needs to be done before th role of Al can be completely understood.

CONCLUSION

In conclusion, the Al-induced compositional changes at the grain boundaries are believed to be the primary reason for the coercivity increase. No major changes were observed in either the crystal structure/lattice parameter of the hard magnetic 2:14:1 phase, nor the general microstructure of the magnet.

ACKNOWLEDGEMENTS

The authors would like to thank Dr. Rodewald of Vacuumschmelze for generously providing the samples. We would also like to thank R. Ramesh and Dr. K. Krishna of the National Center for Electron Microscopy at the Lawrence Berkeley Laboratory for many helpful discussions. This work was supported by the Director, Office of Basic Energy Research, Office of Basic Energy Sciences, Materials and Chemical Sciences Division of the US Department of Energy under contract No. DE-AC03-76SF00098.

REFERENCES

1. M. Sagawa, S. Fujimura, N. Togawa, H. Yamamoto, and Y. Matsuura, J. Appl. Phys. 55, 2083 (1984).
2. Y.C. Yang, W.J. James, X. Li, H. Chen, and L. Xu, IEEE Trans. on Mag. 22, 757 (1986).
3. E. Burzo, A.T. Pedziwiatr, and W.E. Wallace, Sol. State Comm. 61, 57 (1987).
4. R. Mishra, J.K. Chen, and G. Thomas, J. Appl. Phys. 59, 2244 (1986).
5. R. Ramesh, J.K. Chen, and G. Thomas, presented at the 1986 Intermag Conference, Phoenix, AZ. (in press)
6. R. Ramesh and W. Soffa (private communication).

EFFECT OF HYDROGEN ON MAGNETIC PROPERTIES
OF QUATERNARY $Pr_2Fe_{14-x}Co_xB$ COMPOUNDS

F. POURARIAN, S. Y. JIANG AND W. E. WALLACE
MEMS Department and Magnetic Technology Center, Carnegie-Mellon University,
and Mellon Institute, Pittsburgh, PA 15213

The effect of absorbed hydrogen on the structural and magnetic behavior of $Pr_2Fe_{14-x}Co_xB$ has been investigated. Introduction of hydrogen in the host material reduces the average magnetic moment of the 3d element by $\sim$ 4 to 7%. Anisotropy fields for all systems studied are significantly reduced upon hydrogenation. Evidence of spin reorientation at temperatures ranging between $\sim$ 295 K and 496 K for some of the hydrides is observed. $Pr_2Co_{14}B$ and its hydride exhibit uniaxial anisotropy at room temperature and at 77 K.
The role of hydrogen in expanding the lattices of these intermetallics and modifying their magnetic properties is discussed.

INTRODUCTION

Many intermetallic compounds containing rare earths (R = La to Lu) and 3d transition metals (T = Mn,Fe,Co and Ni) absorb large quantities of hydrogen rapidly and reversibly [1]. For example, the compounds with a $CaCu_5$-type structure such as $LaNi_5$ (which is commonly utilized as a hydrogen storage material), the Laves-phase compounds of the formula RT_2 (T = Fe,Co) and the cubic compounds of R_6T_{23} (T = Mn,Fe) are well-known hydrogen absorbers [2-5]. Recently, we have observed that most of the new class of permanent magnet materials belonging to the $R_2Fe_{14}B$ system absorb hydrogen readily and form stable hydrides at room temperature [6]. These observations have been cor-roborated by independent work from other researchers [7].
The changes in magnetic properties of these compounds which occur due to hydrogen absorption are of great fundamental interest. The hydrogen expands the lattice as well as modifies the electron density of the conduc-tion band of the alloy system, thereby modifying the magnitude of the 3d moments, the various exchange interactions and also the magnetocrystalline anisotropy. For instance, it has been observed that, on hydrogen absorption, Th_7Fe_3 transforms from a superconductor to a magnetically-ordered compound [8]; Y_6Mn_{23} from a ferrimagnet to an antiferromagnet [9] and $Y_2Fe_{14}B$ and $Pr_2Fe_{14}B$ from uniaxial to basal plane anisotropy at 77 and 295 K [6]. It also appears that in R-Co or R-Co-B compounds, hydrogen absorption reduces the magnetic moment on cobalt [10,11], while in R-Fe or R-Fe-B compounds either an increase or no change in the iron moment is observed [3,5,12].
In $R_2Fe_{14}B$ structure, the iron atoms occupy six different sites, viz., $16k_1$, $16k_2$, $8j_1$, $8j_2$, 4e and 4c. The rare earth atoms occupy two crystal-lographically inequivalent sites. Neutron diffraction investigation on $Nd_2Fe_{14}B$ material with Fe progressively substituted by Co, revealed that part of the Fe atoms occupy preferentially the $8j_2$ sites while Co atoms show a partial preference for occupying 4e sites [13]. Recent neutron diffraction experiments performed on deuterated $Y_2Fe_{14}BD_{3.5}$ [13] inferred the following sequence of hydrogen filling: first, hydrogen fills the interstitial tetra-hedral 4e sites and then the $8j_1$ or $8j_2$ and $16k_2$ sites are simultaneously and progressively occupied up to H $\sim$ 2.5 g atom H/mole. These tetrahedral sites have R_2-T_2, R_3-T and $R-T_3$ configurations, respectively (where R = rare earth and T = transition metal).
Two main points of interest are crucial for hydrogen absorption to be technologically useful in $R_2Fe_{14}B$ systems: (1) in powder decrepitations (fine powder formation) for magnet fabrication and (2) in raising the mag-netic ordering temperature of $R_2Fe_{14}B$ systems. Another interesting aspect

of the present investigation was to ascertain whether or not the iron moment
in cobalt-rich compositions of $Pr_2(Fe,Co)_{14}B$ hydrides behave the same way as
in the $Pr_2Fe_{14}B$ hydride. The present work was undertaken to examine the
influence of hydrogen on the crystal structure, saturation magnetization,
anisotropy field and the spin-reorientation temperature of $Pr_2Fe_{14-x}Co_xB$
systems.

EXPERIMENTAL DETAIL

Sample Preparation

The $Pr_2Fe_{14-x}Co_xB$ systems (x = 0,8,10,11,12 and 14) were prepared by
melting the constituents Pr (99.95%), Fe and Co (99.99%) and B (99.98%) in
a water-cooled copper boat using rf induction heating. A continuous flow of
titanium-gettered argon was maintained during melting. As-cast samples were
wrapped in a tantalum foil, sealed in quartz tubes, filled with $\sim$ 1/3 atmosphere of argon gas, annealed at $\sim$ 900°C for 14 days. X-ray diffraction
analysis was performed at room temperature on randomly oriented powders with
the use of Rigaku diffractometer and Cr-Kα radiation. All samples investigated were found to be single phase.

Hydrogenation

The hydride samples were obtained by exposing the host materials at room
temperature to approximately 60 atm of pure hydrogen. They were then heated
to 250°C for fast hydrogenation process. After induction of 2 to 3 hours,
the samples absorbed hydrogen and were converted to a fine powder ($\sim$ 20 to
50 µ). A portion of the sample was used for the lattice constant measurements. For this purpose the sample was hydrogenated to saturation and then
quenched in liquid nitrogen. The hydrogen remaining in the vapor state was
removed and the sample was then poisoned with sulfur dioxide gas. The compounds $Pr_2Fe_{14-x}Co_xB$ were hydrogenated to concentrations varying between 3
and 5 g.H per formula unit. X-ray diffraction patterns for these hydrides
showed that they were single phase.

X-ray powder diffraction was also employed to determine the easy direction of magnetization for the hydride powders. The intensity ratios of two
reflections, $I_{hko}/I_{oo\ell}$, were compared for magnetically aligned and unaligned
powders. An increase in the ratio for a sample with field oriented parallel
to the slide was an indication of an easy c-axis, whereas a decrease indicated
that the magnetization vector resides in the basal plane.

Thermomagnetic analyses were performed with a Faraday balance by recording magnetization vs. temperature curves at low external fields ($\sim$ 0.5 to
3 kOe) in the temperature range 4.2-1100 K with samples in the form of rough
chunks. The spin-reorientation temperatures, T_{SR}, were determined from the
M vs T curves. Curie temperatures for the hydrides under investigation were
not determined because of hydrogen loss from the samples at the elevated
temperatures involved. Magnetization measurements were performed at room
temperature and at 77 K with a PAR vibrating sample magnetometer in external
fields up to 20 kOe. From these measurements, saturation moments were calculated using Honda plots. The anisotropy fields were obtain at 295 K and
at 77 K on powders (diameter < 37 µ) aligned in wax by measuring M vs H in
the easy and hard directions and extrapolating the latter to intersect the
former.

RESULTS AND DISCUSSION

It was observed that the $Pr_2Fe_{14-x}Co_xB$ systems, where x = 0,8,10,11,12 and 14, absorb typically 3.5 to 5 hydrogen atoms per formula unit at room temperature and $\sim$ 10 atm pressure. X-ray diffraction studies show that the hydrides retain the tetragonal $Nd_2Fe_{14}B$-type crystal structure. The lattice parameters a, c and unit cell volume of the host material decrease monotonically with the increasing cobalt content [14,15]. The same trend is observed for the hydrides only in the range x = 8 to 14. For lower Co concentration (iron-rich compounds) and at x = 0, there is a larger unit cell volume expansion ($\sim$ 6.2%) upon hydrogenation compared to the values determined for x $\geq$ 8. This effect may partly be a consequence of a larger number of hydrogen atoms observed for $Pr_2Fe_{14}B$, occupying more abundant interstitial sites relative to the cobalt-rich compounds. As cited above, the main interstitial sites filled by hydrogen are 4e, 8j or 16k, as was inferred from the results of neutron diffraction studies on $Y_2Fe_{14}BD_{3.5}$ system [13]. Results of the lattice dimensions are given in Table 1 and shown in Fig. 1. The changes in the c/a ratio of the hydrides are found to be between 1.375 to 1.384. The same trend is observed for the host $Pr_2Fe_{14-x}Co_xB$ compounds [14,15], which may indicate a change in the preferential site occupation of the 3d metal could arise along the series. This effect may have important consequences in changing the observed magnetic anisotropy of the host [15] and also the hydrides.

The magnetic ordering temperatures of the host $Pr_2Fe_{14-x}Co_xB$ compounds were observed to be between 556 K and 986 K [14]. It is believed that 3d-3d exchange interactions in these systems are the strongest and they mainly determine the Curie temperature, T_c, [16]. The interaction between the 3d atoms located at different crystallographic sites with a distance less than 2.45 Å

TABLE I

Lattice Parameter Data of $Pr_2Fe_{14-x}Co_xB$ (x = 0.0,8,10,11,12 and 14) Systems and Their Hydrides

Compound	a (Å)	c (Å)	V_c (Å)^3	$\Delta V/V$ %
$Pr_2Fe_{14}B$	8.814	12.253	951.9	
$Pr_2Fe_{14}BH_5$	9.006	12.464	1010.9	6.2
$Pr_2Fe_6Co_8B$	8.717	12.084	918.2	
$Pr_2Fe_6Co_8BH_{3.6}$	8.765	12.129	931.8	1.5
$Pr_2Fe_4Co_{10}B$	8.749	12.025	909.5	
$Pr_2Fe_4Co_{10}BH_{3.7}$	8.749	12.098	926.0	1.8
$Pr_2Fe_3Co_{11}B$	8.682	11.997	904.3	
$Pr_2Fe_3Co_{11}BH_{3.4}$	8.731	12.050	918.6	1.6
$Pr_2Fe_2Co_{12}B$	8.666	11.960	898.2	
$Pr_2Fe_2Co_{12}BH_{3.4}$	8.724	12.026	915.3	1.9
$Pr_2Co_{14}B$	8.635	11.890	886.5	
$Pr_2Co_{14}B$	8.701	11.966	905.9	2.2

234

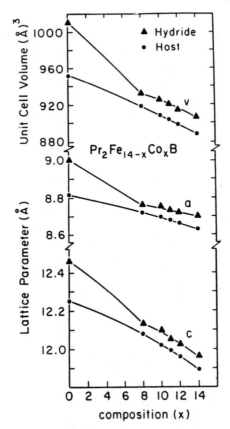

Fig. 1. The variation of the lattice
parameters a and c and the unit
cell volume of the $Pr_2Fe_{14-x}Co_xB$
system, compared to the host
compound.

are determined to be negative and are positive for larger distances [14].
The effect of hydrogen on the magnetic ordering of some $R_2Fe_{14}B$ (or iron-rich
compounds) systems studied in our laboratory [6] revealed an increase of the
T_c by $\sim 50°C$. This increase in T_c is perhaps related to the expansion of the
lattice dimensions and may be attributed to the decrease of the negative exchange interactions. As noted above for Co-rich or $Pr_2Co_{14}B$ hydrides, T_c were
not obtained due to the high temperatures involved. Perhaps a small or no
change in Curie temperature for the latter systems is expected from the nature
of the variation of the magnetization vs temperature curve in the temperature
range investigated.

    Field dependence of magnetization of the $Pr_2Fe_{14-x}Co_xB$ hydrides was mea-
sured for magnetically aligned powders at 295 and 77 K. Representative
results for $Pr_2Co_{14}B$ and its hydride are displayed in Fig. 2(a,b). The data
for the host material were previously obtained in this laboratory and are
included here for comparison. All hydrides studied showed saturation of
magnetization, M_s, along the easy axis beyond ~ 10 kOe. The composition

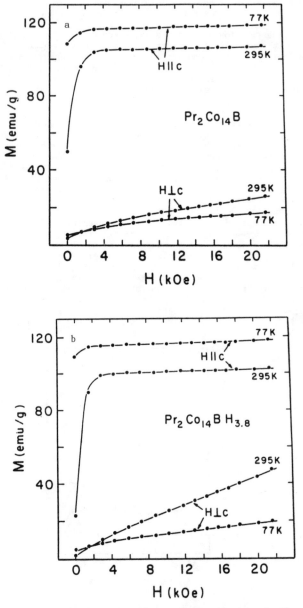

Fig. 2(a,b) Field dependence of magnetization in the
direction parallel and perpendicular to the
external magnetic field for $Pr_2Co_{14}B$ and
the hydride at 77 K and 295 K.

dependence of M_s at 295 K and 77 K is plotted in Fig. 3. Results determined from extrapolation of M vs 1/H to zero and those obtained at 20 kOe are given in Table II.

Partial substitution of Fe by Co increases M_s for $x \simeq 1$ and then decreases it monotonically for larger substitution of Co atoms. In the case of the hydrides, the same trend was observed in the range $x \geq 8$ at 295 K. Hydrogenation lowers the M_s value for the Co-rich compounds but enhances M_s for the compounds with composition $x < 6$. The easy direction of magnetization (D.O.M.) is also sensitive to hydrogenation. The D.O.M. changes from easy c-axis to a basal plane at room temperature (see Table II) for all the samples investigated, except for $Pr_2Co_{14}B$.

Magnetic structure studies observed by powder neutron diffraction of $Pr_2Fe_{14}B$ and pseudoternary alloys of $Nd_2(Fe,Co)_{14}B$ [17,18] revealed a ferromagnetic ordering between the rare earth and the 3d magnetic moment. From the results of the bulk magnetization it appears that the coupling remains the same for the hydrides. The room temperature average Pr moment is determined to be 1.9 μ_B [17]. Assuming a constant Pr moment along the $Pr_2Fe_{14-x}Co_xB$ series and also a collinear ferromagnetic coupling between Pr and the 3d moment, the average 3d transition moments were determined for all the compositions for the host and the hydrides studied. As shown in Fig. 4, upon hydrogen absorption the 3d moment is decreased by $\sim 6\%$ in the range $x \geq 8$, while it increased by the same percentage for $Pr_2Fe_{14}B$ compounds. The same trend of the Fe or Co moment modification was also observed previously in several binary RFe_2 or pseudobinary $RFe_{2-x}Co_x$ hydride systems [10]. This behavior in the hydrides may arise due to the weakening of the R-3d exchange interactions caused by the increase of R-3d separation.

It is believed that the anisotropy of $R_2Fe_{14}B$ arises from both contributions of 4f and 3d sublattices [19]. The Fe magnetization vectors align along the c-axis. For $Pr_2Co_{14}B$ (cobalt-rich) the Co moment favors a planar anisotropy [20,21]. The composition dependence of the anisotropy field, H_A,

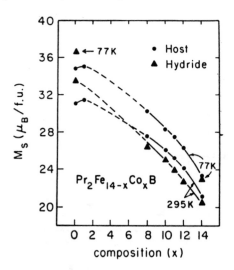

Fig. 3. Dependence of saturation magnetization (M extrapolated to 1/H = 0) on Co content of $Pr_2Fe_{14-x}Co_xB$ hydride, compared to the host compound at 77 K and 295 K.

TABLE II

Magnetic Moment at Saturation M_S and 20 kOe, Direction of Magnetization,
Anisotropy Field H_A and Spin-Reorientation Temperature T_{SR}
of $Pr_2Fe_{14-x}Co_xB$ and Their Hydrides

Compound	$M_S(\mu_B/f.u.)$		$M(\mu_B/f.u.)$ at 20 kOe		D.O.M.		$H_A(kOe)$		T_{SR} (K)
	295 K	77K	295 K	77K	295K	77 K	295K	77K	
$Pr_2Fe_{14}B$	31.0	34.8	30.6	34.4	axis	axis	79	172	–
$Pr_2Fe_{14}BH_5$	33.9	36.6	33.3	36.4	plane	plane	–	–	–
$Pr_2Fe_6Co_8B$	27.5	30.2	27.2	29.7	axis	axis	68	170	–
$Pr_2Fe_6Co_8BH_{3.6}$	26.4	–	25.9	26.8	plane	plane	–	–	–
$Pr_2Fe_4Co_{10}B$	26.0	28.3	25.3	27.7	axis	axis	67	187	760
$Pr_2Fe_4Co_{10}BH_{3.7}$	25.0	–	24.8	23.5	plane	plane	–	–	< 77
$Pr_2Fe_3Co_{11}B$	25.2	27.5	24.5	26.8	axis	axis	72	200	684
$Pr_2Fe_3Co_{11}BH_{3.4}$	23.8	–	23.1	19.9	plane	plane	–	–	< 77
$Pr_2Fe_2Co_{12}B$	24.1	26.3	23.4	25.9	axis	axis	84	232	660
$Pr_2Fe_2Co_{12}BH_{3.4}$	22.6	–	22.0	10.8	plane	axis	–	53	<295
$Pr_2Co_{14}B$	21.0	23.3	20.8	23.0	axis	axis	140	350	668
$Pr_2Co_{14}BH_{3.7}$	20.5	22.9	20.1	21.8	axis	axis	55	217	496

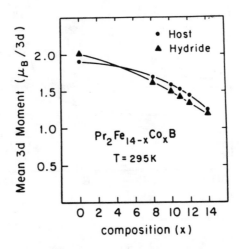

Fig. 4.   Dependence of mean 3d
moment on x in $Pr_2Fe_{14-x}Co_xB$
hydride, compared to the host
compound at 295 K.

observed earlier [14,15] for host compounds showed an anomalous behavior at a composition ratio Co/Fe ≃ 0.67, beyond which H_A values increased drastically. This behavior is suggested to be attributed to the sharp changes in the c/a ratio accompanied by a strong increase in the Pr sublattice anisotropy [15]. Results of the anisotropy field for some of the $Pr_2Fe_{14-x}Co_xB$ hydrides studied at 77 K and 295 K are given in Table II. It can be seen that hydrogenation has a marked influence in modifying the easy direction of magnetization of the host compounds. For the systems with composition x = 0,8,10 and 11, the anisotropy changes from uniaxial to planar, whereas for x = 12 the compound remains axial down to 77 K with an anisotropy field H_A of 53 kOe. For $Pr_2Co_{14}B$ the H_A values reduced two-fold and 1.5-fold at 295 K and 77 K, respectively, upon hydrogenation. It is thus clear that both the rare earth and 3d sublattice anisotropies are strongly influenced by the hydriding process [6,22].

In $Pr_2Fe_{14-x}Co_xB$ host compounds, it was observed that for Co/Fe ratio > 0.67 a spin-reorientation T_{SR} occurs [14]. This is believed to be due to a transition between uniaxial and basal anisotropy. The effect of hydrogenation has eliminated T_{SR} for all compositions studied, except for x = 12 and 14. T_{SR} has been reduced by 172 K for the latter composition. For x = 12, T_{SR} is expected to be between 77 and 295 K. This was evidenced from the axial anisotropy field observed at 77 K. The observed T_{SR} for $Pr_2Co_{14}B$ and its hydride, from the measurements of magnetization versus temperature above 295 K, is displayed in Fig. 5.

The behavior of lowering the T_{SR} upon hydrogen absorption was also observed for the $Nd_2Fe_{14}B$ system [6]. In the host compounds it is believed that a reduction in Pr–3d exchange interaction increases the importance of the Pr crystal field interaction and, as a result of this, T_{SR} occurs. Therefore, the effect of decrease of the T_{SR} by hydriding is a consequence of excessive weakening of R–3d exchange.

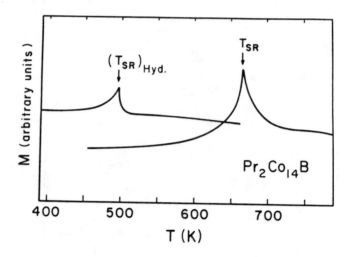

Fig. 5.   Temperature dependence of magnetization near the T_{SR} region, at a constant applied field for $Pr_2Co_{14}B$ and its hydride.

CONCLUSIONS

Hydrogen absorption of $Pr_2Fe_{14-x}Co_xB$ intermetallics results in the modification of magnetic properties and strong reduction of the spin-reorientation temperature. It also resulted in a reduction of the aniostropy field (at 298 and 77 K). The spatial site occupations of the interstitial hydrogen in the lattice may have a significant effect in modifying the above properties. The saturation magnetization of the compositions in the range $x \geq 8$ reduces significantly upon hydrogenation and may be partly attributed to the weakening of the R-3d exchange interaction. It is postulated that such a behavior may arise as a result of the fanning of the praseodymium sublattice moment, analogous to the behavior observed in RFe_2H_x compositions [23].

ACKNOWLEDGEMENTS

Work was supported by the Lawrence Livermore National Laboratory and the National Science Foundation.

REFERENCES

1. W. E. Wallace, R. S. Craig and V. U. S. Rao, Advances in Chem. Series, No. 186, eds. S. L. Holt, J. B. Milstein and M. Robbins (Am. Chem. Soc., New York, 1980), p. 207.
2. J. H. N. Van Vucht, F. A. Kuijpers and H. C. A. M. Bruning, Philips Res. Rep. 25, 133 (1970).
3. F. Pourarian, W. E. Wallace and S. K. Malik, J. Mag. Mag. Mat. 25, 299 (1982).
4. F. Pourarian, E. B. Boltich, W. E. Wallace, R. S. Craig and S. K. Malik, ibid., 21, 128 (1980).
5. E. B. Boltich, F. Pourarian, W. E. Wallace, H. K. Smith and S. K. Malik, Sol. State Commun. 40, 117 (1981).
6. F. Pourarian, M. Q. Huang and W. E. Wallace, J. Less-Common Met. 120, 63 (1986).
7. P. Dalmas de Teotier, D. Fruchart, L. Pontonnier, F. Vaillant, P. Wolfers, A. Yaovanc, J. M. D. Coey, R. Fruchart and PH. L'Heritier, ibid, 129 (1986).
8. S. K. Malik, W. E. Wallace and T. Takeshita, Sol. State Commun. 23, 599 (1977).
9. K. Hardman, J. J. Rhyne, H. K. Smith and W. E. Wallace, J. Appl. Phys. 52, 2070 (1981).
10. F. Pourarian, W. E. Wallace and S. K. Malik, J. Less-Common Met. 83, 95 (1982).
11. L. Y. Zhang, F. Pourarian and W. E. Wallace, to be published.
12. L. P. Ferreira, R. Guillen, P. Vulleit, A. Yaonanc, D. Fruchart, P. Wolfers, P. L'Heritier and R. Fruchart, J. Mag. Mag. Mat. 53, 145 (1985).
13. D. Fruchart, P. Wolfers, P. Vulliet, A. Yaoanc, R. Fruchart and P. l'Heritier, in Nd-Fe Permanent Magnets, Their Present and Future Applications. I. V. Mitchell (editor), Elsevier, Amsterdam, 1985, p. 173.
14. A. Pedziwiatr, S. Y. Jiang and W. E. Wallace, J. Mag. Mag. Mat. 62, 29 (1986).
15. F. Bolzoni, J. M. D. Coey, J. Gavigan, D. Givord, O. Moze, L. Parati and T. Viadieu, ibid., 67, 123 (1987).
16. S. Y. Jiang, F. Pourarian, E. B. Boltich and W. E. Wallace, presented at Intermag Conf., Japan, 1986.
17. J. F. Herbst and W. B. Yelon, J. Appl. Phys. 57, 2343 (1985).
18. J. F. Herbst and W. B. Yelon, ibid., 60, 4224 (1986).

19. E. B. Boltich and W. E. Wallace, Sol. State Commun. 55, 529 (1985).
20. K. H. J. Buschow, D. B. de Mooij, S. Sinama, R. J. Radwanski and
    J. J. M. Franse, J. Mag. Mag. Mat. 51, 211 (1985).
21. F. Pourarian, S. G. Sankar, A. Pedziwiatr, E. B. Boltich and W. E.
    Wallace, Proc. of this Symposium (1987).
22. J. M. Friedt, A. Vasquez, J. P. Sanchez, P. l'Heritier and R. Fruchart,
    J. Phys. F. 16, 651 (1986).
23. J. J. Rhyne, G. E. Fish, S. G. Sankar and W. E. Wallace, J. de
    Physique, Coll. C5, Suppl. No. 5 40, C5-209 (1979).

NEUTRON DIFFRACTION STUDIES OF MAGNETIC MATERIALS

WILLIAM J. JAMES
Department of Chemistry and Graduate Center for Materials Research, University of Missouri-Rolla, Rolla, Missouri 65401

ABSTRACT

The ability of neutron diffraction in determining the nature and extent of magnetic ordering is illustrated for the intermetallic compounds, $Y_6(Fe,Mn)_{23}$ and $ErFe_3$. Substitution with other 3d transition metals influences the Fe-Fe exchange forces such as to alter, sometimes considerably, the magnetic properties, e.g., local site magnetic anisotropies in $Er(Fe,Ni)_3$ and thermal expansion anomalies in the $R_2(Fe,Co)_{14}B$ compounds. When the 3d atoms are near neighbors in the periodic chart, their nuclear scattering lengths for neutrons are sufficiently different to permit the detection of preferential occupation of the several nonequivalent crystallographic 3d metal sites, i.e., atomic ordering, present in the R_6M_{23} and $R_2Fe_{14}B$ structures.

INTRODUCTION

Determination of magnetic structures is not an easy task. Many aspects of the magnetic structure can only be inferred or guessed from conventional magnetization studies. A reliable picture of the complete structure is often elucidated only by combining these studies with data from neutron diffraction experiments. Although neutron diffraction provides a direct probe of magnetic structure and changes therein, it may at times be inadequate as ambiguities can arise in interpretation of the data, particularly for polycrystalline samples. Usually magnetization data, Mossbauer data, or sometimes specific heat data can remove these ambiguities.

It is best if single crystals can be obtained for neutron diffraction experiments as these can allow one to determine absolute directions of magnetic moments. However, for many materials, it is impossible to obtain sufficiently large single crystals for neutron work. Typically, crystal dimensions need to be around 0.3 to 0.5 cm. Thus, for these materials, one must be content to obtain data on a polycrystalline sample where the moment directions can be expressed only in terms of degrees from a unique axis.

There now exists a refinement method which is quite powerful in extracting the maximum amount of information from neutron diffraction patterns. This method was developed by H. M. Rietveld [1] in the late 60's and is designed to allow analysis of entire powder diffraction patterns since it accounts for reflection overlap and background. The older, integrated, intensity method does not do this. The Rietveld approach has made powder diffraction studies nearly as valuable as single crystal work in structure determination.

The nuclear component of scattered neutrons can provide useful information. Elements having nearly equal atomic numbers are practically indistinguishable using x-ray diffraction techniques, but can have quite different scattering lengths for neutrons. Because of this, neutrons are superior to x-rays when studying intermetallic alloys containing metals with nearly equal atomic numbers, e.g., Fe and Co. Also, light atoms are nearly invisible in the presence of heavy metals when probing with x-rays, but can have neutron scattering lengths that compare in magnitude with those of heavy metals. Thus, neutron diffraction is essential for locating hydrogen atoms in metallic and intermetallic hydrides. In practice, deuterium is used instead of hydrogen because hydrogen has a high incoherent neutron scattering cross-section.

For the purpose of illustrating the power of neutron diffraction techniques in probing magnetic structure, I have selected six intermetallic alloy systems, with which I am very familiar and which nicely illustrate the several aspects of application of neutron diffraction referred to above.

The first, Y_6Mn_{23}, illustrates the ability of neutron diffraction to unravel the magnetic structure of a complex compound containing several nonequivalent crystallographic sites. The second system, $Y_6(Fe_{1-x}Mn_x)_{23}$, was chosen because the site preference of the 3d-metal atoms plays a dominant role in its unusual magnetic properties. Furthermore, whereas the scattering factors of Fe and Mn for x-rays are nearly the same, the nuclear scattering lengths of Fe and Mn are considerably different; 0.95×10^{-12} cm for Fe vs. -0.37×10^{-12} cm for Mn.

The third system is $Y_6Mn_{23}D_{23}$, chosen to illustrate the importance of locating the hydrogen (deuterium) atoms as they not only play a major role in determining the stability of these hydrides, but also influence their electronic and magnetic properties.

The fourth and fifth choices are $ErFe_3$ and $Er(Fe_{1-x}Ni_x)_3$. The former exhibits three magnetic phases between 4.2K and T_c, one of which is noncollinear [2,3]. The ability of the neutrons to ascertain simultaneously the long range magnetic ordering and distribution of Er, Fe, and Ni atoms among the different crystallographic sites provides an insight into local and competing anisotropies of the occupied sites, i.e., the point symmetry of nearest neighbor atoms surrounding a particular site dictates the nature of the local anisotropy, e.g., axial or planar.

The final system selected is $Nd_2(Fe_{1-x}Co_x)_{14}B$. The magnetic and atomic structures of one end member $Nd_2Fe_{14}B$ were determined directly from neutron diffraction of a powder [4]. The substitution of small amounts of Co for Fe atoms substantially increases T_c and the site preference of Co atoms clearly explains the anomalous thermal expansion behavior exhibited by $Nd_2Fe_{14}B$ and $Nd_2(Fe_{1-x}Co_x)_{14}B$ [5,6].

NEUTRON DIFFRACTION STUDIES

Magnetic Structure of Y_6Mn_{23}

This compound is isotypical with Th_6Mn_{23} which is f.c.c. (space group Fm3m) and has 116 atoms/unit cell. The structure consists of an octahedral cluster of rare earth atoms surrounded by 50 transition metal atoms on the faces and corners of the octahedron. The transition metal, Fe or Mn, occupies four crystallographically distinct sites, b, d, f_1, and f_2, Fig. 1.

Until 1968, Y_6Mn_{23} was considered to be a ferromagnet with equal Mn moments of $0.4 - 0.5$ μ_B coupled parallel [7-10], although in 1967, Dworak et al. [9] reported a magnetic structure from the analysis of six peaks obtained from polarized neutron diffraction of a powder sample. Polarized neutrons permit the separation of the magnetic contributions to the diffraction peaks from those of the nuclear contributions. At room temperature, Y_6Mn_{23} was said to be weakly ferrimagnetic with the moments on the b and d sites coupled antiparallel to those of the two f sites. However, they concluded that at low temperatures, it should be a ferromagnet.

In 1978, Hardman et al. [11] showed conclusively, on the basis of neutron diffraction data of a powder sample refined by a modified Rietveld technique [12] that Y_6Mn_{23} was a ferrimagnet down to 4.2K. The Mn atoms occupy the four sites, each with its own unique magnetic moment. Whereas the moments are collinear along <111>, the moments on the b and d sites couple antiferromagnetically with those on the f sites, as proposed by Dworak et al. [9], Table I. Note that the best fit placed a small magnetic moment on the S-state Y atom. These powder diffraction results were later confirmed by Delapalme et al. [13] using a single crystal and polarized neutrons.

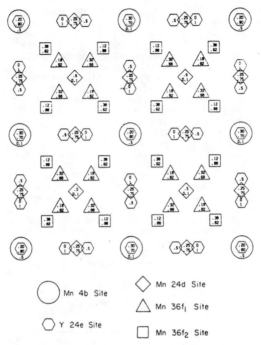

Fig. 1. The Th_6Mn_{23} structure.

Atomic Ordering in and Magnetic Structure of $Y_6(Fe_{1-x}Mn_x)_{23}$

The unusual magnetic behavior of the $Y_6(Fe_{1-x}Mn_x)_{23}$ system was first observed by Kirchmayr [14]. The end members magnetically order at about 485K; Y_6Mn_{23} has a bulk magnetization of 12.4 μ_B/mole and Y_6Fe_{23} a bulk magnetization of 43.1 μ_B/mole. However, as Mn is substituted for Fe in the

Table I. Summary of neutron diffraction results for Y_6Mn_{23} at 295K.

	$Y_6Mn_{23}(1)$	$Y_6Mn_{23}(2)$	$Y_6Mn_{23}(P.N.)$**
$Y(\mu_B$/atom)	-0.36*	-0.18	---
Mn b	-3.60	-2.65	-2.25
Mn d	-2.87	-1.92	-1.72
Mn f_1	2.11	1.66	1.52
Mn f_2	1.87	1.47	1.27
μ_B/f.u.	8.86	9.79	9.75
R_{nuc}	4.5%	4.2%	---
R_{mag}	3.7%	4.1%	6.5%

*Negative sign indicates the direction of the moment.
**Polarized neutron study of a single crystal [13].

latter', or Fe for Mn in the former, both bulk magnetization and Curie temperature decrease markedly, Fig. 2. In fact, when x is between ≈0.4 and
0.7, no magnetic order is observed down to liquid He temperatures. Based on
magnetic measurements of $Y_6(Mn_{0.75}Fe_{0.25})_{23}$, Oesterreicher et al. [15]
explained the absence of magnetic order as the result of a hybrid state
between strong intrinsic magnetic hardness and mictomagnetism, i.e., a
partly disordered magnetic state originating from fluctuations of magnetic
exchange and anisotropy changes from site to site. Bechman et al. [16]
suggested microscopic regions of antiferromagnetically coupled atoms interspersed with ferromagnetically coupled regions, based on a random (stoichiometric) distribution of Fe and Mn atoms among the lattice sites.

Neutron diffraction studies were, therefore, undertaken by Hardman
[17,18] to determine the distribution of Fe and Mn atoms among the four
sites and the magnetic structures. Atomic ordering was clearly evidenced
across the entire compositional range, Fig. 3 [19]. The striking result of
the ordering was the preference of Mn atoms for the f_2 site and that of the
Fe atoms for f_1. Neutron data obtained for $Y_6(Fe_{0.1}Mn_{0.9})_{23}$ showed no
moment for the Fe atoms. Mossbauer spectra obtained for samples with x =
0.8 and 0.9 indicated no internal hyperfine field and hence no spontaneous
long-range magnetic ordering of the Fe atoms from 300 to 1.5K. The same
result was obtained in applied fields of up to 6T at 4.2K [20,21]. The Fe
magnetic moments vs. Mn composition are shown in Fig. 4. A sharp drop
occurs upon introduction of Mn, favoring antiferromagnetic exchange.
Maximum randomization of Fe and Mn atoms and thus competing anti- and

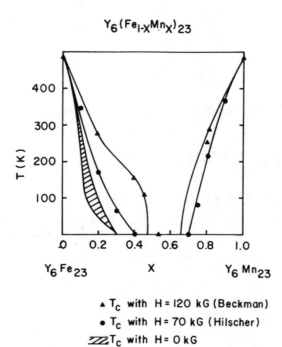

Fig. 2. Magnetic ordering temperatures of $Y_6(Fe_{1-x}Mn_x)_{23}$ as a function of
Mn composition, x. Values deduced from moment data at high fields
are given along with an approximate zero field extrapolation.

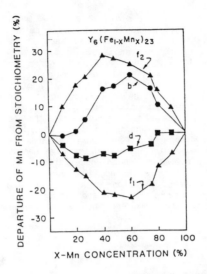

Fig. 3. Preferential site ordering on Mn atoms expressed as a percentage departure from stoichiometric distribution.

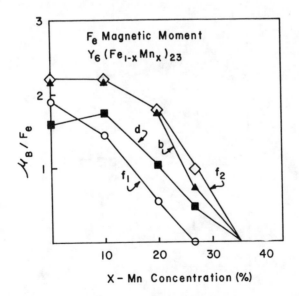

Fig. 4. Fe sublattice magnetic moments as a function of Mn composition. The Mn atoms are not ordered. All iron moments are coupled parallel except for x = 0.27 which has the antiferromagnetic configuration shown.

246

ferromagnetic exchange occur in the range of x = ≈0.4 to 0.7, in the region of the absence of long range magnetic order. The authors concluded that this frustration phenomenon was similar to that found in spin glass systems.

The origin and nature of this anomalous behavior were the objects of several investigations [10,22,23] but no direct and/or conclusive evidence of spin clusters resulted. Small angle neutron scattering (SANS) experiments were carried out on compositions in the region of anomalous behavior [24,25] but evidenced only the absence of ferromagnetic spin clusters. On the premise that antiferromagnetic clusters might be present, Hardman-Rhyne and Rhyne [25] decided to probe the structure for forbidden reflections, i.e., (100), (110), etc. Such clusters would not contribute to small angle scattering but would produce weak diffuse peaks at Bragg angles for a unit cell doubled in one or more directions. Using high resolution scans and a wavelength of 2.4Å, they indeed evidenced the existence of antiferromagnetic short range order in the presence of the (100) superlattice peak, Fig. 5a. The (111) and (200) nuclear reflections reduced 10-fold in intensity are shown in Fig. 5b for comparison. Diffuse scattering is most intense at x = 0.6 and diminishes sharply near the ferrimagnetic boundaries at x ≈ 0.4 and

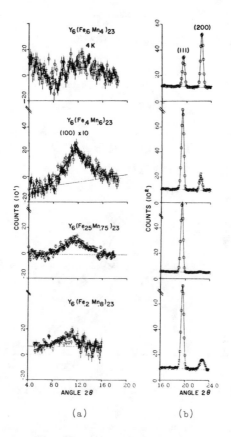

Fig. 5.   (a) Short-range order diffuse (100) peaks for Mn compositions in the region of no long-range order. (b) The nuclear (111) and (200) peaks.

0.8. The (100) peaks were found to be strongly temperature dependent, Fig. 6. Measurements above r.t. to 200°C showed no evidence of diffuse peaks. An average cluster size of 29Å was obtained for the x = 0.6 and 0.75 compounds. The average Mn cluster moment was maximized in the range of x = 0.5-0.6 and decreased as the average Mn-Mn distances became shorter for larger x.

Nuclear and Magnetic Structures of $Y_6Mn_{23}D_{23}$

The crystallographic and magnetic behavior of the isotypes, Y_6Mn_{23} and Th_6Mn_{23}, are dramatically altered upon deuteration. Th_6Mn_{23} is a Pauli paramagnet whereas the deuteride (D_{30}) is a ferrimagnet with T_c = 355K and a bulk magnetization of 18.5 μ_B/f.u. [10,26]. In contrast, $Y_6Mn_{23}D_{23}$ exhibits no net spontaneous magnetization [27,28] while Y_6Mn_{23} is a ferrimagnet.

Neutron diffraction studies by Hardman et al. [29] of $Y_6Mn_{23}D_{23}$ showed the deuterium atoms to prefer the a, f_3, j_1, and k_1 sites at 295K, Table II, in basic agreement with Westlake's geometric model [30]. The temperature factors for deuterium in the a and j_1 sites were quite large. The high thermal factor for deuterium in the octahedral a site indicated little or no bonding to the Y atoms. The high thermal factor for deuterium in the j_1 site indicated an averaging effect of deuterium atoms disordered between two j_1 sites only 0.90Å apart.

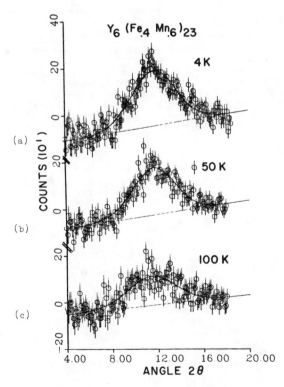

Fig. 6. Temperature dependence of the diffuse (100) peak in $Y_6(Fe_{0.4}Mn_{0.6})_{23}$ at (a) 4K, (b) 50K, and (c) 100K.

Table II. Atomic parameters for $Y_6Mn_{23}D_{23}$ at 295K in the f.c.c. (Fm3m) structure. Refined occupancy is N (maximum occupancy of that site), position is (x,y,z), and B is the Debye-Waller temperature factor.

Atom	Site	N/2 f.u. (max)	x	y	z	B
Y	e	12 (12)	0.205	0	0	0.82
Mn	b	2 (2)	1/2	1/2	1/2	0.47
Mn	d	12 (12)	0	1/4	1/4	0.47
Mn	f_1	16 (16)	0.179	0.179	0.179	0.47
Mn	f_2	16 (16)	0.372	0.372	0.372	0.47
D	a	2 (2)	0	0	0	6.35
D	f_3	10 (16)	0.100	0.100	0.100	1.37
D	j_1	20 (48)	0	0.169	0.373	4.95
D	k_1	14 (48)	0.161	0.161	0.049	2.28

$$a_0 = 12.805\text{\AA} \qquad R_{wp} = 10.64 \qquad \chi = 1.69$$

$$Z = 4 \qquad R_e = 6.29$$

At T < 180K a crystallographic transition from Fm3m to P4/mmm and the onset of antiferromagnetic coupling took place in agreement with magnetic susceptibility measurements by Crowder et al. [28], Fig. 7, and Mossbauer spectroscopy by Stewart et al. [31].

Analysis of the neutron data at 78 and 4K revealed a weak antiferromagnetic coupling of the c-axis moments of Mn on the b and f_2 sites of the original f.c.c. structure. Larger moments on the f_2 (s_2 and t_2 in the P4/mmm structure) sites were found at 78K than for 4K, corresponding to the broad minimum at ~80K in the magnetic susceptibility [28], Fig. 7.

Magnetic Structures of $ErFe_3$

ErFe$_3$ crystallizes in the rhombohedral structure of the PuNi$_3$, space group R$\bar{3}$m. This structure can be derived from the structure of the RM$_5$ compounds by means of appropriate ordered substitutions of rare earth atoms for transition metal atoms [32], e.g.,

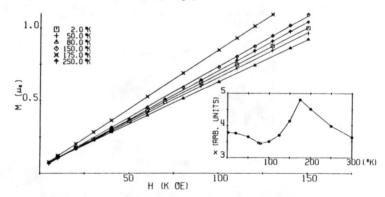

Fig. 7. Magnetization and susceptibility for $Y_6Mn_{23}D_{23}$.

$$2RM_5 - M + R = 3RM_3 = R_I 2R_{II} M_I 2M_{II} 6M_{III}$$

Examination of Fig. 8 reveals that the rare earth atoms are restricted to two sites: the a site wherein R_1 has an environment of nearest neighbor atoms nearly that of the rare earth in the RM_5 structure; and the c site where R_{II} has an environment nearly that of a rare earth atom in the cubic Laves phase. This latter phase can also be derived from the RM_5 structure by the same substitutions which generate the RM_3 compounds, namely

$$RM_5 - M + R = 2RM_2$$

In the Laves phase one crystallographic site of the rare earth has a cubic symmetry corresponding to $\overline{4}3m$ and as a consequence, the nearest neighbor environment of the R_{II} atom in the RM_3 structure is a slightly distorted tetrahedron, with elongation occurring parallel to the $\overline{3}$ axis. Accordingly, any magnetic moment associated with the R_1 (a site) atom would favor axial, $\vec{c}$, symmetry whereas moments associated with R_2 (c site) atoms would favor planar symmetry.

The transition metal atoms are distributed over three sites where two, M_I(d) and M_{II}(c), have axial symmetry, with the third, M_{III}(h), having planar symmetry. M_I has the same environment as the M atoms in the Laves phase and M_2 the same as in the RM_5 phase, Fig. 8.

Magnetization data taken on a small single crystal of $ErFe_3$ (T_c = 555K) at the Neél laboratory in Grenoble, France [2] indicated the easy direction at 4.2K to be the c axis with a spontaneous moment of about 3.74 μ_B/f.u. As the temperature increased, a sharp discontinuity in the magnetization at zero field along $\vec{c}$ was observed at T_{R_1} = 42K, Fig. 9. Above this

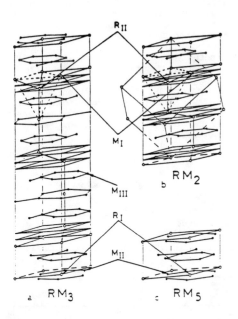

Fig. 8. (a) Crystalline structure of the rhombohedral RM_3 compounds. (b) Crystalline structure of the Laves cubic phase RM_2. (c) Crystalline structure of the hexagonal RM_5 phase.

250

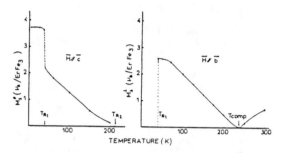

Fig. 9. Thermal variations of the spontaneous moment components $M_s \parallel$ and $M_s \perp$ along $\vec{c}$ and $\vec{b}$.

temperature, the c axis was no longer the easy axis. Above T_R = 210K, the spontaneous component along $\vec{c}$ was zero with equal values along $\vec{a}$ and $\vec{b}$, indicating the basal plane to be the easy plane. Accordingly, between T_R and T_R a reorientation of the moments arose leading to a noncollinear structure as evidenced by the sharp discontinuity of the angle of the spontaneous moment with $\vec{c}$.

Powder neutron diffraction data were collected to establish the exact magnetic configurations of the three phases. At 4.2K the magnetic (00ℓ) peaks were absent indicating the easy axis to be $\vec{c}$ in accord with the magnetization data. At 77K the (003) intensity was considerably increased; the moments were no longer parallel to $\vec{c}$ and the data refined to a noncollinear model with the moments in a plane containing the c axis. At 300K a collinear structure with the magnetic moments in the basal plane gave excellent agreement with the neutron data. The refinement results are given in Table III.

It should be noted that at 77K the angle θ for Er_I was less than that for Er_{II} in keeping with the fact that the environment of the nearest neighbors for Er_{II}, the 6c site, favored a more planar anisotropy than that of Er_I. Therefore, above 210K, the competition among the local anisotropies of the Fe atoms, which by reason of strong exchange interactions, leads to a total (net) anisotropy favoring the plane perpendicular to the c axis. Below 42K the local anisotropies of the Er atoms dominate resulting in a c axis collinear structure.

Table III. Results of the refinement of neutron data for $ErFe_3$ at 4.2, 77, and 300K.

Atoms	4.2K		77K		300K	
	$M(\mu_B)$	θ	$M(\mu_B)$	θ	$M(\mu_B)$	θ
Er_I	9.6±0.5	0°	8.8±0.2	38±4°	3.6±0.3	90°
Er_{II}	8.9±0.3	0°	8.4±0.2	57±4°	4.1±0.2	90°
Fe	1.9±0.2	180°	1.9±0.1	236±4°	1.7±0.1	90°

Magnetic Structures of Er(Fe$_{1-x}$Ni$_x$)$_3$

Neutron diffraction experiments on powders of Er(Fe$_{1-x}$Ni$_x$)$_3$ revealed that atomic ordering of Fe atoms was dominant at the 3b site and that the 18h site was preferred by Ni atoms, Table IV. Neutron diffraction powder patterns of ErFe$_2$Ni were obtained at r.t., 77K, and 4.2K, Fig. 10. The (003) reflection (intensity entirely magnetic) is observed at 300K as well as at 4.2K. However, at 4.2K it is considerably reduced in magnitude and nearly absent at 77K. This indicates that spin orientations occurred as the temperature changed. Stated otherwise, if the (003) reflection were absent, all moments would point along the c axis. At 77K, the Fe moments were found to lie along the c axis, but the Er moments were inclined to the c axis with the moment of Er$_{II}$ inclined at a larger angle θ. In ErFe$_2$Ni, at r.t., the Fe moments were found to be in the basal plane as observed in ErFe$_3$ [2]. However, the substitution of Fe with nonmagnetic Ni decreases the Fe-Er exchange interaction resulting in a noncollinear structure even up to room temperature, Table V. For this composition, Fe atoms prefer both the 3b and 6c sites. Due to the axial local symmetry, Fe atoms on the 3b and 6c sites have an additional orbital moment and, as a consequence, are associated with a more important anisotropy than of the Fe atoms on 18h. Fe atoms on the 3b site have an easy axis whereas those on the 6c site have an easy plane. The preferential occupancy of Fe atoms on 3b thus favors the Fe sublattice moments parallel to $\vec{c}$. This was further confirmed by neutron data for ErFe$_{1.5}$Ni$_{1.5}$ and ErFeNi$_2$. With increasing Ni concentration, a large increase in the preferential occupation of the 3b site and a corresponding decrease of the 6c site were found, accounting for the observed collinear structures along $\vec{c}$ in the temperature range 77-300K, as evidenced by the virtual absence of the (003) reflection for ErFeNi$_2$, Fig. 11.

Atomic Ordering in Nd$_2$(Fe$_{1-x}$Co$_x$)$_{14}$B

The Nd$_2$Fe$_{14}$B structure is tetragonal (space group P4$_2$/mnm, 4 formula units/unit cell) and contains two rare earth sites, f and g, six transition metal sites (k$_1$, k$_2$, j$_1$, j$_2$, e$_1$, and c) and one boron site (g), Fig. 12 [4]. Herbst and Yelon [33] have carried out neutron powder diffraction studies of five Nd$_2$(Fe$_{1-x}$Co$_x$)$_{14}$B compounds, x = 0.1 to 0.9 and the two corresponding end numbers. The results showed a generally random filling of the k$_1$, k$_2$,

Table IV. Occupation of atoms on the crystallographic sites.

	ErFe$_2$Ni	ErFe$_{1.5}$Ni$_{1.5}$	ErFeNi$_2$
	No. of Atoms	No. of Atoms	No. of Atoms
Fe (3b)	2.845	3.000	2.832
Fe (6c)	5.870	2.638	2.651
Fe (18h)	9.285	7.862	3.517
Ni (3b)	0.155	0.000	0.168
Ni (6c)	0.130	3.362	3.349
Ni (18h)	8.715	10.138	14.483
R$_n$	4.30	4.77	4.95
χ^n = R$_w$/R$_{expected}$	2.29	2.87	2.90

Table V. Magnetic moments on the crystallographic sites of ErFe$_2$Ni.

	300K				77K				4.2K			
	K_y	K_z	$M(\mu_B)$	$\theta(M,c)$*	K_y	K_z	$M(\mu_B)$	$\theta(M,c)$	K_y	K_z	$M(\mu_B)$	$\theta(M,c)$
Er (3a)	-1.79	2.22	2.85	38±4°	-1.07	-7.96	8.03	7±1°	-2.15	-8.66	8.93	14±3°
Er (6c)	-2.06	0.23	2.06	83±8°	-2.52	-6.14	6.64	22±3°	-7.50	-4.42	8.71	60±10°
Fe (3b)	2.16	0.00	2.16	270°	0.00	2.30	2.30	180°	0.00	2.45	2.45	180°
Fe (6c)	2.16	0.00	2.16	270°	0.00	2.30	2.30	180°	0.00	2.45	2.45	180°
Fe (18h)	2.16	0.00	2.16	270°	0.00	2.30	2.30	180°	0.00	2.45	2.45	180°
Ni (3b)	0.00	0.00	0.00		0.00	0.00	0.00		0.00	0.00	0.00	
Ni (6c)	0.00	0.00	0.00		0.00	0.00	0.00		0.00	0.00	0.00	
Ni (18h)	0.00	0.00	0.00		0.00	0.00	0.00		0.00	0.00	0.00	
R_n		3.98				4.07				5.95		
R_m		10.91				5.32				7.80		
$X^m = R_w/R_{expected}$		2.55				2.62				3.45		

*$\theta(M,c)$ angle between the c axis and the magnetic moment direction.

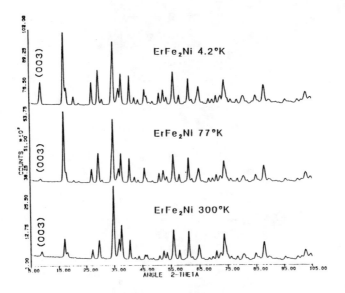

Fig. 10. Neutron diffraction spectra of ErFe$_2$Ni.

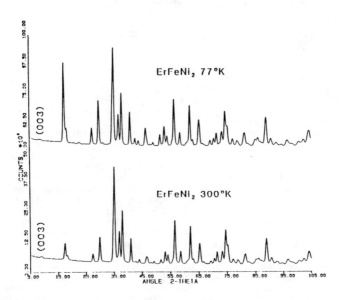

Fig. 11. Neutron diffraction spectra for ErFeNi$_2$.

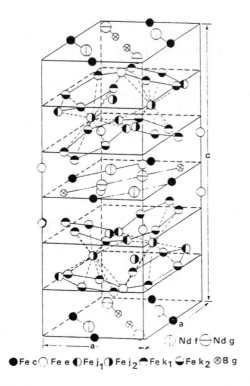

Nd f ⊖ Nd g

● Fe c ◯ Fe e ◖ Fe j₁ ◑ Fe j₂ ◐ Fe k₁ ◒ Fe k₂ ⊗ B g

Fig. 12. Tetragonal unit cell of the $Nd_2Fe_{14}B$ structure.

j_1, and c sites. The j_2 site Co occupancies were, however, considerably below that corresponding to the Co stoichiometry; the Fe atoms exhibited strong preference for the j_2 site.

In contrast, Co preferred the e site, but not to the extent that Fe preferred the j_2 site. The Co deficit of the j_2 site was not fully compensated by its excess on the e site.

It has been pointed out by several investigators [4,34] that the j_2 site in $Nd_2Fe_{14}B$ and the c (or "dumbell") site in Nd_2Fe_{17} are crystallographic as well as magnetic cognates. Each site has the largest number of nearest neighbor Fe atoms and the largest moment. Mossbauer studies [5,35] have also confirmed the strong preference of Fe for the j_2 site. The neutron diffraction results account clearly for the observed large anomalous thermal expansion observed for $Nd_2Fe_{14}B$ [5], $Y_2Fe_{14}B$ [6], and for its behavior with increasing substitution of Co for Fe [6]. As pointed out by Givord et al. in R_2Fe_{17} [36] and $R_2Fe_{14}B$ [37], there are some very short interatomic distances between Fe atoms which tend to decrease the Fe-Fe exchange interaction by favoring antiferromagnetic coupling in a local region of the cell. Since the normal Fe-Fe distances in the remainder of the crystal favor ferromagnetic coupling, the lattice tends to expand in the magnetic ordered state at the expense of the elastic energy. Minimizing the total energy in such a magnetic system results in an Invar-type thermal expansion anomaly below T_c. As Co is substituted for Fe, the anomalous expansion diminishes with increasing Co content, Fig. 13. The Co atoms are nearly excluded from the j_2 sites and distributed randomly among the

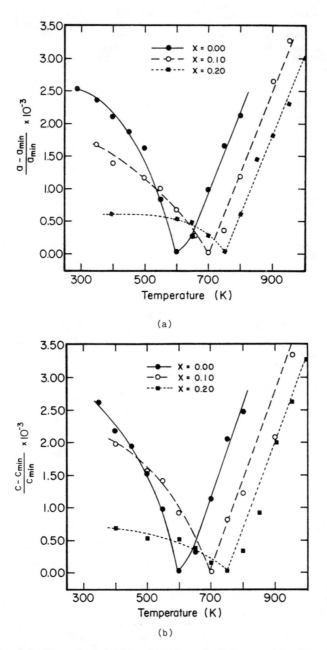

(a)

(b)

Fig. 13. (a) Thermal variation of lattice parameter a of $Nd_2(Fe_{1-x}Co_x)_{14}B$, x = 0.00, 0.10, and 0.20. (b) Thermal variation of lattice parameter c of $Nd_2(Fe_{1-x}Co_x)_{14}B$, x = 0.00, 0.10, and 0.20.

remaining sites. Accordingly, the short Fe-Fe distances existing in the antihexagonal prisms above and below the j_2 sites (12 near neighbors), Fig. 12, are altered by the insertion of the smaller Co atoms, thus reducing the localized magnetic stress arising from the tendency to antiferromagnetic coupling in the short Fe-Fe distances vs. the positive coupling for the longer Fe-Fe distances in the crystal. A compromise between the localized magnetic stress and the elastic energy of the crystal can thus be effected by Co substitution without as large an anomalous lattice expansion as results for $R_2Fe_{14}B$ compounds.

REFERENCES

1. H. M. Rietveld, J. Appl. Cryst., 2, 65 (1969).
2. B. Kebe, W. J. James, J. Deportes, R. Lemaire, W. Yelon, and R. K. Day, J. Appl. Phys., 62 (3), 2052 (1981); R. Ballou, J. Deportes, B. Kebe, and R. Lemaire, J. Magn. Magn. Mater., 54-57, 494 (1986).
3. D. E. Tharp, Y.-C. Yang, W. J. James, W. B. Yelon, D. Xie, and J. Yang, J. Appl. Phys., 61 (8), 4249 (1987).
4. J. F. Herbst, J. J. Croat, and W. B. Yelon, Phys. Rev. B, 29, 4176 (1984); J. Appl. Phys., 57, 4086 (1985).
5. D. E. Tharp, Y.-C. Yang, O. A. Pringle, G. J. Long, and W. J. James, J. Magn. Magn. Mater., in press (1987).
6. B.-P. Cheng, Y.-C. Yang, S.-C. Fu, and W. J. James, ibid, in press (1987).
7. B. F. DeSavage, R. M. Bozorth, F. E. Wang, and E. R. Callen, J. Appl. Phys., 36, 922 (1965).
8. H. R. Kirchmayr, IEEE Trans. Magn., 2, 493 (1966).
9. A. Dworak, H. R. Kirchmayr, and H. Rauch, Z. angew. Phys., 6, 318 (1968).
10. S. K. Malik, T. Takeshita, and W. E. Wallace, Solid State Comm., 23, 599 (1977).
11. K. Hardman, W. J. James, and W. B. Yelon, The Rare Earths in Modern Science and Technology, G. J. McCarthy and J. J. Rhyne (eds.), Vol. 1, Plenum Press, New York (1978), p. 103.
12. Program received from R. B. Van Dreele, Arizona State University (1975).
13. A. Delapalme, J. Deportes, R. Lemaire, K. Hardman, and W. J. James, J. Appl. Phys., 50, 1987 (1979).
14. H. R. Kirchmayr, ibid, 39, 1088 (1968).
15. H. Oesterreicher, H. F. Bettner, and F. T. Parker, Mag. Lett., 1, 89 (1978).
16. C. A. Bechman, K.S.V.L. Narasimhan, W. E. Wallace, R. S. Craig, and R. A. Butera, J. Phys. Chem. Solids, 37, 245 (1976).
17. W. J. James, K. Hardman, W. B. Yelon, and B. Kebe, Journal de Physique, 40 (C5), 206 (1979).
18. K. Hardman, W. J. James, and W. B. Yelon, J. Phys. Chem. Solids, 41, 1105 (1980).
19. K. Hardman, J. J. Rhyne, and W. J. James, J. Appl. Phys., 52 (3), 2049 (1981).
20. G. J. Long, K. Hardman, and W. J. James, Solid State Comm., 34, 253 (1980).
21. K. Hardman, W. J. James, G. J. Long, W. B. Yelon, and B. Kebe, The Rare Earths in Modern Science and Technology, G. J. McCarthy, J. J. Rhyne, and H. B. Silber (eds.), Vol. 2, Plenum Press, New York (1980), p. 316.
22. J. J. Reilly, Z. Phys. Chem. N.F., 117, 5155 (1979).
23. G. K. Shenoy, B. D. Dunlap, P. J. Viccaro, and D. Niarchos, Adv. Chem. Ser., 194, 502 (1981).
24. K. Hardman, Ph.D. Thesis, University of Missouri-Rolla (1979).
25. K. Hardman-Rhyne and J. J. Rhyne, J. Less-Common Met., 94, 23 (1983).
26. K. Hardman-Rhyne, H. K. Smith, and W. E. Wallace, ibid., 94, 95 (1983).

27. M. Commandre, D. Fruchart, A. Rouault, D. Sauvage, C. B. Shoemaker, and D. P. Shoemaker, J. Phys. (Paris) Lett., $\underline{40}$, L639 (1979).
28. C. Crowder, B. Kebe, W. J. James, and W. B. Yelon, The Rare Earths in Modern Science and Technology, G. J. McCarthy, H. B. Silber, and J. J. Rhyne (eds.), Vol. 3, Plenum Press, New York (1982), p. 473.
29. K. Hardman-Rhyne, J. J. Rhyne, E. Prince, C. Crowder, and W. J. James, Phys. Rev. B, $\underline{29}$ (1), 416 (1984).
30. D. G. Westlake, J. Mater. Sci., $\underline{18}$, 605 (1983); Scr. Metall., $\underline{16}$, 1049 (1982).
31. G. A. Stewart, J. Zukrowski, and G. Wortmann, J. Magn. Magn. Mater., $\underline{25}$, 77 (1981).
32. E. Parthe and R. Lemaire, Acta Cryst., $\underline{331}$, 1879 (1975).
33. J. F. Herbst and W. B. Yelon, G. M. Research Laboratories, Research Publication, GMR-5473, July (1986).
34. L. H. Bennett, R. E. Watson, and W. B. Pearson, J. Magn. Magn. Mater., in press (1987).
35. H. M. van Noort and K.H.J. Buschow, J. Less-Common Met., $\underline{113}$, L9 (1985).
36. D. Givord, R. Lemaire, W. J. James, J. M. Moreau, and J. S. Shah, IEEE Trans. Magn., $\underline{\text{MAG-7}}$, 657 (1971).
37. D. Givord, H. S. Li, and F. Tasset, J. Appl. Phys., $\underline{57}$, 4094 (1985).

USE OF PERMANENT MAGNETS IN ACCELERATOR TECHNOLOGY:
PRESENT AND FUTURE*

KLAUS HALBACH
Center for X-Ray Optics, Lawrence Berkeley Laboratory, Berkeley, CA 94720

ABSTRACT

Permanent magnet systems have some generic properties that, under some circumstances, make them not only mildly preferable over electromagnets, but make it possible to do things that can not be done with any other technology. After a general discussion of these generic advantages, some specific permanent magnet systems will be described. Special emphasis will be placed on systems that have now, or are likely to have in the future, a significant impact on how some materials research is conducted.

FOREWARD

Due to a large number of previous commitments, it is impossible for me to find the time to write a paper before the deadline that describes in the standard fashion the content of the talk that I gave at the conference. While Dr. Sankar, one of the chairmen of this symposium, accepted this as a matter of fact, he nevertheless wanted me to contribute something to the proceedings. As a consequence, I proposed to write a paper in a very unconventional manner that reduces the work involved significantly: After an introduction, I will simply reproduce (most of) the viewgraphs I showed in my talk, and give a detailed legend for each figure, but without connecting text, hoping that this will trigger the memory of the audience present at the symposium sufficiently to be useful. Dr. Sankar graciously agreed to this proposal. Since this is an experiment of sorts, I would greatly appreciate blunt comments, both positive and negative, from all readers willing to contribute to this experiment.

INTRODUCTION

Even though it may seem that the purpose of this paper is the description of some very specific devices, that is not the intent. It is the purpose of this paper to first explain some generic properties and advantages of a class of permanent magnet materials, and then to explain a number of very general design concepts that happen to be applied to components that are used in accelerators. I will first describe general purpose components (bend magnets, lenses) that are essential for the operation of an accelerator or storage ring, illustrated with specific examples. I will then follow that up with an explanation of the functioning, and specific examples, of special devices (undulators/wigglers) that allow accelerator technology to serve other disciplines, among them materials research.

* This work was supported by the Division of Advanced Energy Projects, U.S. Department of Energy.

ADVANTAGES OF PM SYSTEMS

· Strongest fields when small

· Compact

· Immersible in other fields

·"Analytical" material

· No power supplies ⎤

· No cooling   ⎬— · Reliability

· No power bill   ⎦   · Convenience

Fig. 1 Of the many advantages of PM systems, the one listed on top of this figure is the one that is usually the most exciting. The reason for the advantage listed on top is the following: When one scales an electromagnet in all dimensions while keeping the magnetic field at equivalent locations fixed, it is easy to see that the current density in the coils scales inversely proportional to the linear dimensions L of the magnet. Since superconductors have an upper limit for the current density j that can be carried, and dissipative coils have an upper limit for j due to the need to remove the dissipated power, j needs to be reduced below that prescribed by simple scaling when L reaches a certain small value that depends, of course, on many details of the magnet design. When j is reduced, the field in a magnet that does not use iron obviously is also reduced, even if the total Ampere turns are maintained by increasing the coil size. The same is also true for a magnet using iron, since an increase of the coil size invariably leads to a field reduction due to increased saturation of the iron. PM, on the other hand, can be scaled to any size without any loss in field strength. From this follows that when it is necessary that a magnetically significant dimension of a magnet is very small, a permanent magnet will always produce higher fields than an electromagnet. This means that with permanent magnets one can reach regions of parameter space that are not accessible with any other technology. The critical size below which the PM out-performs the electromagnet depends of course on a great many details of both the desired field strength and configuration as well as the properties of the readily available PM materials. In the region of the parameter space that is accessible to both technologies, the choice of one technology over the other will be made on the grounds of cost or convenience, (main specifics: power supplies, power needed to run the system, equipment associated with cooling) and in this arena permanent magnet systems are often also preferable, but less so the larger the smallest magnetically relevant dimension becomes.

The other advantages listed are self-explanatory, except for the meaning of "analytical" material. What I mean by that is the fact that the magnetic properties of charge (current) sheet equivalent materials (see Fig. 4) are so simple that they can be described by very simple analytical expressions, which in turn makes the design of very complex systems surprisingly simple.

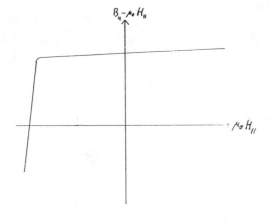

Fig. 2 After a powder, with grains equal in size to ferromagnetic domains, has been oriented in a magnetic field, sintered, and cooled to room temperature, it is exposed to a strong magnetic field parallel to the crystalline axis of the grains. Fig. 2 shows the resulting magnetization curve. It is assumed throughout, that the material is never driven beyond the knee of that curve.

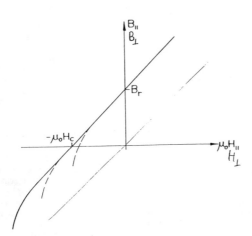

Fig. 3 Shows the resulting B(H) curve both in the direction parallel and perpendicular to the preferred direction, commonly called the easy axis. The remanent field for rare earth materials varies between .8 and 1.2 T, and for various ferrites between .3 and .35 T. The differential permeabilities are as small as 1.03 for some materials, and rarely larger than 1.15.

$$B_{\shortparallel} = \mu_0 \mu_{\shortparallel} H_{\shortparallel} + B_r$$
$$B_\perp = \mu_0 \mu_\perp H_\perp$$
$$\left.\begin{array}{l}\end{array}\right\}$$
$$\vec{B} = \mu_0 \,\hat{\mu} \times \vec{H} + \vec{B}_r$$
$$\vec{H} = \hat{\gamma} \times \vec{B} - \vec{H}_c$$

this $\nearrow$ into $\quad curl \; \vec{H} = 0 \quad or \quad div \, \vec{B} = 0 :$

$curl \left(\hat{\gamma} \times \vec{B} \right) = curl \, \vec{H}_c = \vec{j}_{eq} \, ,$ or $div \left(\mu_0 \hat{\mu} \times \vec{H} \right) = -div \, \vec{B}_r = \rho_{eq}$

This represents passive material $\left(\hat{\gamma}, \hat{\mu} \right)$
with active terms/properties $\left(\vec{j}_{eq}, \, \rho_{eq} \right).$

Homogeneous magnetization:

$\vec{j}_{eq} \longrightarrow$ current sheet $\longrightarrow$ $CSEM$
$\rho_{eq} \longrightarrow$ charge sheet $\longrightarrow$

Fig. 4 Shows that magnet properties can be described, in addition to "frozen magnetization" (Fig. 2) and B(H) curve (Fig. 3), by an anisotropic permeability very close to 1, plus either an impressed current or charge distribution, which, for the usual case of homogeneously magnetized material, corresponds to current or charge sheets on the surface.

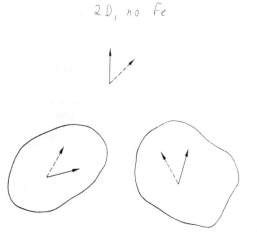

2D, no Fe

Fig. 5 Demonstrates the easy axis rotation theorem (Ref.1), which says: in an iron-free, two dimensional system, rotation of the easy axis of all CSEM of the system by any given angle keeps the magnetic field strength everywhere outside the material fixed, but rotates the field direction by the same angle in the direction opposite to the rotation direction of the easy axis. This theorem gives a lot of insight that can be used to great advantage in the design of systems.

a

b $E = v \times B \longrightarrow$ Magnets if $v/c > .03$

2 D

6b Magnets give more
focusing than elec-
tric fields that are
limited by breakdown
when v/c >.03.

c $V(r,\varphi) = const. \cdot r^n \cdot \cos(n\varphi) \rightarrow B \sim r^{n-1}$

6c 2D scalar poten-
tial for useful
field distributions.

6d Classification of
magnet types.

d $\begin{cases} n & \text{Use of magnet} \\ 1 & \text{bend magn.} \qquad dipole \\ 2 & \text{lin. focus. (A G)} \quad quadrupole \\ 3 & \begin{cases} \text{correction of aberrations} \\ \end{cases} \\ 4 \\ \vdots \end{cases}$

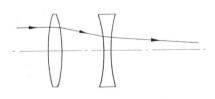

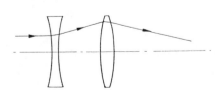

Fig. 7 A doublet
assembled from
lenses of equal
focusing and defo-
cusing strengths can
always be made focu-
sing in both direc-
tions.

264

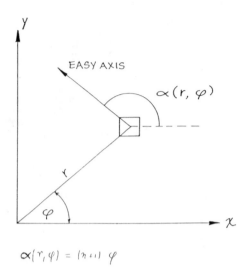

$$\alpha(r,\varphi) = (n+1)\,\varphi$$

Fig. 8 Using "frozen magnetization," and with it the linear superposition principle, the question "what is the optimum orientation of the easy axis of material anywhere to produce a given field distribution?" is a good question that has a very simple answer in the case of 2D iron-free multipoles (Ref.1).

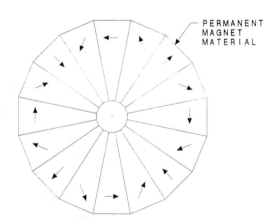

PERMANENT MAGNET MATERIAL

Fig. 9 Material with continuously changing easy axis orientation is very difficult to make. Therefore one uses segmented multipoles, (Fig. 9 shows a quadrupole.) Material in each block is homogeneously oriented and magnetized in the ideal direction for the centerline of block.

2 D QUADRUPOLE FIELD

$$B_x - i B_y = B_r \cdot \frac{X + iY}{r_1} \cdot 2 \cdot (1 - \frac{r_1}{r_2}) \cdot \frac{\sin (2\pi/M)}{2\pi/M} \cdot \cos(\bar{\pi}/M)$$

Possible Harmonics: $n = 2 + \nu \cdot M; \ \nu = 0, (1), 2 \ldots$

$2D \ dipole$

$$B = B_r \cdot \ln(r_2/r_1) \cdot \frac{\sin 2\bar{\pi}/M}{2\bar{\pi}/M}$$

Fig. 10 Gives the field for a 2D iron-free segmented quadrapole and dipole. M is the number of blocks in the magnet, and r_1 and r_2 are the inside and outside radii of the magnet (Ref. 1). Both magnets can produce fields larger than B_r. When M is reasonably large, the loss of field strength due to segmentation is very small.

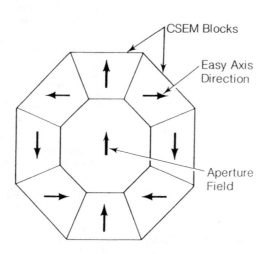

CSEM Blocks

Easy Axis Direction

Aperture Field

Fig. 11 Segmented dipole. This type of magnet can also be used to provide the main field for magnetic resonance imaging.

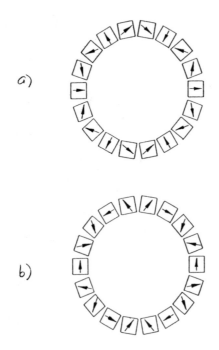

a)

b)

Fig.12a,b These are
two sextupoles that
can be used for
electron cyclotron
resonance ion
sources. The easy
axis rotation
theorem tells us
that both configur-
ations have essen-
tially the same
fields. Configura-
tion b is preferable
because some plasma
will escape along
radial field lines,
leading in case b to
heating of the
vacuum chamber
between permanent
magnet blocks, and
not right under them
as is the case with
configuration a.

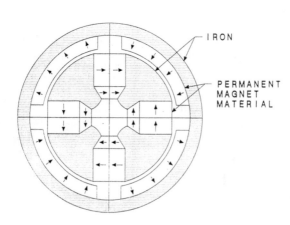

IRON

PERMANENT
MAGNET
MATERIAL

Fig. 13
Shows schematically
a PM quadrapole with
adjustable field
strength. In this
hybrid magnet, the
dotted material is
soft iron. The
field distribution
is controlled by the
iron surface, thus
it is very insens-
itive to material
properties as long
as the permeability
of the iron is large
compared to 1. The
arrows indicate the
easy axis of the
CSEM. By rotating
the outer ring with
the attached CSEM
one can clearly
change the field
strength in the central region. I design such systems with a design code
that consists of an appropriately linked set of analytical formulas. This
is possible only because the material is "analytical."

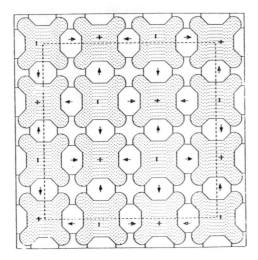

Fig. 14 Shows an arrangement of soft iron and CSEM to serve as a multiple beam focusing system.

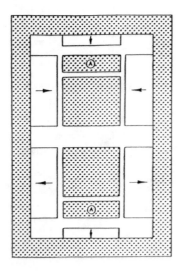

Fig. 15
Shows schematically a hybrid dipole. By drilling a hole through the whole magnet along the vertical center line, one can produce a strong magnet that can, under some circumstances, replace a solenoid. However, for this kind of solenoidal field magnet, the line integral of H along the whole vertical centerline is, because of Ampere's Law, always identically zero.

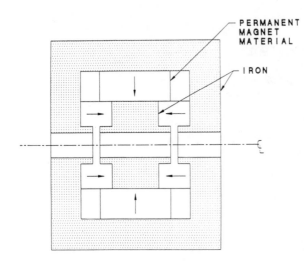

PERMANENT MAGNET MATERIAL

IRON

Fig. 16 Shows a solenoidal field doublet.

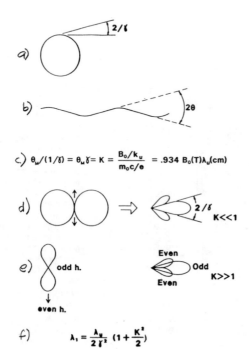

a) $2/\gamma$

b) 2θ

c) $\theta_M/(1/\gamma) = \theta_M \gamma = K = \dfrac{B_0/k_u}{m_0 c/e} = .934\ B_0(T)\lambda_u(cm)$

d) $\Rightarrow$ $2/\gamma$ $K \ll 1$

e) odd h. $\quad$ even h. $\quad$ Even Odd Even $K \gg 1$

f) $\lambda_1 = \dfrac{\lambda_u}{2\gamma^2}\left(1+\dfrac{K^2}{2}\right)$

Fig. 17 Is a very brief synchrotron radiation tutorial. 17a Shows schematically the radiation emitted by an electron circulating in a homogeneous field which is equivalent to the field in the bend magnet of a synchrotron or an electron storage ring. At any instant, the radiation is emitted essentially in a cone of opening angle 2/γ centered along the forward direction of the motion of the electron. γ= energy of the electron divided by its rest energy, ≃ 2000 for an energy of 1 GeV.

17b By providing a field that periodically changes direction, one can get much more radiation, and one can change the field strength in this kind of magnet called an undulator (U) or wiggler (W) without affecting the operation of the storage ring, thus allowing one to change the properties of the radiation.

17c The ratio of the angle between the extreme trajectory directions and the natural radiation cone angle is called the deflection parameter K.

17d To understand synchrotron radiation better, it is useful to describe the physics phenomena first in a coordinate system that moves with a constant velocity that equals the average velocity of the electron in the U or W. Assuming first a very small magnetic field, the electron experiences a weak electric field, leading to a linear motion, which in turn causes radiation to be emitted in the manner shown. Observed from the laboratory frame, one expects to see the pattern shown on the right.

17e When the field becomes sufficiently large, the electron will not move linearly any more, but executes a figure 8-like motion, leading to harmonics which, when K is sufficiently large (K>3), leads to so many overlapping harmonics that one is dealing with an essentially continuous spectral distribution. In this case, the magnet is called a wiggler.

17f When K is much smaller than 1, (as described in Fig. 17d), only the fundamental is excited, and it is quantitatively given by Fig. 17f. $\lambda\mu$ is the undulator period length, and θ is, in this case, not the extreme trajectory angle, but the direction of observation.

- *Element sensitive biology*

- *Interface, thin films, and material sciences*

- *Chemical kinetics and photochemistry*

- *Atomic and molecular science*

- *X-ray Lithography*

Fig. 18 This is a partial list of applications of synchrotron radiation. Even though U and W are used now in synchrotron light sources, the storage rings were not specifically designed to take advantage of the light that can be produced with U and W. A new generation of synchrotron light sources is being designed (with construction to start very soon) that will take advantage of the properties of modern permanent magnet U and W.

Trace Element Profiles Along a Sample Are Measured in Air with Part-per-Billion and Femtogram Sensitivities ⅊

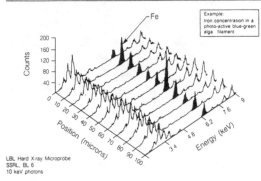

Example:
Iron concentration in a photo-active blue-green alga filament

Fe

Counts

200
160
120
80
40

0 10 20 30 40 50 60 70 80 90 100
Position (microns)

3.4 4.8 6.2 7.6 9
Energy (keV)

LBL Hard X-ray Microprobe
SSRL, BL 6
10 keV photons

Fig. 19 Shows one example of the application of synchrotron radiation: a series of element sensitive scans of an alga filament.

X-ray Microprobe With Multilayer Mirrors for Materials and Biological Studies ⅊

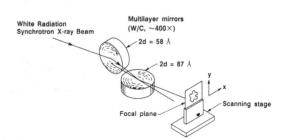

White Radiation
Synchrotron X-ray Beam

Multilayer mirrors
(W/C, ~400×)

2d = 58 Å

2d = 87 Å

y
x

Focal plane

Scanning stage

Fig. 20 Shows an experimental setup typical for the kind of research described in Fig. 19.

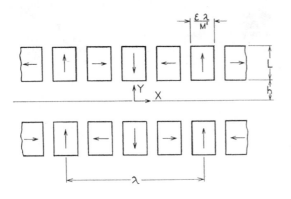

Fig. 21 Shows an iron-free U/W. The performance is obtained by taking the performances of a multipole, and letting the radius go to infinity while keeping the distance between poles fixed, thus leading to one linear array. The complete U/W consists of 2 such arrays and its performance is given in Ref. 2. The easy axis rotation theorem again tells us that blocks with

their easy axis parallel to the symmetry plane contributes as much to the total field as the other blocks do.

Hybrid Insertion Device configuration
with field tuning capability.

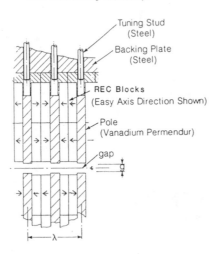

Fig. 22 Shows schematically the crossection of an U/W that uses both soft iron and CSEM. Again, this magnet is not nearly as sensitive to material properties as the iron-free U/W.

272

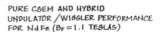

PURE CSEM AND HYBRID
UNDULATOR/WIGGLER PERFORMANCE
FOR NdFe (Br = 1.1 TESLAS)

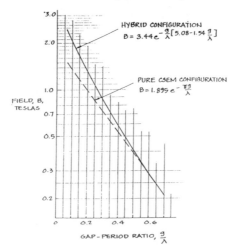

HYBRID CONFIGURATION
$B = 3.44 e^{-\frac{g}{\lambda}[5.08 - 1.54\frac{g}{\lambda}]}$

PURE CSEM CONFIGURATION
$B = 1.895 e^{-\frac{\pi g}{\lambda}}$

FIELD, B,
TESLAS

GAP-PERIOD RATIO, $\frac{g}{\lambda}$

Fig. 23 Gives the achievable fields for the two types of U/W described so far.

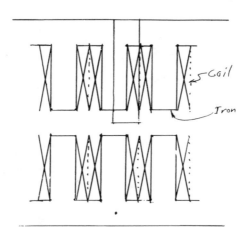

Coil

Iron

Fig. 24 In some types of free electron lasers (i.e. synchrotron radiation produced by large arrays of "fat" electrons, i.e., bunched with light wave period-icity) a substantial fraction of the energy of the electrons is converted into light, requiring an adjustment of the resonance condition (see Fig. 17f) due to the changing value of γ. One way to do that is adjustment of K through B. Since one needs to have a reasonable adjustment range for B, one would like to use an electromagnet (em). This requires a quantitative understanding of the field limitation in an em U/W. From this Figure, showing a periodic em, it is clear that one needs to analyze only 1/4 of a period in one half of the magnet.

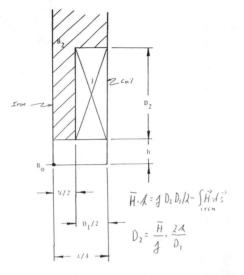

Fig. 25
Shows 1/4 of a
period. Taking
the line integral
over H along the
path going from
the lower left
corner up into the
iron along the
left problem
boundary to a
point higher than
the coil, to the
right boundary,
down the right
problem boundary,
to the bottom
boundary, and then
left to the start-
ing point, one
obtains the upper
of the two equa-
tions. Ignoring
the saturation term

$$\bar{H} \cdot \lambda = j\, D_2 D_1/2 - \int_{iron} \vec{H} \cdot d\vec{s}$$

$$D_2 = \frac{\bar{H}}{j} \cdot \frac{2\lambda}{D_1}$$

(if it's significant, the design is not useful), one obtains the expression
shown for the height D_2 of the coil. For a given desired field H and a
maximum "coolable" current density j, one obtains a D_2 that is independent of the absolute size of the device. Since the flux going into the top
of the pole is proportional to both H and D_2, the iron at the top will
saturate very strongly if the width of the pole at the top (scaling with
the period length) becomes too small.

Plan view of iron poles of U/W

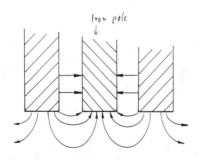

Iron pole

Fig. 26 In
reality, satu-
ration will be
worse because of
the lateral width
of the pole, as is
clear from this
figure, showing a
plan view of the
U/W (with the beam
going from left to
right), with
fieldlines also
shown schemati-
cally. This
Figure also
suggests a solu-
tion: attach CSEM
to the poles, with
the easy axis ori-
ented in such a
way as to put flux
into the pole in
the direction
opposite to the
flux generated by
the coil.

Plan view of PM assisted em U/W

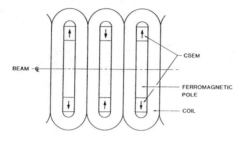

CSEM

BEAM — ℄

FERROMAGNETIC POLE

COIL

Fig. 27 Shows schematically the complete arrangement.

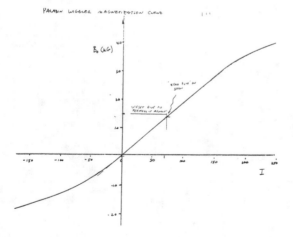

PALADIN WIGGLER MAGNETIZATION CURVE

B_0 (kG)

I

Fig. 28 Since the CSEM pre-loads the pole with flux (i.e. with the power supply off, there is flux going through the pole because of the CSEM) one expects an asymmetric saturation curve that allows one to reach a higher peak field than is possible without CSEM, and this saturation curve, taken from a log book, shows asymmetry as expected.

High-Field Electromagnetic Wiggler

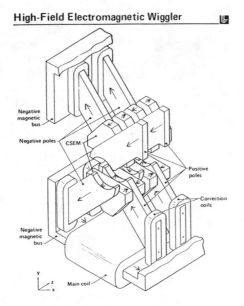

Fig. 29 Shows
schematically an
U/W when the con-
cepts outlined
above are carried
to an extreme,
leading to very
high field levels.

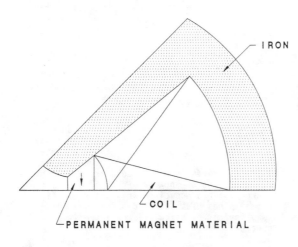

Fig. 30 Shows a
similar way to use
CSEM in an em, in
this case a quad-
rupole. Shown is
1/8 of the com-
plete magnet, with
CSEM close to the
aperture region.

276

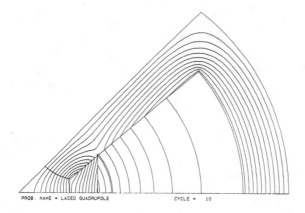

PROB. NAME = LACED QUADRUPOLE   CYCLE =   10

Fig. 31 Shows a field line pattern in the magnet depicted in Fig. 29. It is clearly visible how the CSEM reduces the flux density in the pole, making it possible to produce higher fields in the aperture, to reduce the size of the magnet, and making it less sensitive to iron properties.

Variably Polarized Radiation Can Be Generated with
Crossed Undulators In Low Emittance Storage Rings

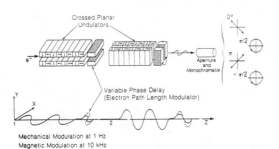

Mechanical Modulation at 1 Hz
Magnetic Modulation at 10 kHz

Fig. 32 For some applications it is desirable to produce circularly polarized synchrotron radiation, and the system shown here is a very clever and elegant way to do that; see Ref.3. Looking at the light emitted by one electron at a location just in front of the monochromator, one will find two wave trains with orthogonal linear polarization, separated in time, and each wave train containing the same number of optical periods as the number of periods in each undulator. A monochromator of sufficient resolution will increase the number of periods in each wave train, making them overlap and thus produce circularly polarized light, if the delay between the wave trains has been adjusted properly. By adjusting that delay in the way indicated, circularly polarized light of either helicity can be produced.

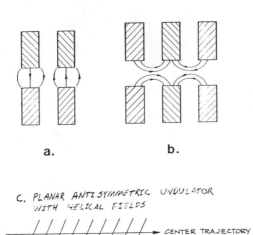

a. b.

c. PLANAR ANTISYMMETRIC UNDULATOR
WITH HELICAL FIELDS

CENTER TRAJECTORY

STEEL POLES

Fig. 33 Another method to produce circularly polarized light is to produce synchrotron radiation in a helical magnetic field. A conceptually simple way to produce such fields is to use short segmented dipoles of the kind described in Fig. 11, with each dipole rotated by a given angle relative to its next upstream neighbor (Ref. 2).A still different method is to use the basic hybrid U design, but excite the poles not with the standard symmetry pattern shown in Fig. 33a, but the antisymmetric pattern shown in Fig 33b. If, in addition, one arranges the poles not perpendicular to the center electron trajectory but at an angle as shown in Fig. 33c, such a magnet has helical fields of opposite helicity above and below the midplane (Ref. 4) that could be used by stored electron beams since they are always very small in the vertical direction.

REFERENCES

K. Halbach: Design of Permanent Multipole Magnets with Oriented Rare Earth Cobalt Material, NIM 169

K. Halbach: Physical and Optical Properties of Rare Earth Cobalt Permanent Magnets, NIM 187, 109 (1981).

K.J. Kim, A Synchrotron Radiation Source with Arbitrary Adjustable Elliptical Polarization, NIM, 219, 425 (1984).

K. Halbach: Some New Ideas About Undulators, published in the proceedings of the First International Conference on Insertion Devices for Synchrotron Radiation Sources, SPIE 582, 68, 1985.

(Acknowledgement)
I want to thank Steve Marks for his assistance in preparing this paper in what must be a record short time.

IMPACT OF THE HIGH-ENERGY PRODUCT MATERIALS
ON MAGNETIC CIRCUIT DESIGN

HERBERT A. LEUPOLD, ERNEST POTENZIANI II, JOHN P. CLARKE, AND DOUGLAS J.
BASARAB
Electronics Technology and Devices Laboratory, Fort Monmouth, NJ 07703-5000

ABSTRACT

    Many devices that employ magnetic fields are encumbered by massive sole-
noids with their equally bulky power supplies or by inefficient permanent-
magnet structures designed for use with obsolescent magnet materials. This
paper describes how the high-energy product materials are employed in several
structures to afford mass and bulk reductions of an order of magnitude or
more. Also discussed are novel designs that are not attainable with the
older materials such as the alnicos. Substitution of solenoids with permanent magnets also eliminates considerable energy consumption and the attendant problems arising from generation of heat. In many cases, all this is
accomplished within leakage-free systems. Among the designs described are:
nuclear magnetic resonance imagers; cylindrical solenoidal field structures
for klystrons and nonperiodic field TWT's; cylindrical field structures with
arbitrary axial gradients for advanced gyrating beam sources; annular field
sources for high harmonic gyrotrons; helical transverse field sources for
circularly polarized radiation sources; miniature periodic permanent-magnet
configurations; and clad permanent-magnet circuits for biasing fields in
millimeter-wave filters. Where alternate designs exist, they are compared
with regard to performance, bulk, and economy. All of the structures are in
various stages of design and construction, and, for completed structures, comparisons are made between theoretical projections and actual performance.

INTRODUCTION

    While the rare-earth permanent magnets (REPM) afford greatly enhanced
performance when substituted for older magnets in conventional devices, such
employment amounts to mere improvements in degree, rather than the revolution
in kind of which these remarkable materials are capable. Not only do their
high-energy products enable them to supply higher fields to greater volumes
with less material, but their extreme magnetic rigidity under the influence
of large antiparallel (coercivity) and orthogonal (anisotropy) fields greatly
simplifies magnet design and fabrication. It also makes possible efficient
permanent-magnet configurations that are unviable with conventional permanent
magnets. A number of such structures have been designed at the U. S. Army
Electronics Technology and Devices Laboratory (ETDL) for use in a variety of
light, compact, millimeter-wave, microwave, and optical devices for electron-
beam lenses; Faraday rotators; radiation sources; amplifiers; filters; iso-
lators; and circulators. A selection of these structures is described in
this paper to illustrate the design methods appropriate for high-energy product magnet materials, to suggest further uses and development of such devices
and to highlight future research on permanent magnets needed to further expand their efficacy and range of usefulness.

THE MAGNETIC ANALOGUE TO OHM's LAW

    Particularly helpful in the conception and analysis of new magnetic cir-
cuits is the magnetic analogue to Ohm's law in electrical circuit theory.
The derivation and operational details of this approach have been extensively

published elsewhere[1-4] and only a brief summary of its principal features is given here.

CIRCUIT ANALOGUES

MAGNETIC			ELECTRIC		
MAGNETOMOTIVE FORCE	$\Delta U, F$	GILBERTS	ELECTROMOTIVE FORCE	$\Delta V, E$	VOLTS
MAGNETIC POTENTIAL	U	GILBERTS	ELECTRIC POTENTIAL	V	VOLTS
MAGNETIC FLUX	Ø	MAXWELLS	ELECTRIC CURRENT	I	AMPERES
PERMEANCE	P	G-CM/OE	CONDUCTANCE	G	SIEMENS
RELUCTANCE	R	OE/G-CM	RESISTANCE	R	OHMS
MAGNETIC FIELD	H	OERSTEDS	ELECTRIC FIELD	·E	VOLTS/M
FLUX DENSITY	B	GAUSS	CURRENT DENSITY	J	COULOMB/MS
PERMEABILITY	μ	GAUSS/OERSTEDS	CONDUCTIVITY	σ	SIEMENS/CM
MAGNETIC FIELD ENERGY	W	JOULES	ELECTRIC POWER	P	WATTS
MMF OF A MAGNET	$L_M H_C$	GILBERTS	EMF OF A BATTERY	V_B	VOLTS

FIG. 1. Magnetic circuit quantities and their electrical counterparts.

Figure 1 shows the important magnetic quantities and their electrical analogues. Because the mathematical relationship between the magnetic quantities is formally identical to those governing the electrical ones, all the usual theorems and algorithms (Ohms's law, Helmholtz's laws, Maxwellian loops, etc.) used in the analysis of electrical circuits are applicable to magnetic ones as well. Thus, we may write for the magnetic Ohms's law

$$\phi = PF$$

or

$$\phi = F/R, \tag{1}$$

where ϕ is the magnetic flux, P the permeance, F the magnetomotive force (mmf), and R the reluctance in respective correspondence with the electrical current I, the conductance G, the electromotive force V, and the resistance R of the electrical case. Other pertinent magnetic quantities are listed with their magnetic analogues in Fig. 1. Because magnetic flux is not confined to constrained, tractable paths such as the wires of an electrical circuit, the permeances are estimated by dividing space around the magnetic circuit into somewhat arbitrary flux paths of which the boundaries are planes, cylindrical arcs, or spherical segments that emanate normally from the circuit surfaces to connect points of different magnetic potential. The permeances of these simplified paths can, then, be calculated through standard formulae and, if the division has been judiciously made, surprisingly good approximations to the true fluxes can often be obtained. Figure 2 shows some standard flux paths, together with their respective permeance formulae.

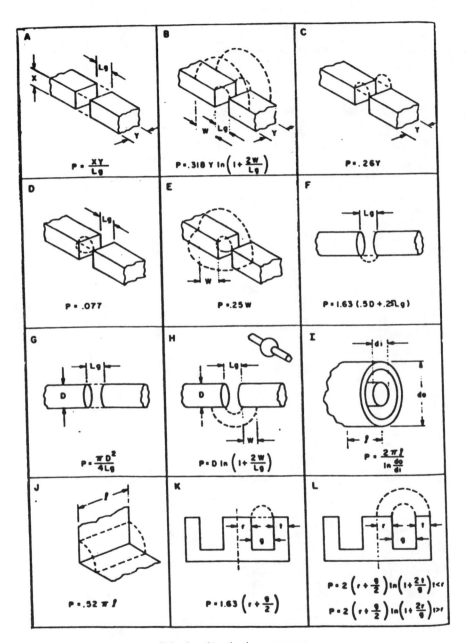

FIG. 2. Standard permeances.

A rare-earth permanent magnet (REPM) behaves like a magnetic battery in that the total mmf F that it produces around the circuit in which it is placed[3,4] is independent of the permeance of that circuit and is given by

$$F = {}_BH_c \; \ell, \qquad (2)$$

where ℓ is the length of the magnet in the direction of its orientation. This is not true of the older magnets such as the alnicos. Figure 3 illustrates the origin of the difference. A magnet's total mmf is proportional to the H axis intersection of the reversible recoil line at its circuit operating point.[4] For a REPM as shown in Figs. 3a and 3b, this intersection will always be at ${}_BH_c$, since its demagnetization and recoil lines are identical. This identity is a consequence of its square hysteresis loop in $4\pi M$, that is, its magnetic rigidity. For an alnico magnet, a change in operating point from A to B results in a change of recoil line from L_1 to L_2 and, consequently, a change in H intercept from H_1 to H_2. As can be seen from the $4\pi M$ curve for alnico, this result comes about because the change in operating point has resulted in a demagnetization from M_1 to M_2.

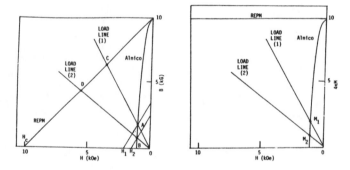

FIG. 3. Comparison of REPM and alnico behavior when operating load line is changed from (1) to (2), operating point of alnico goes from A to B, and total mmf from $H_1\ell$ to $H_2\ell$. In contrast, the mmf of REPM on going from C to D remains unchanged at ${}_BH_c\ell$.

Hence, we see that no unique mmf can be assigned to a conventional permanent magnet, and, if the useful magnetic field is to be modulated by variation of air-gap length or electrically generated mmf's, the magnet mmf will always be that corresponding to the lowest point on the demagnetization curve reached in the course of the modulation cycles. For this reason, circuit design with traditional magnets has been as cumbersome and difficult as electrical circuit design would be if the emf's of batteries were dependent upon the circuits in which they are placed. Therefore, because they are magnetic batteries of constant mmf, REPM's afford greatly simplified magnetic circuit analysis and design.

MAGNETIC CLADDING

A particularly useful design option afforded by REPM's is that of magnetic cladding,[5-9] which entails the strategic placement of auxiliary magnets
to confine the flux produced by the field supply magnets to the desired working space. This approach is suggested by analogy with electrical circuit
theory, wherein current in a particular branch of an electrical circuit can
be reduced or suppressed by inserting a source of emf in opposition to it.
In illustration, a typical horseshoe magnet circuit with its electrical analogue is shown in Fig. 4A. Magnetic flux is desired only in the gap between
the opposing pole pieces P_W, but, in fact, much flux leaks through the other
permeance paths P_L, collectively denoted G_L in the analogue. Figure 4B
shows how the unwanted flux and current are each eliminated in the magnetic
circuit and its electrical analogue: by cladding of the supply magnets with
radially oriented magnets of appropriate mmf in the former, and by insertion of batteries of the correct emf in the latter.

On the principal of cladding are based several of the permanent-magnet
field sources recently designed at ETDL, several of which will be described
here.

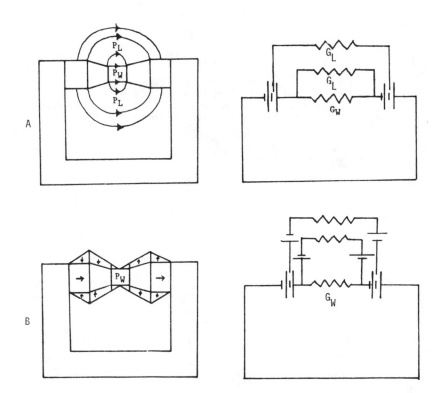

FIG. 4. A - Horseshoe magnet (left) and its electrical analogue (right).
B - Clad version of magnet (left) and analogue (right).

284

The 58 GHz Filter Magnet

This structure is pictured in Fig. 5, and is similar to the example of
the previous section except that the cladding about the working gap has been
deleted to provide access for a waveguide. The omission allows some lateral
flux leakage through the cladding, but only a small fraction of what would
occur without cladding.

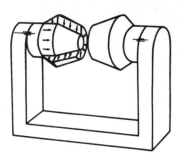

FIG. 5.   Partial section of 58 GHz
filter magnet.

If a field of H_G is desired in a circular space of cross-sectional
area A_G and length L_G, and if the magnet cross section is limited to area
A_M, the design parameters to be determined are the combined supply magnet
length L_M and the maximum thickness of the cladding magnets t_M. These are
found by first pretending that the design objectives have already been accomplished, so that all of the flux furnished by the supply magnets passes
through the gap and all fields and magnetization in the gap and magnets are
uniform. Then, the circuital form of Ampere's law demands that

$$H_M L_M + H_G L_G = 0,\qquad(3)$$

where H_M is the field in the magnet. Conservation of flux requires

$$B_M A_M = B_G A_G = H_G A_G,\qquad(4)$$

and the equation for magnet's second quadrant demagnetization curve is

$$B_M = H_M + B_r,\qquad(5)$$

where B_r is the magnet remanence. These relations determine the desired
quantities H_M and L_M
For no leakage to occur, every point on the exterior surface of the
structure must be at the same magnetic potential so that, according to the
magnetic Ohm's law, no streaming of flux between exterior points is possible.
If we take our reference potential to be that of the iron return path, the
potential of a point along the magnet axis at distance X from the yoke is
given by

$$F_x = \int_0^X H_M \cdot dx$$

$$F_x = H_M X. \tag{6}$$

If one considers the radial path from X to what is to be the outer surface of the cladding, and H_C is the radial field in the cladding, we must have

$$F_x + \int_0^{t_c(x)} H_C \, dr = 0 \tag{7}$$

for the outer surface point t_c to be at the same potential as the yoke. Since, by hypothesis, no flux passes through the cladding, B_C must be zero and from the demagnetization curve

$$H_C = {}_BH_C \tag{8}$$

and

$$\int_0^{t_c(x)} H_C \, dr = {}_BH_C \, t_c. \tag{9}$$

Finally, Eqs. (6), (7), and (9) yield

$$t_c^{(x)} = \frac{H_M X}{{}_BH_C}, \tag{10}$$

which reaches a maximum at $X = L_M/2$ and

$$t_c^{max} = H_M L_M/2{}_BH_C. \tag{11}$$

The same thickness should extend over the pole pieces with the magnet orientation normal to the pole piece surface. In practice, this is approximated by cladding with the technologically easier normal-to-axis orientation and with radial thickness $t_c^{max}/\cos \Theta$, where Θ is the angle of taper of the pole pieces.

The structure of Fig. 5 was actually constructed for ETDL by Electron Energy Corporation for use as a biasing field source for a 58 GHz wave filter. It produces 20.5 kOe in a 0.51 cm gap at a weight of 9.0 kg. An electromagnet and power supply of similar performance weighs hundreds of kilograms.

Permanent-Magnet Solenoids

Solenoidal magnetic fields have an expansive range of application and usually require heavy electrical solenoids and equally bulky power supplies. They also require considerable energy for their operations if kilooersted fields are desired in spaces of any size. Therefore, for many operations requiring static solenoidal fields, permanent-magnet structures would be highly desirable. Again, we resort to cladding to produce structures that are unviable with alnico.

A plausible approach is suggested by a configuration conceived by

Neugebauer and Branch[10] for a double-chambered klystron designed for the Air Force. (See Fig. 6.) Magnetic flux is supplied to the cylindrical working spaces w by the axially oriented supply magnets A via the permalloy pole pieces B, so that any circular cross section of w is an equipotential surface. The conical magnets M with which the supply magnets are clad are shaped to ensure confinement of all the flux produced by the supply magnets to the working space. In this manner, opposed solenoidal fields are set up in tandem in the adjacent chambers, as indicated in Fig. 6 by the large arrows.

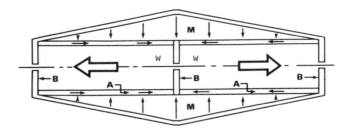

FIG. 6. Clad double-chambered klystron magnet designed by Neugebauer and Branch.

The theoretical rationale for such structures is similar to that given for the filter magnet and is presented elsewhere.[7] Hence, only the resulting formulae for the dimensions R_s and t_c are given here, viz.,

$$R_s = R_w \left[1 + (1 + B_r/H_w)^{-1} \right]^{1/2} \tag{12}$$

$$t_c = H_w L/_BH_c, \tag{13}$$

where R_s is the outer radius of the supply magnets A, t_c is the maximum thickness of the conical shell magnets M, R_w is the radius of the working spaces, L is their length, H_w is the specified working field, B_r is the remanence of the magnets, and $_BH_c$ is their coercivity. If only a single chamber is desired, the end of high magnetic potential must be clad as shown in Fig. 7.

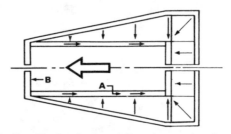

FIG. 7. Single-chambered Neugebauer structure.

287

Although Neugebauer cladding confines flux, it is unnecessarily bulky and heavy. If, instead of taking the reference potential at an end of the working space, we take it halfway between the ends, we can use cladding of inward orientation to reduce all of the surface to the right to reference potential.[6] Similarly, with outwardly oriented cladding, we raise all of the surface to the left to reference potential. Because the potential differences between the ends and the reference at the center is only half that in the Neugebauer structure, the cladding now need be only half as thick, and a reduction of more than 50 percent in the structural mass results. Figure 8 summarizes these results; in Fig. 9 is a sectional view of the ETDL design with the reference potential at the center.

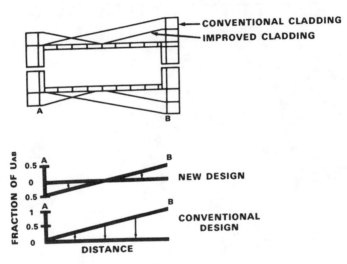

FIG. 8. Comparison of the actions of the new ETDL cladding with the conventional Neugebauer cladding.

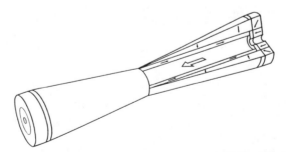

FIG. 9. Sectional view of ETDL version of REPM solenoid.

288

    The field uniformity is illustrated by the computer plot of Fig. 10 and
the graph of Fig. 11A. Because of the impracticability of having proper fan-
shaped cladding in the corners at the ends of the structure, some lateral
leakage through the conical cladding results, as does a lessening of axial
field uniformity. Figure 11B shows that the latter can be amended by small
adjustments in the diameters of the cladding magnets at the ends, so that the
axial field uniformity becomes comparable to the sectional uniformity. Figure 12 shows the operating points of the supply and cladding magnets on their
demagnetization curve, demonstrating the unviability of alnico magnets for
this purpose.

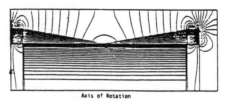

Axis of Rotation

FIG. 10. Flux plot of REPM solenoid from finite element
analysis. The lines in the working space do not appear
evenly spaced because of the cylindrical symmetry, i.e.,
each line represents the total flux in the annulus in
which it occurs.

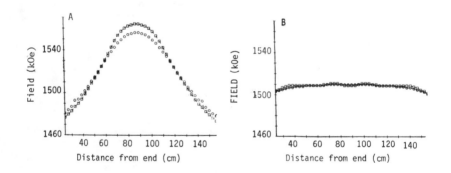

FIG. 11. Magnetic field profile along the axis of the REPM of Figs. 9 and 10.
A - unaltered solenoid, B - with quasi-optimized cladding magnets.   o - on-
axis fields, □ are one-half radius off-axis fields.

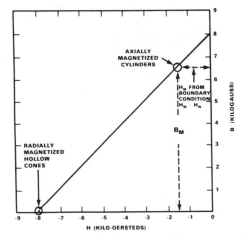

FIG. 12. Operating points for the magnets in a permanent-magnet solenoid. The desired value of H_W is not on the alnico curve, so alnico cannot be used as a supply magnet. If used as a cladding magnet, alnico is limited to an mmf of 800 Oe/cm.

Solenoids with Axial Field Variation

Solenoids can also be made to produce axially tapered fields by suitable adjustment of either the ring thicknesses of the supply and cladding magnets or of their magnetic remanences and coercivities.[7] The former procedure is called "geometric tapering" and the latter "parametric tapering." The two methods are illustrated in Fig. 13. In practice, the variation of the cladding magnets is usually geometric and that of the supply magnets is parametric. The choice for the cladding magnet is made to minimize weight. The supply magnet is tapered parametrically because its thinness and fragility would make a geometric taper impracticable in most cases. Such a hybrid embodiment was constructed by Electron Energy Corporation after an ETDL design to produce a linear field variation from about 1.0 kOe to 0.4 kOe over a device interaction region. The structure and its calculated and measured field profiles are shown in Fig. 14. The distortions outside of the interaction region are due to the rather large access holes through the pole pieces and the cladding magnets at the ends, but should not adversely affect the operation of the inserted device. The calculated curves of Fig. 15 show the remarkable cross-sectional field uniformity, which seems to be characteristic of all of the permanent-magnet solenoids investigated thus far.[6-9]

Annular Magnetic Fields

Some micro/millimeter-wave devices employ hollow electron beams. Examples are high-harmonic gyrotrons[10-11] and hollow-beam traveling-wave tubes.[12] In such applications, it is sometimes advantageous to confine the magnetic field to the region occupied by the interacting beam, that is, in a peripheral annular ring.[13] This can be accomplished by cladding both the inner and outer surfaces of a cylindrical shell, as shown in Fig. 16. There need not be two supply magnets as shown; the flux may be provided by either a

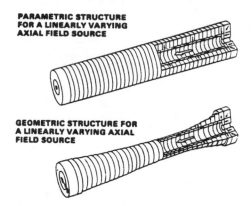

PARAMETRIC STRUCTURE FOR A LINEARLY VARYING AXIAL FIELD SOURCE

GEOMETRIC STRUCTURE FOR A LINEARLY VARYING AXIAL FIELD SOURCE

FIG. 13. Alternative structures for producing solenoidal magnetic fields with a linear taper.

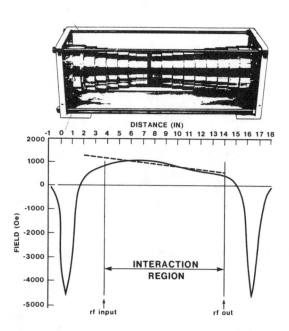

FIG. 14. On-axis field of CEBA magnet. The dashed line is the calculated field for a perfect structure without access ports. The solid curve is the measured field.

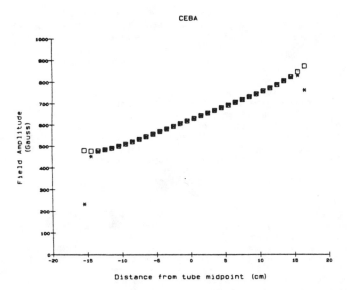

FIG. 15. Magnetic field as a function of distance from tube midpoint for on-axis points (crosses) and for points 2 cm from axis (squares).

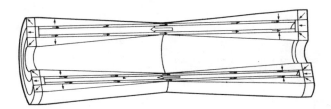

FIG. 16. Annular magnetic field from a permanent-magnet structure.

single inner shell or central cylinder. In the case of the central cylinder, no inner cladding is necessary. This may sometimes save weight by making the inner and outer radii of the cladding magnets smaller.

Transverse Rectangular Fields

All of the structures discussed thus far produce longitudinal fields of cylindrical symmetry. The same ideas used to design these structures can also be employed to obtain transverse rectangular fields.[9] (See Fig. 17.) Again, the appropriate fan-shaped orientations at the corners of the cladding magnets are approximated by single directions of magnetization along the diagonals. This, again, results in slight flux leakage near the corners and inhomogeneities in the vicinities of the supply magnets shown in Fig. 18. As in the case of the constant field solenoid, the small, off-center inhomogeneities can be greatly reduced with adjustments in the cladding.

Structures of this type are potentially useful whenever transverse biasing magnets are required.

One of the most interesting of these is for a NMR field source for magnetic resonance imaging in medical diagnostics.[9] (At present, such sources employ superconducting magnets.) It possesses the advantage of stability, simplicity, lower mass, lack of electrical currents with attendant power supplies and energy consumption, and no necessity for expensive and onerous liquid helium cooling. A REPM structure of the type shown in Fig. 17 would weigh about four tons, as opposed to the 11 tons of a conventional permanent-magnet circuit previously designed.[14]

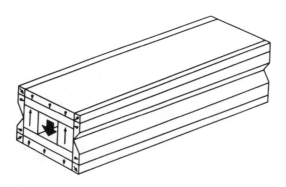

FIG. 17. Uniform transverse field from a clad REPM structure.

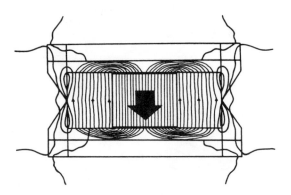

FIG. 18. Flux plot of structure in Fig. 17.

Periodic Structures

Figure 19 is a schematic cross section of a conventional periodic permanent magnet (PPM) stack such as those used for electron-beam focusing in
traveling wave tubes (TWT). The orientations of the permanent magnets are
denoted by the small arrows. The large arrows show the field directions in
the bore. The flux is conducted from the permanent magnets into the bore via
the interspersed pole pieces. Unfortunately, flux is conducted to the outside as well. This leakage can be curtailed by cladding each section of the
PPM in the manner of the permanent-magnet solenoid as in Fig. 20.[15] The
cladding results in either higher fields in a given configuration or, alternatively, in smaller structural masses and bulks at the same fields. Figure
21 compares an unclad PPM with its clad equivalent with regard to structural
mass dependence on attained axial field amplitude. It is clear that mass
savings can be considerable. For example, at the typical field of 4 kOe,
cladding results in a mass reduction of about 60 percent from that of the
simple unclad structure. Comparisons of cladding with other methods of mass
reduction of PPM's are given in detail in reference 16.

Improvement of permanent-magnet materials is particularly important for
TWT field sources. Figures 22 and 23 illustrate the benefits from further
increases in magnetic remanence, especially for the easy-to-fabricate, conventional type of Fig. 19. While improvement of remanence does not eliminate
the weight advantage held by less conventional designs, it greatly reduces
it, thereby increasing the viability of the less expensive, conventional design for application where the very lightest weights are not needed.

Also of great importance is the PPM stack capability to withstand temperatures to 500 K or higher. With increasing miniaturization and power densities, operating temperatures inevitably rise to the detriment of magnetic
remanence and coercivity. This can result in defocusing of the electron beam,
reduced performance, and shorter tube life. Tube cost is also increased by
the added difficulty of shunting a magnet stack in manufacturing. In addition to the cost of applying shunts, which can exceed ten percent of the
total tube cost, reduced manufacturing yield and reduced tube life can be
related directly to inadequate temperature-compensated magnet structures.

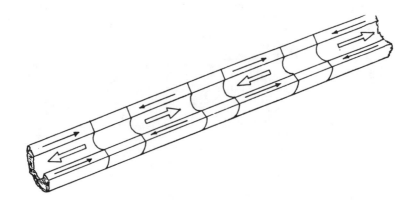

FIG. 19.   Periodic permanent-magnet structure.

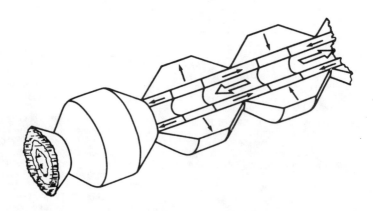

FIG. 20.   Clad periodic permanent-magnet structure.

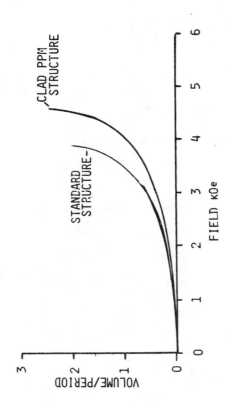

FIG. 21. Bulk advantage of clad versus unclad periodic permanent magnet.

296

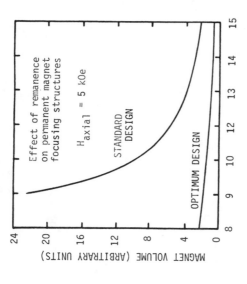

FIG. 23. Comparison of magnet volume dependence on remanence for two different TWT magnet stacks when both supply an on-axis field amplitude of 5 kOe.

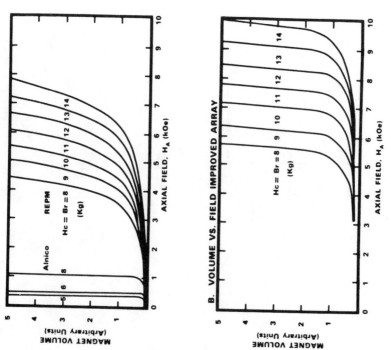

FIG. 22. Comparison of the volume dependence on an axial field amplitude of two different TWT magnet stacks.

STRUCTURES WITHOUT IRON FLUX PATHS

Sometimes, when the very highest fields obtainable from permanent magnets are desired, it is advantageous to think of permanent magnets as rigid collections of poles and to try to place them so that fields are optimized where desired. Sufficiently judicious arrangements can sometimes result in fields much higher than the remanences and coercivities $_BH_C$ of the source magnets. Examples are the multipolar cylindrical configurations conceived by Halbach[17] for beam-focusing in particle accelerators. The dipolar, or "Magic Ring," is a special case of this class of structure.

The Magic Ring

Figure 24 illustrates the magnetic orientation in a magic ring. The magnetization vector is of constant magnitude and in a direction given by

$$\alpha = 2\phi + \pi/2, \tag{14}$$

where ϕ is the cylindrical angular coordinate. The fields within the central cavity are easily found by noting that the surface and volume pole densities engendered by the variation in orientation are given, respectively, by

$$\sigma = \vec{M} \cdot \hat{m} \tag{15}$$

$$\rho = \nabla \cdot \vec{M}, \tag{16}$$

where $\vec{M}$ is the magnetization and $\hat{m}$ the unit vector normal to the surface at any point. These expressions can, then, be inserted into Coulomb's law and integrated over the entire structure. The resulting on-axis field is given by

$$H = B_r \ln R_o/R_i, \tag{17}$$

so that, in principle, there is no limit to the magnitude attainable for H. However, because of the logarithmic dependence of H on the outer radius, the structure would be prohibitively large for fields much greater than about $2 B_r$. Nevertheless, if a material of B_r of 10 kG is used, a 15 kOe field in a one-inch space can be obtained by a structure only 4.5 inches in diameter.[18] The field is also quite uniform and very forgiving of rather coarse approximations to the ideal structure. For example, in Fig. 24C is pictured an octagonal approximation of only eight different magnetic orientations. Uniformity is still seen to be quite good, and field strength is still 90 percent of that of the ideal structure.

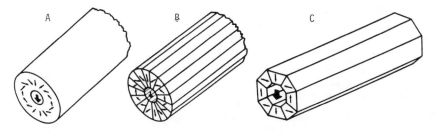

FIG. 24. A - magic ring, B - approximation to magic ring with 16 uniformly magnetized segments, C - octagonal approximation to a magic ring.

Fabrication of even the octagonal approximation entails some difficulty because of the multiple orientations required. One can minimize the effects of this difficulty by observing that a uniformly magnetized annular shell contains every possible orientation with respect to the local radius. Thus, one need only apply a transverse orienting field to a cylindrical shell during manufacture, magnetize it after fabrication, and, then, cut it to the desired number of segments as shown in Fig. 25A. The resulting segments are then interchanged, as in Fig. 25B, and glued together, thereby forming the magic ring of Fig. 25C. The prescription for the interchange of segments is

$$S (\phi) \rightleftharpoons S (-\phi). \tag{18}$$

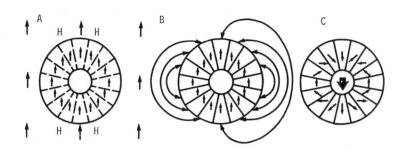

FIG. 25. Fabrication of a magic ring. A - Cylinder is aligned in a field H and divided into 16 segments. B - The segments are interchanged as shown, forming C, the completed magic ring.

With rare-earth magnets of remanence $\geq$ 10 kG, such structures would provide compact, simple, easily portable sources of uniform transverse fields of 15 or 20 kOe. Application would be for biasing fields in millimeter-wave filters and other reciprocal devices or for nuclear magnetic resonance (NMR) imagers. Also, by slicing a magic ring transversely to its axis and then rotating each successive segment by the same angle, with respect to its predecessor, one can make a twister stack for the production of circularly polarized millimeter-wave radiation.[17,18]

All of these structures have the advantages of being much lighter and more compact than equivalent electric solenoids, and of not requiring energy sources for their operation.

The Magic Sphere

There are applications that require high biasing fields for only a centimeter or two, where a magic ring would have to be wastefully long to ensure the requisite field uniformity. A possible solution would be to make a spherical structure based on the same principle as the magic ring, but without the field inhomogeneities induced by end effects.[19,20] We refer to such a structure as a "magic sphere."

We arrive at the magic sphere by way of the same considerations used to derive the magic ring. If a uniform bipolar axial field is desired in a spherical cavity, the magnetic orientation in the surrounding spherical permanent-magnet shell should vary as

$$\alpha = 2\,\theta + \pi/2, \tag{19}$$

where θ is the polar angle. (See Fig. 26.) The resulting field in the cavity is

$$H = (4/3)\,B_r\,\ln\,(R_o/R_i), \tag{20}$$

so, it is apparent that the field in the cavity of a magic sphere is 4/3 that of a magic ring with the same outer and inner diameters. If a field of 20 kOe is needed in 1.0 cm, and if we use a NdFeB magnet with a remanence of 12 kG, the whole structure would weigh less than 0.15 kg, a very small mass for so large a field in that sized space.

Figure 27 shows the effect of access holes drilled axially through the poles on the field profile. It is clear that the structure is relatively insensitive to such mutilation for holes up to one-fourth the working space diameter with only slight distortions of field profile and a reduction of only a few percent in peak field.

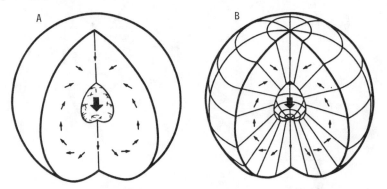

FIG. 26. A - Sectioned magic sphere. B - Approximation to magic sphere built with uniformly magnetized segments.

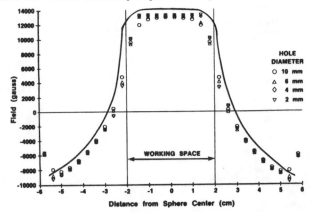

FIG. 27. On-axis field profile of magic sphere with axial polar holes of different diameters.

The Magic Igloo

This structure consists of either the northern or southern hemisphere of a magic sphere placed on a plane of iron so that its antimirror image in the iron completes the magic sphere.[20] (See Fig. 28.) Although it provides only half the working space of a full magic sphere with the same inner and outer diameters, it has the advantage that access holes can be drilled through the iron slab, rather than through the expensive magnets, as would be necessary for a full sphere. Figure 29 shows that the field profile of the magic igloo also exhibits high tolerance for such access holes.

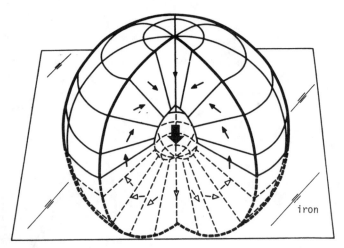

FIG. 28. Magic igloo. A hemisphere of magic sphere rests on an iron plane, as shown. The paramagnetic, or anti-, image in the plane is shown in dashed lines and unblackened arrows.

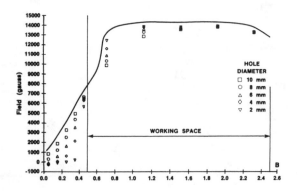

FIG. 29. Axial field profile for a magic igloo for various sized axial holes in its iron plate anti-mirror.

TWISTERS

The species of free-electron laser, known as the ubitron,[21] produces circularly polarized radiation and requires a constant magnetic field, the orientation of which is transverse to a cylindrical space and which varies linearly with distance along the axis, viz.,

$$\phi = kZ, \qquad (21)$$

where ϕ is the cylindrical angular coordinate, Z is the distance along the axis, and k is the pitch, or twist. Such a field is usually produced by a bifilar electrical solenoid and is, therefore, subject to all the attendant inconveniences of electrical sources: bulk, heating, etc. The Naval Research Laboratory is constructing a ubitron that, in prototype, employs a bifilar solenoid, but is to use, eventually, a permanent-magnet structure designed by ETDL.[22] Three design alternatives were considered.[23] They are pictured in the third column of Fig. 30. The first and second columns, respectively, show the untwisted structural bases and the ideal embodiments of the twisted structures.

The top row pertains to the simplest and least expensive alternative. Its basis is a rectangular bar transversely magnetized, as shown, and with a cylindrical hole bored along its axis. The field arises from the poles on the surface of the hole. It is somewhat reduced by a counter field generated by poles on the narrow walls of the bar.

When the bar is twisted about its long axis, as in the second column of Fig. 30, the desired field arises. The transverse field will be reduced by an amount dependent on the rate of twist. Since no actual magnetic material can be twisted in this manner, the desired structure is approximated by slicing the basis into thin cross-sectional slabs and rotating each by an angle $2\pi/n$, where n is the number of slabs per period.

The middle row of Fig. 30 shows a twister based on the octagonal approximation to the magic ring. It produces more field per unit mass of structure, but is, by far, more expensive and difficult to manufacture.

The bottom row of Fig. 30 shows the most complex and expensive structure of all. It is based on the clad rectangular structure discussed in the previous section. It produces a lower maximum field than either of the other structures, and is intermediate between the others with regard to field per unit mass at the lower fields. It would seem that its only advantage lies in its iron pole pieces, which would tend to smooth out the permanent-magnet inhomogeneities that can be quite troublesome in undulator free-electron lasers.

Figure 31 shows the comparison of the three twisters with regard to maximum, practically obtainable fields as a function of twist. Although the magic-ring structure is best in this regard, the bar magnet was chosen for embodiment in a prototype because of its low price, simplicity, and mechanical robustness. Figure 32 shows an expanded view of how the magnets are to be mounted in brass discs, with peripheral holes for pitch variation.

Figure 33 shows the insensitivity of the magic ring twister to the angular increments between magnet sections. Fortunately, even increments as large as $\pi/5$ (10 slabs per period) result in a negligible field loss for the 2.5 cm twist used. The other structures display similar insensitivity in this regard.

302

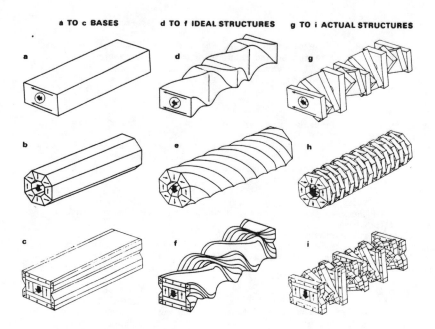

a TO c BASES   d TO f IDEAL STRUCTURES   g TO i ACTUAL STRUCTURES

FIG. 30. Three twister structures with their bases and actual physical embodiments.

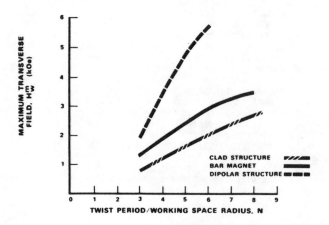

FIG. 31. Dependence of maximum practical fields of the three twisters on twist period.

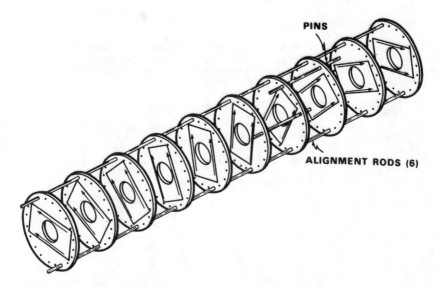

FIG. 32. Expanded view of twister structure built for NRL. The alignment rod holes afford variable twist.

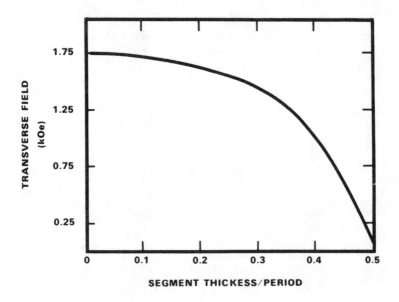

FIG. 33. Effect on segment thickness on transverse field.

SUMMARY AND CONCLUSIONS

Although much progress has been made in REPM technology over the past decade, considerable work remains to be done if many of the projected novel devices are to reach fruition. In this connection, achievement of the following would be very useful and, in some cases, critical.

(1) Better performance of temperature-compensated magnets. This includes further reduction of the already low-temperature coefficients of energy product and their extension to broader temperature ranges. Also important is the increase in energy products at all temperatures. Increases of the order of 25 percent would be attained by the squaring of the hysteresis loops alone. These developments are absolutely essential for the attainment of higher fields in smaller structures operating at greater power densities in TWT's projected for use in future lightweight, airborne devices. Raising of the coercivities and squaring of the loops of the currently best 2-17 temperature-compensated magnets might be a solution. Compensation of the NdFeB compounds might be an even better one.

(2) Full magnet alignment in novel magnet geometries by getting sufficient flux where it is needed during the fabrication and magnetization processes. This can be very difficult in the case of conical shell magnets aligned normal to the cone surface, which would be very useful for the cladding of pole pieces in horseshoe-type magnetic circuits. Also difficult, for the same reason, is the radial magnetization of toroids with small holes and broad annuli.

(3) Improvement of fabrication and assembly techniques through the development and use of better solders, glues, epoxies, molds, presses, dies, etc. Routinization of details of fabrication of very unforgiving materials for better industry-wide standardization.

(4) A better variety of square-looped materials with low-energy products is needed for widespread implementation of parametric field shaping. Possible approaches would be the development of very high coercivity ferrites, misch metals, magnetically "knocked down" $SmCo_5$, SmCo's alloyed with heavy rare earths, and composites of REPM's with solders or epoxies. Heat-resistant forms of the dilutants may also become very useful in the near future.

(5) Not directly connected with REPM technology, but potentially very important for the intensification and broadening of their technological exploitation, are the development of innovative methods of heat handling. Included are both schemes for leading heat away from the thermally sensitive regions of devices, as well as improved resistance of materials to increased temperatures arising from miniaturization and increased power levels. Both retention and constancy of magnetic properties at high temperature are important.

(6) Investigation of deposition of REPM materials in the form of films on surfaces of different geometries should continue to be pursued. Energy products must be raised, hysteresis loops squared, and mechanical and thermal ruggedness assured. Especially promising is the possibility of depositing films of in-surface and normal-to-surface orientations upon each other. Success in such a venture could lead to an entire generation of small electronic devices that are unattainable at present.

(7) A keener awareness in the magnetic and industrial communities that the REPM's are more than just very powerful alnicos. They afford not only the possibility of doing conventional tasks better, but offer a cornucopia of design options that were impossible before their advent. Still prevalent mind-sets acquired in the course of design experience with conventional magnets must be discarded when REPM's are used, if the technological riches they offer are to be fully and speedily realized.

References

[1]H. C. Roters, Electromagnetic Devices, Chapters 4 & 5, (Wiley, New York, (1941).

[2]R. J. Parker and R. J. Studders, Permanent Magnets and Their Application, Chapters 4 & 5, (Wiley, New York, London, 1962).

[3]H. A. Leupold, in Proceedings of the Fifth International Workshop on Rare-Earth-Cobalt Permanent Magnets and Their Applications, Roanoke, VA, June 7-10 1981, p. 270.

[4]H. A. Leupold, "A Magnetic Design Primer," Part V, summer short course, Modern Permanent Magnets, Applications and Design, May 19-23, 1986, p. 140.

[5]W. Neugebauer and E. M. Branch, "Applications of Cobalt Samarium Magnets to Microwave Tubes," Technical Report, Microwave Tube Operation, General Electric (15 March 1972).

[6]F. Rothwarf, H. A. Leupold and L. J. Jasper, Jr., Proc. of the Third International Workshop on Rare-Earth Magnets and Their Applications, p. 255, Ed. K. J. Strnat (1978).

[7]A. Tauber, H. A. Leupold and F. Rothwarf, Proc. of the Eighth International Workshop on Rare Earth Magnets and Their Applications, p. 103, Ed. K. J. Strnat (1985).

[8]J. P. Clarke and H. A. Leupold, IEEE Transactions on Magnetics, MAG 22, No. 5, p. 1063 (1986).

[9]E. Potenziani II and H. A. Leupold, IEEE Transactions on Magnetics, MAG 22, No. 5, p. 1078 (1986).

[10]D. S. Furuno, D. B. McDermott, and N. C. Luhmann, Jr., Proceedings of IEDM, p. 346, Los Angeles, CA, December 7-10 1986.

[11]D. B. McDermott, N. C. Luhmann, Jr., D. S. Furuno, A. Kupiszewski, and H. R. Jory, Int. J. I. R. and Millimeter Waves 4, 639 (1983).

[12]N. J. Dionne, Proceedings of IEDM, p. 513, Los Angeles, CA, Dec. 7-10, (1986).

[13]J. P. Clarke, E. Potenziani II, and H. A. Leupold, 31st Annual Conference on Magnetism and Magnetic Materials, Paper BP-17, Baltimore, MD, Nov. 17-20 (1986). (To be published in J. A. P.)

[14]J. H. Battacletti and T. A. Knox, IEEE Trans. Magn., MAG 21 (5), 1874 (1985).

[15]H. A. Leupold, J. P. Clarke, W. E. Lockwood, and D. J. Meyers, Proceedings of IEDM, p. 693, Los Angeles, CA, Dec. 7-10 (1986).

[16]H. A. Leupold and J. P. Clarke, submitted for publication in Electron Devices Society Transactions; Part II, Vacuum Electron Devices.

[17]Klaus Halbach, Proceedings of the Eighth International Workshop on Rare-Earth Magnets and their Applications, p. 128, Dayton, OH, May 1985.

[18]A. B. C. Morcos, University of Dayton, School of Engineering, M.S. Thesis, November, 1985.

[19]H. A. Leupold and E. Potenziani II, Intermag Paper GH-12, Tokyo, Japan, April 14-17, 1987.

306

[20]H. A. Leupold and E. Potenziani, IEEE Trans., Magnetics, (to appear September, 1987).

[21]R. M. Phillips, IRE ED-7, p. 231 (1960).

[22]R. H. Jackson and D. E. Pershing, Proceedings of IEDM, p. 86, Los Angeles, CA, Dec. 7-10, 1986.

[23]A. B. C. Morcos, E. Potenziani II, and H. A. Leupold, IEEE Transactions on Magnetics, MAG 22, No. 5, p. 1066, 1986.

STATIC DEVICES WITH NEW PERMANENT MAGNETS

J. CHAVANNE, J. LAFOREST and R. PAUTHENET
Laboratoire Louis Néel, C.N.R.S., 166X, 38042 Grenoble cedex, France

ABSTRACT

The high remanence and coercivity of the new permanent magnet materials are of special interest in the static applications. High ordering temperature and large uniaxial anisotropy at the origin of their good permanent magnet properties are obtained in rare earth-transition metal compounds. Binary $SmCo_5$ and Sm_2Co_{17} and ternary $Nd_2Fe_{14}B$ compounds are the basis materials of the best permanent magnets. New concepts of calculations of static devices with these magnets can be applied : the magnetization can be considered as rigid, the density of the surface Amperian current is constant, the relative permeability is approximately 1 and the induction calculations are linear. Examples of hexapoles with Sm-Co and NdFeB magnets are described and the performances are compared. The problems of temperature behaviour and corrosion resistance are underlined.

INTRODUCTION

New static devices with specific magnetic field configurations can be designed and constructed with new permanent magnets. The permanent magnets are ferromagnets or ferrimagnets in which ordering temperature above room temperature, e.g. large exchange interactions and high spontaneous magnetization are quested. The specific property of a permanent magnet is the ability to stay in a metastable state of saturated magnetization. This can be obtained through the phenomenon of coercivity which is intrinsically linked to magnetic anisotropy of uniaxial character. In rare earth-transition metal rich compounds, Curie temperatures above 300 K are observed. Some have large uniaxial anisotropy associated with a uniaxial crystallographic structure. New permanent magnets such as $SmCo_5$, Sm_2Co_{17} and Nd-Fe--types with very high characteristics are now manufactured. This has led to an important development of static devices for charged particle beam applications using these permanent magnets, such as in travelling wave tubes, undulators/wigglers and in Electron Cyclotron Resonance (ECR) multi-charged ion sources.

We summarize in the first part some physical aspects of rare earth-transition metal alloys and the second part emphasizes the interest of the rare earth magnets. In the third part we present a method of calculation of the magnetic induction in static devices made with these magnets. The fourth part is devoted to the hexapolar magnetic field configuration with these magnets describing the evolution during the last few years and new improvements due to the use of Nd-Fe-B magnets.

RARE EARTH-TRANSITION METAL COMPOUNDS FOR PERMANENT MAGNETS

A permanent magnet material requires simultaneously, at room temperature, a large atomic magnetic moment and a large anisotropy. The first of these properties is characteristic of the 3d elements : Fe, Co or Ni and the second one is characteristic of the 4f elements : rare earths. Alloys between these two series of elements allow the combination of large values of both magnetic moment and anisotropy. Due to the difference of the electronegativity and the size of the atoms of the rare earth metals and of iron, cobalt or nickel, the binary phase diagrams present a large number of intermetallic compounds [1,2]. Their magnetic properties result in the combined effects of the magnetism of the two types of constituting atoms. The conduction electrons of the rare earth metals have a higher chemical potential than the 3d ones, they fill up the empty 3d orbitals and then reduce the 3d magnetism. This effect is crucial in nickel or cobalt compounds : no nickel binary compounds have Curie temperature above room temperature. Only for cobalt-rich compounds, R_2Co_{17} and RCo_5 the magnetism is strong enough to give rise to useful permanent magnet properties. In iron binary intermetallic compounds this effect is not so pronounced, however, either Curie temperature (for R_2Fe_{17}) or magnetocrystalline anisotropy (RFe_3 and RFe_2) are too weak. On the contrary, the ternary compounds, $R_2Fe_{14}B$ show magnetic properties very close to that of RCo_5.

Structure

The RCo_5 compounds crystallize in the hexagonal $CaCu_5$-type structure [3]. This structure may be described as a stacking of atomic planes of two kinds. The planes of the first kind are occupied by R atoms in a site, 1a, of hexagonal symmetry and by cobalt atoms in a site, 2c, of symmetry $\bar{3}$/m. Only cobalt atoms occupy the other planes or a 3g site of symmetry mmm (figure 1). In the R_2Co_{17} phases pairs of cobalt atoms (dumbbells) are substituted for 1/3 of R atoms [4,5]. It has been shown that the RCo_5 phase is actually metastable at room temperature and tends to decompose in the neighbouring phases R_2Co_{17} and R_2Co_7 [6]. The structure of $R_2Fe_{14}B$ has also a uniaxial character : it is tetragonal and can be described as a sequence of different kinds of layers. In the layer containing R and B atoms, atomic arrangement reminiscent of the $CaCu_5$-type structure occurs (figure 2) [7,8].

Exchange and anisotropy

In RCo_5, R_2Co_{17} and $R_2Fe_{14}B$, the distances between cobalt or iron atoms are of the same order of magnitude as in the corresponding pure metals. The magnetic exchange interactions between these atoms, preponderant and positive, lead to ferromagnetic arrangement of the 3d moments. As in pure metals, the conduction electrons are negatively polarized by the spins of the 3d atoms. This leads to an antiparallel coupling of the spins of the R atoms and the spins of the 3d atoms. The coupling between the orbital moment and the spin moment of the R atoms depending on the filling of the 4f shell, the antiparallel coupling between 4f and 3d spins gives rise to parallel coupling of the magnetic moments (ferromagnetism) when the 4f shell is less than half-filled (light rare earths) and to antiparallel coupling (ferrimagnetism) for the heavy rare earths. Then, it is not possible to add a strong 4f magnetization to the 3d magnetization. Only the intermetallic compounds with light rare earths (Nd, Pr, Sm) or non magnetic (La, Lu, Y) show enough magnetization to give rise to useful permanent magnet properties at room temperature.

Due to the spin-orbit coupling, a strong anisotropy appears in the uniaxial rare earth-transition metal compounds. In YCo_5, the value of the uniaxial anisotropy energy is $5.8 \ 10^7$ erg/cm^3 (5.8 MJ/m^3) at room temperature [9] ; the easy axis of magnetization is a sixfold axis. At room temperature in $Y_2Fe_{14}B$, the fourfold axis is the easy axis of magnetization, the value of the anisotropy energy is $1.0 \ 10^7$ erg/cm^3 (1.0 MJ/m^3) [10]. When the rare earth atoms are magnetic, the electronic density of the 4f shell is also anisotropic. The direction of the easy magnetization axis results from the combined effects of the local anisotropies of rare earth and transition metals through the exchange coupling between the two types of atoms. In

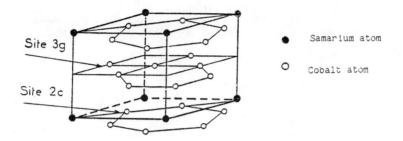

Site 3g

Site 2c

● Samarium atom

○ Cobalt atom

<u>Figure 1</u> : SmCo$_5$ crystallographic structure.

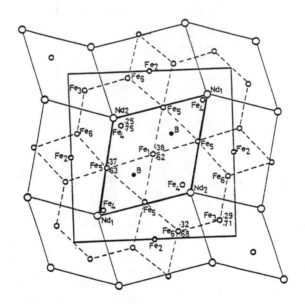

<u>Figure 2</u> : Nd$_2$Fe$_{14}$B crystallographic structure. The rhombic in heavy line
schematizes the unit cell of the CaCu$_5$-type structure [8].

Sm$_2$Co$_{17}$ the cobalt has a small anisotropy ; the resultant anisotropy is
essentially due to the samarium atoms. On the contrary, in SmCo$_5$ and
Nd$_2$Fe$_{14}$B the anisotropies of the rare earth on one hand and of Co or Fe on
the other hand are of the same order of magnitude and they add, to give
rise to very high value of the macroscopic anisotropy energy, respectively
12.0 and 4.4 10^7 erg/cm^3 (12.0 and 4.4 MJ/m^3) (Table 1).

310

Table 1 : Comparison of the saturation magnetization, $\mu_0 J_s$, the anisotropy
energy, K, at room temperature and Curie temperature, T_c, in rare
earth-cobalt and rare earth-iron materials.

	$\mu_0 J_s$ Tesla	K MJ/m^3	T_c (°C)
YCo_5	1.06	5.8	648
$SmCo_5$	0.97	12	724
Sm_2Co_{17}	1.20	3.3	920
$Y_2Fe_{14}B$	1.28	1.0	290
$Nd_2Fe_{14}B$	1.57	4.4	310

RARE EARTH PERMANENT MAGNET MATERIALS

Coercivity

When the magnetization in a fine particle ferromagnet (size of the
order of a micron) is saturated, an energy barrier, equal to the anisotropy
energy, should be overcome in order to rotate the magnetization in the
opposite direction. This gives a theoretical maximum value for the coercive
field which is, for instance in $SmCo_5$, 400 kOe (32 MA/m) at 300 K. The
actual value of coercive field in bulk specimens is reduced by several
orders of magnitude. Indeed, magnetic domains, with magnetization direction
opposed to that of the bulk specimen, are nucleated at structural defects
and a subsequent free displacement of domain walls may occur. Metallurgical
processes are thus needed in order to recover a significant value for the
coercivity. A general purpose is to obtain an inhomogeneous material in which
the metastable state of saturated magnetization may be more easily kept.
Powder metallurgy is used for elaboration of $SmCo_5$ [11] and $R_2Fe_{14}B$ magnets
[12]. In the magnets obtained after sintering, domain walls cannot propagate
from grain to grain. Thus magnetization reversal involves nucleation of
domains at grain boundaries. In some grains, nucleation may occur very
easily at structural defects, but the associated change of magnetization
involves magnetization reversal in individual grains : the coercive field
may reach a significant fraction of the anisotropy field. In a second
approach, advantage is taken of the existence of a miscibility gap occurring
in the phase diagram of rare earth alloys containing cobalt and iron together
with copper. Appropriate thermal treatments in bulk specimens lead to a
separation into two phases with different magnetic properties. Domain walls
tend to be pinned in the less-magnetic phase where their energy is reduced.
This gives rise to two kinds of initial magnetization curves observed in
$SmCo_5$ and $Sm(CoCu)_5$ permanent magnets (figure 3). The origin of the
coercivity in the different types of rare-earth permanent magnets is still
a subject of active discussions [13].

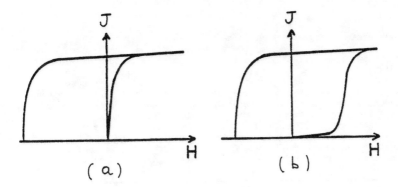

Figure 3 : Schematic first magnetization curves for (a) SmCo$_5$ and (b) SmCoCu$_5$
(a) : nucleation type   ;   (b) : pinning type.

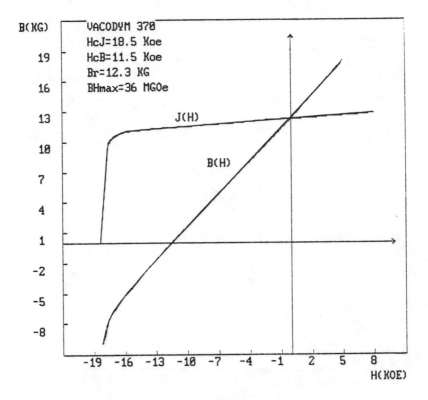

Figure 4 : Demagnetization curves for a NdFeB permanent magnet
(Vacuumschmelze) at room temperature.

312

Hysteresis

All permanent magnet materials have two main properties. The induction, B, is a consequence of the alignement of the atomic moments and the coercivity which reflects the ability of the material to stay in a meta-stable state. This leads to an hysteretic behaviour of the variation of the magnetization intensity or magnetization per volume unit, J, versus internal field, H, represented in the figure 4 for a NdFeB permanent magnet. J_S, the saturation magnetization is an intrinsic property of the material directly related to the magnetic moments of the consistuents and J_R, is the remanent magnetization. The field, H_{cJ}, for which the intensity of magnetization becomes zero is the intrinsic coercivity. The variation of the induction versus internal field for the new rare earth magnets represented in figure 5 emphasizes the improved magnetic properties in

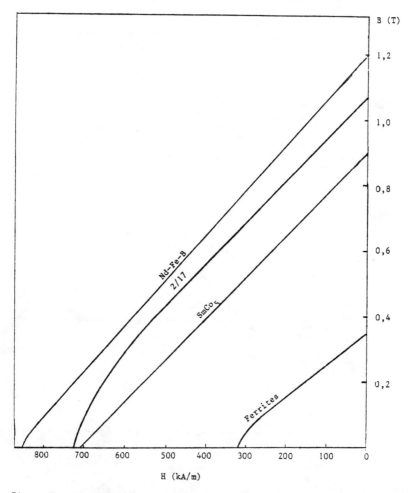

Figure 5 : Comparison of demagnetization curve B(H) for different types of permanent magnets at room temperature.

comparison to the hard ferrites such as BaO, $6Fe_2O_3$. The density of magneto-
static energy available is proportional to BH, its maximum value, $(BH)_{max}$,
corresponds to a particular working point determined by the geometry of the
system. At this point, a minimum volume of permanent magnet is required to
produce a given B value. Table 2 summarizes all the magnetic properties of
the main new permanent magnet materials in comparison with hard ferrites.

Table 2 : Typical values at room temperature of the magnetic properties of
different permanent magnet materials.

Material	B_R	H_{cJ}	H_{cB}	$(BH)_{max}$
Hard ferrites	0.4 T	160 kA/m	150 kA/m	26 kJ/m^3
	4000 G	1950 Oe	1850 Oe	3.3 MG.Oe
$SmCo_5$-type	0.95 T	1200 kA/m	680 kA/m	180 kJ/m^3
	9500 G	15000 Oe	8500 Oe	22 MG.Oe
Sm_2Co_{17}-type	1.05 T	880 kA/m	720 kA/m	210 kJ/m^3
	10500 G	11000 Oe	9000 Oe	26 MG.Oe
NdFeB type	1.2 T	1400 kA/m	880 kA/m	280 kJ/m^3
	12000 G	13500 Oe	11000 Oe	35 MG.Oe

As shown in figure 4, we can consider that the value of J remains
constant until internal field $|\vec{H_i}| < |\vec{H_{cJ}}|$. This important fact leads us to
consider the magnetization vector $\vec{J}$ as rigid. For materials for which
$\mu_0 H_{cJ} > J_R$, the (B,H) curve can be considered as linear, $H_{cB} = B_R/\mu_0$, the
relative permeabilities $\mu_R//$ and μ_R are equal to 1 within an error of
about 5 % for materials such as $SmCo_5$, Sm_2Co_{17} and $Nd_2Fe_{14}B$-type magnets.
As a consequence for the superimposed induction, the magnet material behaves
like vacuum with permeability μ_0. In practice this means that, for two
magnets with their directions of polarization at 90°, these directions are
unchanged. The superimposition of the induction of both permanent magnets
is a linear superimposition of the induction produced by each of them. This
rigidity of the magnetization and the linearity property of the induction
greatly simplify the computation of the magnetic induction in devices with
rare earth permanent magnets.

METHODS OF CALCULATION OF MAGNETIC INDUCTION

Most of the devices are made of prismatic elementary magnets. Inside each of them, the magnetization is assumed to be uniform and equal to the remanent magnetization J_R. The field configuration created by the magnet is the same as the configuration of a distribution of surface Amperian currents with density $\vec{i}_S = \vec{J}_R \wedge \vec{n}$, $\vec{n}$ is a unit vector perpendicular to the surface. On the face of a rectangular parallelepiped of rare earth magnets polarized parallel to the long side of the face, the surface Amperian current density is respectively equal to about 7000 A/cm, 8500 A/cm and 10000 A/cm for $SmCo_5$, Sm_2Co_{17} and NdFeB-type permanent magnets. This method allows us to calculate the induction, B, inside and outside the permanent magnet. The field configuration is also equivalent to that produced by a distribution of magnetic charges with surface density $\sigma = \vec{J}_R . \vec{n}$. Using methods of electrostatics, the calculation gives directly the values of the field H inside and outside the permanent magnet. Such methods have been used to calculate the induction produced by a system of permanent magnets in the case of hexapolar structure [14].

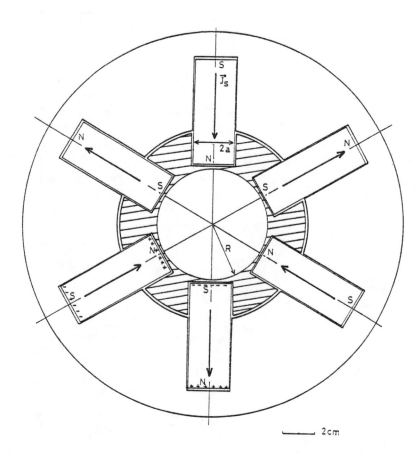

_____ 2cm

Figure 6 : Hexapole with $SmCo_5$ permanent magnets.

DESCRIPTION OF SOME HEXAPOLES

K. Halbach [15] has developed a general theory for the 2n-poles configuration with rare earth permanent magnets assuming a 2-dimensional cylindrical symmetry. He demonstrated that if ϕ designates the azimuth of a radius, the exact 2n-pole configuration is realized when the magnetization vector of the material at any point on the radius makes an angle equal to $n\phi$ with this radius. Such a continuous rotation of the magnetization is difficult to realize in practice ; the decomposition into several sectors uniformly polarized is a good approximation. A hexapolar configuration in a magnetic lens of an Electron Cyclotron Resonance (E.C.R.) multicharged ion sources has been constructed in 1980 [16]. It consists of 6 poles made with parallelepiped SmCo$_5$ permanent magnets (figure 6). The maximum value of the magnetic induction is 0.42 Tesla, the flux distribution and the radial variation of which is represented in figure 7. The properties of the

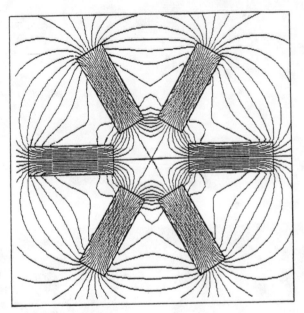

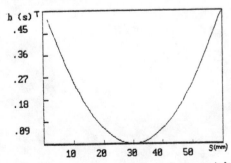

Figure 7 : Flux distribution and variation of the modulus of the induction inside the SmCo$_5$ hexapole.

theoretical hexapolar configuration are observed with an accuracy of 3 %
[14]. A new sketch has been proposed using a magnetic flux concentrator
which consists of prismatic permanent magnets adjacent to the main poles.
The figure 8 shows a hexapolar pancake made of six main parallelepiped
poles respectively north (N) and south (S) between which prisms of samarium-
cobalt permanent magnets are symmetrically placed in such a way that the
north pole of a prism is in contact with the lateral face of a main north
pole and likewise for a main south pole (figure 9). The advantage of this

Figure 8 : A hexapole pancake of SmCo$_5$ permanent magnets.

Figure 9 : Basic element of the hexapole using adjacent permanent magnets
as flux concentrator.

arrangement is that magnetic charges are put on the lateral faces of the main poles, which are particularly efficient in the contribution to the magnitude of the induction. The flux distribution and the variation of the hexapolar induction are plotted in figure 10a. Its maximum value reaches 0.8 Tesla with this magnet configuration [17,18]. For NdFeB magnets the same design leads to a maximum value of 1.2 Tesla (figure 10b).

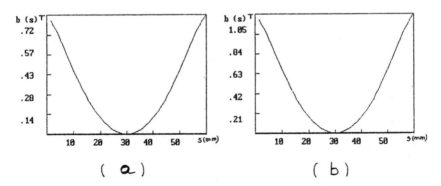

(a)   (b)

Figure 10 : Flux distribution in rare earth-permanent magnet hexapole and variation of the modulus of the induction inside the hexapole
(a) : SmCo₅ magnets   ;   (b) : Nd-Fe-B magnets.

The magnetic field configuration for the ECR ion sources results from the superimposition of such hexapolar structure in a plane perpendicular to a longitudinal configuration produced by solenoids : this leads to a "magnetic bottle". To increase the performances of such heavy ion sources the frequency of the high frequency wave must be increased, leading to an increase of the magnetic fields : radial and axial. ECR sources working at 10 GHz have been constructed with the simple hexapolar design. With the flux concentration system the frequency has been increased to 16 GHz. The NdFeB permanent magnets with their high induction remanence offer a good opportunity to design and construct more powerful ECR multicharged ion sources. However this is only possible with the new grades of NdFeB (i.e. containing dysprosium) which have an improved intrinsic coercivity, H_{cJ}, better than 18 kOe (1440 kA/m). Such a coercivity is needed for ECR ion sources at 28 GHz because permanent magnets in the hexapole are placed in a perpendicular magnetic field of about 13 kOe (1040 kA/m) due to the axial component of the magnetic bottle configuration.

Other static applications of NdFeB permanent magnets are now studied. In nuclear magnetic resonance for medical applications their use will lead to a reduction of weight. In undulators/wigglers the high remanence of NdFeB magnets will increase the domain of use of permanent magnets. For example, in the European Synchrotron Radiation Facility (E.S.R.F.), which will be constructed in Grenoble, the majority of undulators/wigglers will be made with such permanent magnets. However, some problems, such as temperature effects or corrosion could place limits in the use of NdFeB magnets. They have to be taken into account.

Indeed, the magnetic properties vary rapidly with temperature due to their relatively low Curie temperature (310°C). We have compared in Table 3 the different reversible temperature coefficients and the maximum operating temperatures for rare earth magnets and hard ferrites. In NdFeB static systems, where the induction must be stable within 1 %, some caution can be observed with the temperature especially when the magnets are working in the vicinity of the intrinsic coercivity, H_{cJ}.

Table 3 : Comparison of characteristic temperatures and temperature coefficients.

Material	T_c (°C)	$\Delta B_R / B_R$ % per °C	$\Delta H_{cJ} / H_{cJ}$ % per °C	Maximum operating temperature (°C)
$SmCo_5$-type	720	− 0.05	− 0.2	250
Sm_2Co_{17}-type	820	− 0.03	_ 0.2	350
NdFeB-type	310	− 0.1	− 0.6	150
Hard ferrite type	450	− 0.2	+ 0.3	300

The phenomenon of corrosion is particularly crucial in wet and salted atmosphere : without any protection they are corroded within a few hours. One of the solutions is to coat the magnets. Very active work is being done but very few results have till now been published.

CONCLUSION

The rare earth permanent magnet family, with its high intrinsic coercivity closely related to high magnetocrystalline anisotropy has led to new permanent magnet applications. New concepts have been applied to calculate static systems. As the magnetization, $\vec{J}$, can be considered as rigid, permanent magnets with different $\vec{J}$ can be assembled ; the resultant induction is then a linear superimposition of the induction produced by each of them. This property allows the realization of flux concentrator. This associated with high remanent induction of NdFeB magnets has permitted the reduction of weight of the static devices and a range of new applications. A recent development is the use of such magnets associated with soft magnetic materials. However, new computation techniques, such as 3d finite element methods can be used. Indeed, the non-linearity of the soft magnetic materials must be taken into account, ; large value of the induction can then be obtained.

ACKNOWLEDGEMENTS

The authors thank D. Givord, R. Lemaire, R. Geller and P. Sortais for helpful discussions and acknowledge the support of the Commission of the European Communities within the C.E.A.M. project of the Stimulation Program. One of the authors (J.C.) is partly supported by a fellowship from Aimants Ugimag S.A., Saint Pierre d'Allevard, France.

REFERENCES

[1] H.R. Kirchmayr and C.A. Poldy in Handbook on the Phys. and Chem. of Rare Earths, Eds. K.A. Gschneidner and L. Eyring (North-Holland Publishing Co, 1979), p. 55.

[2] K.H.J. Buschow, Rep. Prog. Phys., 40, 1179 (1977).

[3] J.H. Wernick and S. Geller, Acta Cryst., 12, 662 (1959).

[4] J.H. Florio, N.C. Baenziger and R.E. Rundle, Acta Cryst., 9, 371 (1956).

[5] F.S. Makrov and S.P. Vinogradov, Kristallografiya, 1, 634 (1956).

[6] F.J.A. den Broeder and K.H.J. Buschow, J. Less Comm. Metals, 29, 65 (1972).

[7] J.T. Herbst, J.J. Croat, F.E. Pinkerton and W.B. Yelon, Phys. Rev., B 29, 4176 (1984).

[8] D. Givord, H.S. Li and J.M. Moreau, Sol. Stat. Commun., 50, 477 (1984).

[9] J.M. Alameda, J. Déportes, D. Givord, R. Lemaire and Q. Lu, J. Magn. Magn. Mat., 17, 663 (1980).

[10] D. Givord, H.S. Li and R. Perrier de la Bâthie, Sol. Stat. Commun., 51, 857 (1984).

[11] M.G. Benz and D.L. Martin, Appl. Phys. Lett., 17, 176 (1970).

320

[12] M. Sagawa, S. Fujimura, M. Togawa, H. Yamamoto and Y. Matsuura, J. Appl. Phys., $\underline{55}$, 2083 (1984).

[13] D. Givord, J. Laforest, H.S. Li, A. Liénard, R. Perrier de la Bâthie and P. Tenaud, J. de Physique, $\underline{C6}$, 213 (1985).

[14] R. Pauthenet, J. de Physique, $\underline{C1}$, 285 (1984).

[15] K. Halbach, Nuclear Instr. and Methods, $\underline{169}$, 1 (1980).

[16] R. Geller, B. Jacquot and R. Pauthenet, Rev. Physique Appl., $\underline{15}$, 995 (1980).

[17] R. Pauthenet, R. Geller, B. jacquot, M. Lamy and J. Debernardi, Proc. 4th Intern. Workshop on E.C.R. ion sources (C.E.N. Grenoble, 1982).

[18] R. Geller, B. Jacquot and P. Sortais, Nuclear Instr. and Methods, $\underline{A243}$, 244 (1986).

RADIATION EFFECTS IN RARE-EARTH PERMANENT MAGNETS

J. R. Cost[*], R. D. Brown[*], A. L. Giorgi[*], and J. T. Stanley[**]
[*]Los Alamos National Laboratory, Los Alamos, NM 87545
[**]Chemical and Materials Science Engineering Dept., Arizona State University, Tempe, AZ 85287

ABSTRACT

Nd-Fe-B and Sm-Co permanent magnets have been irradiated with fission neutrons and gamma rays. Irradiated samples were periodically removed for room temperature measurements of the open-circuit remanence. Hysteresis loops were measured before and after irradiation. For neutron irradiation, two Nd-Fe-B magnets showed a rapid loss of remanence, while a third magnet from another manufacturer decayed more slowly, suggesting that the radiation hardness of Nd-Fe-B magnets may depend on microstructural details. Irradiation in the Omega West Reactor at Los Alamos with fast neutrons caused the fast-decay samples to have an initial loss of remanence of 1% for irradiation at 350 K to a fluence of 10^{15} n/cm^2. Both SmCo$_5$ and Sm$_2$Co$_{17}$ magnets showed excellent resistance to radiation-induced loss of remanence for neutron irradiation to a fluence of 2.6×10^{18} n/cm^2. Results for gamma irradiation showed no loss of remanence for a dose of about 49 Mrad using a ^{60}Co source. Possible mechanisms for radiation-induced loss of magnetic properties are discussed.

INTRODUCTION

The use of permanent magnets in accelerators allows reduction of the weight compared to using electromagnets, both in the magnet itself and in the elimination of the power supply. To be used successfully in a location where scattered radiation is present, the permanent magnets must be capable of resisting decay in remanence due to this radiation. Radiation produced by proton accelerators operating at energies of 50 to 100 MeV is expected to consist of neutrons and gammas produced when the fringes of the accelerated beam strike structures located physically near the beam line. A rough calculation of the total fluence of neutrons expected for a 50 MeV accelerator gives a fluence of about 10^{15} n/cm^2 over the operating lifetime (approximately 200 h) of the accelerator.

Previous applications of permanent magnets in dipole and quadrupole magnets have specified the use of Sm-Co material. Previous work [1-4] has shown that Sm$_2$Co$_{17}$ magnets are the most resistant to radiation-induced decay of the remanence, and that SmCo$_5$ magnets are also very resistant to doses in the 10^{15} n/cm^2 range. The recently developed Nd-Fe-B magnet family is of interest to accelerator designers due to the high remanence and intrinsic coercivity of these materials. The sensitivity of these magnets to temperature-produced changes in magnetic properties has led to concern over their stability in a radiation environment. As of 1986, the one study which had been done [4] on radiation effects to Nd-Fe-B magnets showed them to be orders of magnitude more sensitive to charged particle radiation than the Sm-Co magnets. We have previously reported some initial results of irradiations of Nd-Fe-B magnets [5]. We found that while the decay of remanence is more rapid than that of the Sm-Co family, the decay for irradiation at 350 K is only about 1% at a fluence of 10^{15} n/cm^2. The present paper is a further investigation of radiation effects on Nd-Fe-B magnets, together with results for Sm-Co magnets irradiated in the same neutron environment.

METHOD

Magnets

The Nd-Fe-B magnets were obtained from three different manufacturers: Crucible Materials Corporation, I. G. Technologies, and Hitachi Magnetics Corporation. The identification and alloy system of the magnets from each manufacturer are given in Table I along with the pre-irradiation values for the residual induction, and intrinsic coercivity specified by the company. The magnets from all three manufacturers were prepared using powder metallurgy combined with thermal treatments unique to each manufacturer.

Two types of Sm-Co magnets were studied, $SmCo_5$, and Sm_2Co_{17}. Both were obtained from I. G. Technologies and were fabricated from powder starting material. The magnetic properties are also given in Table I.

Table I

Pre-Irradiation Properties of Rare-Earth Magnets Used in This Radiation Damage Study

Alloy	Manufacturer	Material	Remanence (Gauss)	Intrinsic Coercive Force (Oersted)
Nd-Fe-B	Crucible Materials	CRUMAX 282	10,800	>17,000
Nd-Fe-B	I. G. Technologies	NeIGT 27H	10,200	>17,000
Nd-Fe-B	Hitachi Magnetics	HICOREX 94EB	11,500	15,000
$SmCo_5$	I. G. Technologies	INCOR 18	8,800	>16,000
Sm_2Co_{17}	I. G. Technologies	INCOR 22HE	9,600	>16,000

Sample Preparation

The materials were received from the manufacturer in the form of blocks with the largest dimension of roughly 20 mm. Small samples required for the irradiation were cut from these blocks using a low-speed diamond saw. Final samples were typically 7 mm long with a square cross-section from 1 mm to 2 mm on an edge. The magnetization was in the long direction.

After being cut, the samples which were not yet magnetized to saturation were magnetized, either in a magnetic charging unit with charging fields up to 100 kOe, or by measuring the complete hysteresis loop using a vibrating sample magnetometer at 250 K (-23°C). A superconducting magnet applied a final field of at least 55 kOe. Saturation, as determined by the hysteresis loops, typically required a field of 40 kOe or less.

Irradiation

Neutron irradiations were done at the Omega West Reactor at Los Alamos using the epithermal cadmium port. This port is located immediately adjacent to the core fuel rod array and is shielded with cadmium in order to attenuate the thermal neutron flux by roughly 10^4. The flux of fast neutrons (E $\geq$ 0.1 MeV) was roughly 2×10^{13} n/cm^2·s as measured by dosimetry foils. The neutron energy spectrum for a port symetrically located on the other side of the core, but without cadmium shielding, has been reported [6] and is in good agreement with our measurements for this facility.

For the irradiation, two samples were placed in a polyethylene rabbit in which they were separated by plastic spacers so that they did not interact magnetically. The rabbit was lowered into position in the reactor port with a chromel-alumel thermocouple, the junction of which was attached to a dummy sample for temperature measurement. This arrangement allowed easy insertion and removal of the rabbit from next to the reactor core. Sample temperature during irradiation was kept close to that of the cooling water of the reactor, roughly 350 K (77°C), by introducing a continuous flow of helium gas. Several samples were also irradiated without helium at a temperature of approximately 426 K (153°C). Samples were held in the reactor for the desired irradiation times and then removed for measurement of the remanence. After each measurement they were reinserted for continuation of the irradiation.

To determine whether gamma irradiation by itself could produce an observable decay in the remanence, several of the samples were irradiated in a gamma cell using ^{60}Co. The dose rate in the gamma cell was approximately 16 rad/s. A measurement of the temperature using a thermocouple indicated no change in temperature compared to the ambient temperature outside the cell. As with the neutron irradiated samples, the same samples were irradiated for increasing periods of time, with the remanence being measured at intervals during the irradiation.

Measurement

Remanence was measured by mechanically moving the sample through a 500-turn pickup coil and monitoring the voltage output of the coil [5]. This voltage was integrated by computer to obtain the area under the voltage-time curve from which a value proportional to the open circuit remanence was calculated. These measured values were normalized to the pre-irradiation value. Typically each reported value was obtained from an average of 20 or more separate readings. This resulted in a fractional uncertainty which was usually less than 0.02.

RESULTS AND DISCUSSION

Fig. 1 shows the relative decay of remanence as a function of irradiation time and neutron fluence for the samples neutron irradiated at 350 K (77°C). For Nd-Fe-B magnets we observe that the CRUMAX 282 and NeIGT 27H samples are sensitive to irradiation by fast neutrons while the HICOREX 94EB material is relatively insensitive. This finding that some Nd-Fe-B magnets can withstand irradiation is particularly important since previous investigations have only reported that these magnets are highly sensitive to neutron irradiation [4,5]. This new knowledge suggests that metallurgical processing can markedly affect the microstructure, which is important in determining the radiation-resistance.

In order to better understand the decay of Nd-Fe-B magnets, the effects of temperature, modification of the neutron energy spectrum, and irradiation with gamma rays were investigated. A comparison was also made between radiation effects on the Nd-Fe-B and Sm-Co magnets. Results for all irradiations are summarized in Table II.

First, non-irradiated samples of Nd-Fe-B magnets prepared from the powder starting material were annealed at 423 K (150°C) to determine if any part of the remanence loss was due to instability at the irradiation temperature. These samples showed an initial 1.5% drop in remanence after heating to this temperature, but no further decay of remanence following periodic measurements for times up to 15,300 seconds. Thus we deduce that the remanence loss is due solely to the irradiation.

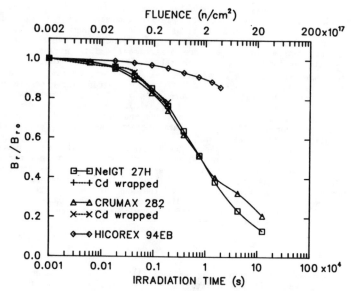

Fig. 1. Retained fraction of pre-irradiation remanence vs time and fluence for Nd-Fe-B magnets irradiated with fast neutrons at 350 K (70°C). Some samples were wrapped in cadmium to further attenuate thermal neutrons.

Table II

Results of Irradiations of Rare-Earth Permanent Magnets

Sample	Type of Irradiation	Irradiation Temperature (K)	Irradiation Time (s)	Total Fluence (n/cm^2)	Normalized Remanence (B_r/B_{ro})
CRUMAX 282	neutron	350	1.26×10^5	2.5×10^{18}	0.209
NeIGT 27H	neutron	350	1.26×10^5	2.5×10^{18}	0.132
HICOREX 94EB	neutron	350	1.92×10^4	3.8×10^{17}	0.860
CRUMAX 282	neutron Cd shielded	350	1.86×10^3	3.7×10^{16}	0.763
NeIGT 27H	neutron Cd shielded	350	1.86×10^3	3.7×10^{16}	0.784
CRUMAX 282	gamma	300	3.05×10^6	48.8*	1.008
NeIGT 27H	gamma	300	3.05×10^6	48.8*	1.003
INCOR 18	neutron	350	1.29×10^5	2.6×10^{18}	1.003
INCOR 22HE	neutron	350	1.29×10^5	2.6×10^{18}	0.998

*Fluence in Mrad.

Because of the high cross section of boron for thermal neutrons and the relatively high energy of the alpha particles emitted by the reaction, it was desirable to check whether thermal neutrons were contributing a large fraction of the atomic displacement damage. To check this we repeated the irradiation of the two samples which had shown decay, but on two similar samples which were enclosed with 0.4 mm of cadmium which attenuated thermal neutrons by an additional factor of 10^4. The results of this are shown both in Table II and in Fig. 1. Fig. 1 shows that there is a negligible difference between irradiation-induced decay of remanence with and without the extra cadmium shielding. Thus we eliminate the reaction between boron and thermal neutrons as being important to the results for Nd-Fe-B magnets irradiated in our cadmium wrapped rabbit tube.

It was desirable to know the extent to which Nd-Fe-B magnets are sensitive to other kinds of irradiation. Thus we gamma irradiated (using a ^{60}Co source) samples of the same magnets which showed remanence decay due to fast neutrons. Following irradiation for approximately 35 days, the samples reached a dose of 49 Mrad, and showed no decrease in remanence. Reactor irradiation exposes the samples to both gamma and neutron irradiation. For comparison, reactor irradiation for roughly 1000 seconds would result in a gamma dose of 50 Mrad together with a neutron dose of about 0.2 Mrad. At this dose the remanence of a NeIGT 27H sample would drop by about 20%. The gamma dose of 49 Mrad corresponds approximately to 2×10^{-9} displacements per atom. The calculated dpa value is highly sensitive to the assumed threshold energy for atomic displacement, which for this calculation was 25 eV. Since gamma irradiation only displaces individual atoms from their lattice sites, while a single neutron irradiation event results in a cascade of displaced atoms, we expect the latter form of irradiation to produce greater decay of the remanence.

In order to provide further understanding of radiation damage effects we investigated Sm-Co permanent magnets. From Table II we see that Sm-Co magnets, both $SmCo_5$ and Sm_2Co_{17}, are quite insensitive to radiation damage by fast neutrons. The remanence of both of these magnets was only minimally changed after 41 hours of irradiation to a fluence of 2.6×10^{18} n/cm^2. Sm-Co magnets would be the obvious choice for high-irradiation environments where the highest stability is required.

We have analyzed the full decay curves in Fig. 1 to determine whether the process of irradiation-induced loss of magnetization fits any of several different models for the reaction kinetics. The predicted fits for all of these models were poor. The reaction clearly does not follow simple first-order kinetics with a single relaxation time. However, first-order kinetics could be involved if there were a fairly broad distribution of relaxation times. From the slope of Fig. 1 we calculate that if this distribution of relaxation times is lognormal, then it has a width parameter, $\beta=2.5$, which means that the distribution is slightly less than two decades wide at half maximum.

For some of the samples hysteresis loops were measured before and after neutron irradiation. Fig. 2 shows such a pair of loops for the Nd-Fe-B magnet NeIGT 27H irradiated at 426 K (153°C) and measured at 250 K (-23°C). Note that the remanence, which decreased roughly 80% during the irradiation (see the start of the solid curve in the second quadrant), recovered to the pre-irradiation value after being magnetically saturated. This would seem to indicate that the same magnetic structure is developed after irradiation and thus that irradiation does not appreciably change the metallurgical structure. Interestingly, the coercive force as shown in Fig. 2 was increased roughly 20% by the irradiation. A possible explanation for this is that the damage introduces extra pinning sites for the magnetic domain walls making them more difficult to move.

An estimate of the rate for atomic displacements due to neutron irradiation has been made and compared with the decay rate of the magnetization [5]. For a flux of 2×10^{13} n/cm$^2 \cdot$s, an elastic scattering cross section of 3×10^{-24} cm^2, and assuming 500 to 5000 displaced atoms per primary knock-on event the sample will have roughly 10^{-7} displaced atoms per second. Since appreciable remanence decay occurs in the first 1000 seconds,

326

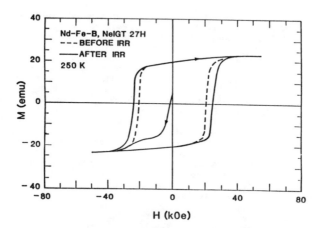

Fig. 2. Hysteresis loops at 250 K (-23°C) for a
NeIGT 27H sample before and after irradiation.

the loss is occuring even though only a small fraction of the atoms is being
displaced. A mechanism which agrees with this high loss rate per atomic
displacement is radiation-induced nucleation of reverse domains. Such
nucleation requires sufficient energy to overcome the domain nucleation
energy barrier. When a 1 MeV neutron collides with an iron atom, it
transfers a maximum energy of 17 keV to the iron atom. This energy is
quickly transferred to other atoms and electrons in the immediate region of
the primary knock-on. The fraction of the shared energy which goes into
creation of displaced atoms is most likely not available for reverse domain
nucleation. Hence the calculation of the dpa level is no more relevant to
the decay of remanence than is the calculation of the total absorbed energy
in rads. The relevant parameter for domain nucleation should be the number
of events which deposit energy in excess of the domain nucleation energy
barrier.

   This mechanism can be expected to be extremely sensitive to the
structure of the magnetic domains. This, in turn, will be determined by the
microstructure of the phases. A study of both of these kinds of structures
in Nd-Fe-B magnets fabricated from powder starting material [7] has
identified the major phase, $Fe_{14}Nd_2B$, as being responsible for the hard
magnetic properties. More recently, Hadjipanayis, et. al. [8] have
investigated the microstructure of this phase in Nd-Fe-B magnets from
different manufacturers and found variations in grain size for this magnetic
phase of more than a factor of one thousand within a given magnet; they also
found large differences between magnets of the different manufacturers.
Such wide differences in microstructure could easily account for the marked
differences in resistance to radiation damage which are observed in Fig. 1.

   At present our understanding of how coercive properties depend upon
structure is not fully clear. We do know, however, that differences in
grain size and grain boundary structure can cause different coercivity
mechanisms to operate [9]. The determination of how these mechanisms affect
the resistance to radiation damage needs further work. Certainly it will be
of value to determine how the microstructure of Hitachi HICOREX 94EB differs
from that of the other magnets, since it resists neutron irradiation better
than its counterparts by a factor of 25. If, as we might expect, a fine
grain structure is one of the key factors, then it will be important to
investigate magnets fabricated from melt-spun starting material since they
have an extremely fine grain size (25 nm) as well as a thin layer of
amorphous material at the grain boundaries to promote pinning of domain
walls [9].

ACKNOWLEDGEMENT

The authors acknowledge the excellent advice and assistance from the staff and supervisors at the Omega West Reactor at Los Alamos. Their help allowed us to utilize that facility to best advantage. We would also like to acknowledge Robert Wolf of Crucible Materials Corporation and John Kaste of I. G. Technologies for providing magnet materials.

REFERENCES

[1] R. D. Brown, E. D. Bush Jr., and W. T. Hunter, Los Alamos National Laboratory report LA-9437-MS, July 1982.
[2] F. Coninckx, W. Naegele, M. Reinharz, H. Schoenbacher, and P. Seraphin, CERN/SPS, TIS-RP/IR/83-07, Feb. 1983.
[3] H. Spitzer and A. Weller, Kernsforschungsanlage Jülich Report SNQ 1 N/BH 22/05/84, May 1984.
[4] E. W. Blackmore, IEEE Trans. Nucl. Sci., NS-32, 3669-3671, Oct. 1985.
[5] J. Cost, R. Brown, A. Giorgi, and J. Stanley, Los Alamos National Laboratory report LA-UR-86-4245. Submitted for publication to IEEE Trans. Magn., Dec. 1986.
[6] W. A. Coghlan, F. W. Clinard Jr., N. Itoh, and L. R. Greenwood, J. Nucl. Mater., 141-143 , 382-386, 1986.
[7] J. F. Herbst, J. J. Croat, F. D. Pinkerton, and W. B. Yelon, Phys. Rev. B, 29, 4176, 1984.
[8] G. C. Hadjipanayis, Y. F. Tao, and K. R. Lawless, IEEE Trans. Magn., MAG-22, 6, 1845, 1986.
[9] R. K. Mishra, J. Magn. and Magn. Mater., 54-57, 450, 1986.

COST AND PERFORMANCE OF LOW FIELD MAGNETIC RESONANCE IMAGING

LAWRENCE E. CROOKS*, CHING YAO**, MITSUAKI ARAKAWA*, JAMES D. HALE*,
JOSEPH W. CARLSON*, PAUL LICATO**, JOHN HOENNINGER*, JEFFREY WATTS*
AND LEON KAUFMAN*
* University of California, San Francisco - Radiologic Imaging Laboratory,
South San Francisco, CA
** Diasonics, Inc., South San Francisco, CA

ABSTRACT

Magnetic Resonance Imaging technology has advanced on many fronts.
Two important areas are the improvement of signal-to-noise levels and the
understanding of imaging pulse sequence optimization. Based on these
advances, low cost imagers using very low field permanent magnets are
providing images of diagnostic quality.

INTRODUCTION

Magnetic Resonance Imaging (MRI) has changed the way diagnostic
radiology is practiced. The accurate depiction of anatomy and
non-invasiveness of the technique make it the modality of choice in an
increasing number of applications. Nevertheless, cost is a continuing
impediment to wider diffusion of the modality. The major component of cost
is the magnet, both as a direct cost and because of its impact on siting
costs. Today, superconducting magnets are used in most of the commercial
systems. The cost of these units is dependent somewhat on field strength,
and siting costs are strongly dependent on field strength [1]. The most
effective way to reduce the cost of a superconducting magnet is to make it
smaller, since this reduces the cost of the cryostat. A smaller magnet
also reduces siting costs because the field is more contained. In any
event, cryostat costs limit the lowest achievable cost with any
superconducting magnet, and physical size can only be reduced to a point
consistent with the need to place a body in the aperture. Other components
are also significant in determining system cost, and these include
computers, gradient coils and power supplies, RF electronics, installation
and warranty. Thus, if drastic cost reduction is desired, subsystems other
than the magnet also need to be considered.
As operating field strength changes, so does the magnet technology
that provides the lowest cost alternative. Roughly, permanent and air core
resistives are the lowest cost options below about 1 Kgauss, air core
resistives are to be favored between 1 and 2 Kgauss and superconducting
magnets are lowest above 2-2.5 Kgauss. From the point of view of
installation, permanent magnets (if not too heavy) are attractive in that
they do not require electrical power and cooling water, services needed for
resistive magnets. In an intergrated approach to cost reduction, the
effects of lower field strengths cascade through the system. Reducing
field strength impacts not just available magnet types, but also gradient
systems and siting. If some speed is sacrificed in computer data
flow times, then further savings are possible. An associated benefit of
low frequency operation is that increased coil efficiency reduces the power
needed to excite the nuclei. This reduces the cost of the RF transmitter.
We are investigating the use of very low field MRI as a low cost
alternative to existing superconductive technology. The magnet used is a
Sumitomo four-post permanent magnet with a 500 gauss field, 65 cm aperture
and 130 cm pole tip. The magnetic material is ferrite, and the total
weight is 6000 Kg. The magnet was shimmed passively by the use of 5
adjusting points on the upper pole tip mount and by the use of iron shims.
With these we were able to improve on the 200 ppm specification even though

the permanent magnet is situated in the 9 gauss field of a 3.5 Kgauss
superconductive magnet and the 2 gauss field of an orthogonally placed 20
Kgauss superconductive magnet, a demanding environment unlikely to be found
in clinical situations. At this time the magnet itself is not insulated
from ambient temperature changes.

DESIGN CONSIDERATIONS

Signal-to-Noise (S/N) Levels

There exists a considerable disparity on the claims made for the
behavior of S/N as field strength changes. Based on theoretical models,
proponents of high field operation (15-20 Kgauss) argue that S/N changes
linearly with field strength [2]. Since the high field units operate with
S/N levels of 25 to 60, it would be hopeless to expect to get useful
diagnostic information from a 500 Gauss magnet that should yield S/N values
of 1 to 2. Fortunately, when considering S/N, field strength cannot be
considered separately from other variables in the imaging experiment [3-5].
These include sequence parameter optimization to account for changes in the
T1 relaxation time with field strength, imaging sequence optimization,
chemical shift artifacts, RF power deposition, and RF coil design. When
these factors are taken into account, we find weak dependences of resolving
power (a numerical measure of diagnostic confidence given by Contrast x
S/N) with field strength [6,7]. Although there is a penalty in terms of
some loss of "intrinsic" signal-to-noise (S/N) levels as field decreases,
this is partially compensated for by other favorably behaved variables.
The T1 relaxation time of most tissues decrease to about one half of those
at 3.5 Kgauss and a quarter of those at 15 Kgauss. As a consequence, for
constant imaging time, signal averaging can be used to improve S/N.
Background field inhomogeneities favor operation at lower field strengths.
A homogeneity of 200 ppm at 500 gauss translates into an absolute
inhomogeneity of 0.1 gauss. To achieve the same inhomogeneity at 3.5
Kgauss the field would need to be specified at 30 ppm and 7 ppm at 15
Kgauss. Because the permanent magnet can be shimmed to better than 200
ppm, this permits weak gradients for signal readout and consequently
improved S/N.

Another area of improvement involves the design of RF coils for signal
pickup. Permanent magnets present a field oriented perpendicular to the
subject's long axis, whereas in air core magnets the field is aligned along
the patient axis. The longitudinal field permits use of saddle-shaped
coils. When such a pair is configured in an orthogonal a arrangement and
the signals detected in quadrature (QD coil), a 40% improvement in S/N over
a single saddle coil is obtained. With the transverse field either a
single saddle or a solenoidal coil can be used. If the patient is the
major source of noise both achieve equivalent performance, as we have
demonstrated at 3.5 Kgauss [8], but, at the lower fields, coil losses may
dominate. In this case the solenoid would become the preferred
configuration. For the permanent magnet we have evaluated both
single-saddle and solenoidal coils. We have found that even at 500 gauss
high Q coils can be built, so that patient loading becomes the major noise
source. For head imaging we find that the solenoidal coil has a 30-40%
advantage in S/N over a single saddle coil. This compensates for the
inherent advantage of a longitudinal field magnet, which makes it easy to
use QD coils. Incorporating QD coils in transverse field magnets is a task
that is conceptually possible but not yet demonstrated.

Although the technology necessary for optimizing S/N at low magnetic
fields is evolving, today we can obtain S/N levels that are 2.5-3 times
lower than those of state-of-the-art superconductive units (Figures 1-3).

These levels are an order of magnitude higher than those predicted by simplistic considerations based on field strength as a separable parameter, and permit a wide range of diagnostic procedures.

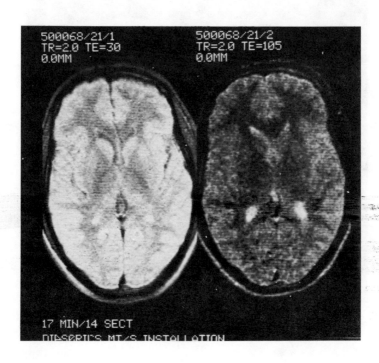

Figure 1. Transverse section of the head, imaged in 17 min for 14 sections. Two echos per section were obtained, at 30 and 105 msec. Resolution is 1.1 x 1.1 x 10 mm.

332

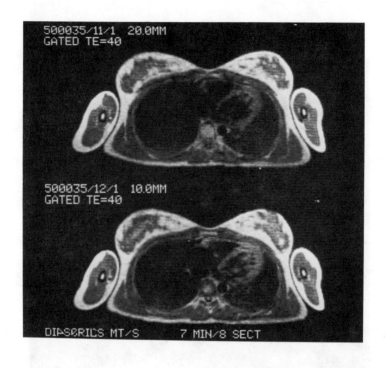

Figure 2. Gated heart, transverse section. Eight sections were imaged in 7 min. Resolution is 2.2 x 2.2 x 10 mm.

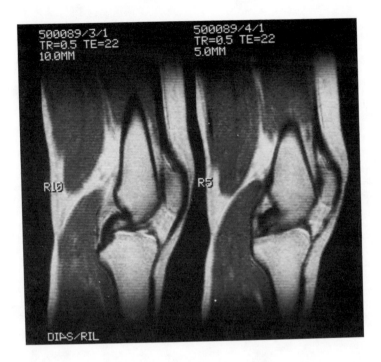

Figure 3. Sagittal section of the knee, imaged in 17 min (8 sections). Resolution is 1.1 x 1.1 x 5 mm.

Sequencing for Low Field Operation

A major aspect of the design of sequences for low field operation is to take advantage of the opportunities afforded by shortened T1 values. An extensive discussion of the process of optimizing sequences is beyond the scope of this presentation, but analytical tools to do this are well established [9].

As described above, the shortening of T1 relaxation times permits sequence optimization by maximizing S/N per unit time. For example, in the head, a TR= 1 sec at 3.5 Kgauss. Consequently, in the same time, twice as much data can be acquired, so that signal averaging compensates by 40% the S/N losses of the lower field unit for the same gray/white matter contrast. If a longer TR is desired for more coverage, then, at TR= 2 sec the signal averaging advantage is lost, but contrast increased by about 45%. Since resolving power is the product of contrast and S/N, the relative advantage of the low field unit is retained. Other as of yet unexploited advantages of short T1 times can be explored with 3-D acquisition techniques and partial flip imaging [10]. What is important to note is that short T1 values and low background field inhomogeneities change the character of sequences for low field operation. Therefore, optimization by sequence matching to magnet characteristics is an essential step in achieving a diagnostically efficacious low field instrument.

334

DISCUSSION

Because of the investment in technology that permits superconductive medium field units to operate with their intrinsic advantages (shorter T1 values, low RF power deposition, easier siting, lower investment and operating costs, and reduced hazards), and at the same time with high S/N

levels, MRI technology has advanced to the point that many of the diagnostic procedures performed with superconductive imagers can be also carried out at very low fields by relatively inexpensive units based on permanent magnets. While today these units cannot reproduce all of the studies done at medium and high field strengths, because of their lower costs they actually present an opportunity to perform most of the same as well as a different set of studies where patient charges have been the main limiting factor. Therefore, low field MRI presents both limitations and opportunities. The generalized acceptance of this mode of MRI practice will depend not only on clinical efficacy, but on more difficult to predict factors such as the willingness to accept adequate as opposed to "the best" in image quality, reimbursement policies, referral patterns, legal issues and siting limitations. Outside the U.S., other considerations include liquid helium cost and availability as well as governmental procurement policies.

ACKNOWLEDGEMENTS

These investigations are supported in part by Diasonics, Inc. The authors wish to acknowledge the contributions of Mr. Charles Peters of Diasonics, of Mr. Hiroo Hayashi of Sumitomo Special Metals and Mr. Paul S. Kimura of Sumitomo Corporation of America.

REFERENCES

1. Kaufman L, Crooks LE, Margulis AR and Proseus JO. Siting, In Clinical Magnetic Resonance Imaging. Edited by Margulis A, Higgins C, Kaufman L and Crooks L. (University of California Press, San Francisco, CA 1983) Chapter 20.

2. Edelstein WA, Bottemley PA and Pfeifer LM. A signal-to-noise calibration procedure for NMR imaging systems. Medical Physics 11, 180 (1984).

3. Hoult DI, Chen C-N, and Sank VJ. The Field Dependence of NMR Imaging. Magnetic Resonance in Medicine 3, 730 (1986).

4. Kaufman L and Crooks LE. Technical Advances in Magnetic Resonance Imaging. In Contemporary Imaging, Goldberg H, Higgins CB and Ring EJ, eds. (University of California Press, San Francisco, CA, 1985) p. 7.

5. Kaufman L. Optimal Field Strength in Proton MRI. Toshiba Medical Review, International Edition, No 16, March 1986, pp. 5-9.

6. Crooks LE, Arakawa M, Hoenninger JC, McCarten B, Watts J and Kaufman L. Magnetic Resonance Imaging: Effects of Magnetic Field Strength. Radiology 151, 127 (1984).

7. Posin JP, Arakawa M, Crooks LE, Feinberg DA, Hoenninger JC, Watts JC, Mills CM and Kaufman L. Hydrogen Magnetic Resonance Imaging of the Head at 0.35 and 0.7 Tesla: Effects of Magnetic Field Strength. Radiology 157, 679 (1985).
8. Arakawa M, Crooks LE, McCarten B, Hoenninger JC, Watts JC and Kaufman L. A Comparison of Saddle-Shaped and Solenoidal Coils for Magnetic Resonance Imaging. Radiology 154, 227 (1985).

9. Ortendahl DA, Hylton NM, Kaufman L, Watts JC, Crooks LE, Mills CM and Stark D. Analytical Tools for MRI. Radiology 153, 479 (1984).

10. Mills TC, Ortendahl DA, Hylton NM, Crooks LE, Carlson JW, and Kaufman L. Partial Flip Angle MRI. Radiology 162, 531 (1987)

Author Index

Subject Index

Formula Index